THE FOREST

OF

PHYSICS

Travis Norsen

ISBN: 978-1-7345280-1-5

Cover photo (looking down from Mount Monadnock in Southern New Hampshire) by Travis Norsen.

Send inquiries to info@pigpugpress.com.

Pig
Pug
Press

https://pigpugpress.com

"I fully agree with you about the significance and educational value of ... history and philosophy of science. So many people today – and even professional scientists – seem to me like somebody who has seen thousands of trees but has never seen a forest. A knowledge of the historic and philosophical background gives that kind of independence from [the] prejudices of his generation from which most scientists are suffering. This independence created by philosophical insight is – in my opinion – the mark of distinction between a mere artisan or specialist and a real seeker after truth."

- Albert Einstein

Contents

Preface

THIS book is the text for a new "Advanced Introductory Physics" course which I had the opportunity to design and teach at Smith College starting in 2019. The intended audience for this course is incoming students with unusually strong high school physics backgrounds (e.g., two years of calculus-based AP physics). Such students have seen most of the mechanics and E&M *content* that we cover in our regular two-semester introductory sequence, but typically have not developed all of the *skills* (associated with designing and running experiments, modeling, dealing with data and uncertainty, setting up integrals, taking appropriately expert-like approaches to solving problems, and running computer simulations) that our college-level introductory courses emphasize.

Prior to the advent of the new Advanced Introductory Physics course, these unusually-well-prepared students either took one of the regular intro courses to develop their skills (but then often felt soul-crushing boredom because the content was so familiar) or skipped the intro courses (and either took no physics at all, or jumped into higher-level physics courses where they sometimes struggled because they lacked some of the prerequisite skills). Either way, they unfortunately tended to drift away from physics.

So the main goal of the new course was simply to give these students an opportunity to develop their skills, but on a fresh, challenging, and inspiring body of content– with the hope that they'd stick with this beautiful, fundamental, and highly practical subject.

As the designer and first instructor for the course, I decided to find the needed fresh and challenging content by looking to the historical development of physics. The idea that emerged was to try to tell the stories, in one admittedly fast-paced semester, of what I think of as the three great pillars of classical (pre-20th-century) physics: Newton's theory of universal gravitation, Maxwell's theory of electricity and magnetism, and the atomic theory of matter. The book thus has three main Parts (of four or five chapters each), one for each of these three theories. (And there is also a stand-alone opening Chapter that just gets some ideas about measurement and uncertainty out on the table.)

Part I begins with a little Ancient Greek astronomy and works its way up through Galileo and Kepler to Newton's laws for particles, interacting systems, and rotational motion. This Part focuses on developing skills including modeling, data analysis (often using Python), and setting up integrals, and has an astronomical theme that includes applications to several more contemporary discoveries such as extra-solar planets.

Part II also begins with a qualitative tour of more ancient observations of electrical and magnetic phenomena, but quickly moves through the 19th century consolidation into Maxwell's unified theory of electrodynamics. Important conceptual themes include the notion of "fields" and their role in energy transport and radiation, and the practical application of electromagnetic principles to both DC and AC electric circuits. Parallels between electromagnetic and gravitational phenomena are explored and applied, for example, to the LIGO group's recent observation of gravitational radiation from a binary black hole merger.

Finally, Part III develops the atomic theory of matter from Dalton's chemistry-based speculations at the beginning of the 19th century, to the head-waters of quantum theory at the beginning of the 20th. The story of atoms is highly entangled with the ideas of heat and thermodynamics and, to understand these connections, students will develop a facility with mathematical concepts from probability and statistics.

Each Part, as I have said, attempts to "tell the story" of how certain big ideas developed and interacted across the centuries. But the presentation is only historical in a sense that is so rough and crude that it will undoubtedly drive any actual historian of science who looks at it completely crazy with frustration. Furthermore, as you can probably tell from the above Part-descriptions, any student who thinks a historically-structured physics course will be light on equations and other technical details is also in for a rude awakening. This book, in short, is not a history of physics. It is rather a physics textbook – and indeed a rather hard-core one for the introductory level – which just happens to take a somewhat more historical perspective than is typical for physics textbooks.

The point – the value – of the book's historical structure is, in my opinion, well captured by the Einstein quote I put on an earlier page. Still today, many professional scientists – and even moreso, many science students – are very good at making calculations and solving problems, but seem to lack the sort of big-picture perspective that is needed to decide what questions are interesting to pursue, needed to critically judge work from outside their own narrow sub-field, and needed to understand how and why science is unique as a way of trying to navigate the world. So my hope is that by cramming the entire long history of physics (at least, prior to the 20th century) into this one course, students will be able to rise above the individual trees and come away with a real appreciation for the amazingly coherent and beautiful whole that is the *forest* of physics.

That, at any rate, is the big long-term abstract goal. But perhaps it is also worth stressing that, in the shorter term, students will learn a lot of really challenging and thought-provoking ideas, develop a bunch of amazing practical skills, and generally have a fantastic time along the way.

I am making this book publicly available with the hope that fellow instructors, self-motivated physics students, and physics-interested auto-didacts may find value in it. Readers from any of those (overlapping!) categories should be aware, however, that the book is only really the framing skeleton for the course. Each chapter includes (conceptual) Questions and (calculational) Exercises designed to help students stay actively engaged with what they have read. But the real "guts" of the course are to be found in the end-of-chapter Projects. These are what students in my class spend most of their time working on and they are where most of the intended learning goals are achieved. The Projects range from hands-on laboratory experiments to theoretical and computational-modeling explorations. In the Projects, students develop comfort estimating and propagating uncertainties; they use python code to visualize data, perform curve-fits, and explore numerical simulations; and they extend and strengthen their abilities to understand and apply complex theoretical ideas. In short, the Projects are where students learn to do physics well by, well, *doing physics*.

Students using the book as part of a course they are enrolled in can look forward to learning from the Projects during the semester. To make this crucial aspect of the course as accessible as possible to other readers, I have made Jupyter notebook versions of (most of) the Projects publicly available. Instructors (who are encouraged to borrow and adapt these materials for their own classrooms) and other readers (who can get at least some of the Projects' added value by working/reading through them as they work through the book) can find this rich set of bonus resources by following the link from `https://pigpugpress.com/forest/`.

Enjoy!

- Travis Norsen (July, 2021)

CHAPTER 0

Uncertainty

BOTH in the popular media and in textbooks, scientific conclusions are often presented in an unfortunately dogmatic way: you are told what is (supposedly) true, but the process by which that truth was established – often by the gradual accumulation of evidence supporting one of a number of competing reasonable hypotheses – remains obscure. One of the goals of this book is to give you a taste of a more realistic understanding of how science really works, by examining not just what contemporary physicists believe, but how these beliefs were built up over the centuries.

One of the important lessons of the physics we'll study is that, in any particular period of time, many things remain unknown and are often mired in (warranted and appropriate) controversy. Does the earth go around the Sun, or vice versa? Do the phenomena involving static electricity and electrical currents imply that there is one kind of electric fluid – or, instead, two? Magnets (and electric charges and celestial bodies) seem to exert forces on one another even though they do not touch. But how is this possible? How can a thing act where it is not? Is the familiar matter (of which tables, trees, people, and planets are made) composed, at the microscopic scale, of tiny indivisible particles, as some ancient Greek philosophers had speculated, or is it instead smooth and continuous like a fluid or field?

As we will see, for each of these questions (and many others), there was a time when the answer was uncertain, and there was a process through which new discoveries and evidence brought clarity.

Uncertainty, in this sense, will be a recurring theme throughout the book. But that is not what this preliminary chapter is about! Instead, in this pre-amble to our tour of the development of some of the most important ideas in physics, we will be discussing the narrower kind of uncertainty that is associated with measurements: why does every measurement have an associated uncertainty, how do we estimate the size of that uncertainty, how do we propagate it through calculations, and how do we take uncertainties into account when we compare measured quantities to each other and to theoretical predictions?

This chapter will help you build up a toolkit of skills, for dealing with uncertainty, by addressing each of these questions. There is a lot of material here, and much of it will probably be new and unfamiliar. But don't worry. You're not expected to fully master all of these topics in one week. We'll be revisiting these things, as we explore various experimental measurements, over the entire course of the semester. But it will be helpful to get all these ideas out on the table up front, and spend some time practicing some of them now, so we can focus on more interesting developments later on. So... let's jump in!

§ 0.1 Terminology and Other Preliminaries

Any time you *measure* something, there will be some uncertainty associated with the result. For example, news organizations often conduct polls to find out what fraction of the voting population would vote for a given candidate if the election were held today. They might report that the fraction, F, planning to vote for Candidate X is $F = 37\%$ and then note that the poll has a margin of error of 3%. To put this in a more standard form, they are saying that

$$F = 37\% \pm 3\% \tag{0.1}$$

which is just a simple way of expressing that they are pretty sure that the true value of F (which we might find out if, for example, the election really is held today!) lies somewhere in the range between $37\% - 3\% = 34\%$ and $37\% + 3\% = 40\%$.

And of course one sees this all the time in the sciences as well. For example, the currently accepted value for the mass m_e of the electron is

$$m_e = 9.10938356 \times 10^{-31} \text{ kg} \pm 0.00000011 \times 10^{-31} \text{ kg} \tag{0.2}$$

which just means that physicists are pretty sure that the true value of the mass (which, unfortunately, is not given to us in the back of any book!) lies bewteen $9.10938345 \times 10^{-31}$ kg and $9.10938367 \times 10^{-31}$ kg.

Let's first clarify a bit of terminology. A generic measured quantity – call it "A" – can be reported in the following form:

$$A = A_{best} \pm \delta A. \tag{0.3}$$

"A_{best}" is called "the best value of A", but this name is slightly misleading. It is really just the value which happens to lie in the *center* of the *range* that we are saying the true value of A probably lies in. So the "best value" is really no better, no more likely to be the true value, than any other value in that range – it just happens to be in the center.

The quantity "δA" that appears after the "$\pm$" sign is, as we have seen, "the uncertainty in A". To be a little bit more precise, δA is often called the "absolute uncertainty", to distinguish it from something called the "fractional uncertainty" that one also sometimes uses. The fractional uncertainty is defined as

$$f_A = \frac{\delta A}{A_{best}}, \tag{0.4}$$

i.e., it is just the ratio of the absolute uncertainty to the best value. (Sometimes this is reported as a percentage and called the "percentage uncertainty" but this is not really a different thing, any more than 0.17 and 17% are different things.)

It is important that A_{best} and δA must have the same *units* – otherwise it would be impossible/meaningless to add/subtract them. So, for example, the best value and (absolute) uncertainty of the mass of the electron, m_e, are both some number *of kilograms* (or some other appropriate unit in terms of which it is possible to measure mass, e.g., grams, slugs, or the mass of a proton m_p).

This implies that the fractional uncertainty f_A of any quantity is a pure (unitless) number, no matter what units A might have.

Using units to help check for mistakes is always a good idea, and it is no different when working with uncertainties. For example, suppose you have measured the mass m and acceleration a of a cart, and are going to calculate the force F that was causing this acceleration, using Newton's second law, $F = ma$. There is of course some uncertainty δm associated with your measurement of the mass, as well as some uncertainty δa associated with your measurement of the acceleration. What is the formula for the uncertainty δF on the force?

We will discuss this sort of question – how to propagate uncertainty through calculations – in more detail shortly. The point for now is that you don't have to even know anything about that topic to know that, for example,

$$\delta F = \delta m + \delta a \tag{0.5}$$

is *NOT* the correct formula! The right hand side asks you to add two quantities, one of which is some number of kilograms (or some other mass unit) and the other of which is some number of meters per second squared (or some other acceleration unit). But these quantities are incommensurable; they cannot be meaningfully added. So the above formula cannot be correct. On the other hand, another candidate formula

$$\delta F = m_{best}\, \delta a + \delta m\, a_{best} \tag{0.6}$$

does have sensible units: the two terms that are added on the right hand side are both some number of $kg \cdot m/s^2$ (the same thing as a "Newton"), so they can be added, and the result is a quantity with appropriate units to represent the uncertainty in a force. (We will see later whether or not this alternative formula is *correct*; the point for now is simply that it is not *obviously incorrect* in the way that Equation (0.5) is.)

One should always remember that something of the form

$$A = A_{best} \pm \delta A \tag{0.7}$$

is just an elegant way of expressing that A almost certainly lies between $A_{best} - \delta A$ and $A_{best} + \delta A$. Sometimes it is helpful to be able to refer to these "edge" values more directly, so let's give them official names:

$$A_{min} = A_{best} - \delta A \tag{0.8}$$

and

$$A_{max} = A_{best} + \delta A. \tag{0.9}$$

These relationships can also be inverted, giving

$$A_{best} = \frac{A_{min} + A_{max}}{2} \tag{0.10}$$

and

$$\delta A = \frac{A_{max} - A_{min}}{2}. \tag{0.11}$$

§ 0.2 Random Error

So far we've just been talking about how to talk about uncertainty. But why is there uncertainty in the first place, and what determines how big it is? There are a number of different possible answers, some of which are more or less important/relevant in different situations. Let's just review a few of these possibilities briefly to get the idea.

First, it could be that the quantity you are measuring is not really a fixed thing at all, but is rather something that intrinsically varies. For example, if you flip a coin 100 times, how many of those 100 flips will come up heads? There is no single number which is "the correct answer". (Note: it's not that we don't *know* "the correct answer"; rather, we know that there isn't one!) If you flip a coin 100 times today, you might get 47 heads, but then if you flip the same coin 100 times again tomorrow you might get 54 heads. It's just going to be different each time. And so, the best you can do, by way of answering the question, is going to be to report some kind of range. For example: the number of heads (when you flip a coin 100 times) is usually between 40 and 60, i.e.,

$$H = 50 \pm 10. \tag{0.12}$$

Similar questions that might similarly only be answerable by reporting a *range* of values include things like this:

- How much does it rain in Northampton in May?

- How many offspring does a queen bee produce during her lifetime?

- How many physics majors graduate from Smith College in one year?

OK, so one reason why it can be necessary to report not just a single value but instead a *range* of values – which is what one is doing when one reports an uncertainty – is that the thing one is measuring is something that happens multiple times, and happens at least somewhat differently each time.

But often there is uncertainty in a measurement even if the thing we are measuring is a non-repeating, stable thing. For example, the width of a certain table (let's assume) just is what it is. It doesn't fluctuate from one moment to the next. But suppose you measure it with a yardstick: you carefully line up one end of the yardstick with one edge of the table, then look over at where the other edge of the table lines up with the numbered marks. Maybe the yardstick has little tick marks for each tenth of an inch, and the other edge is about halfway between the "30.1 inch" and "30.2 inch" tick marks, like this:

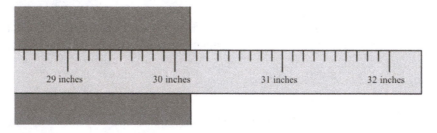

You are tempted to just say

$$W = 30.15 \, \text{inches}. \tag{0.13}$$

But in a situation like this – where you are interpolating, by eye, between the smallest marks on a measuring instrument, there is some uncertainty associated with that interpolation. You can tell, say, by looking, that the width is not 30.12 inches, but the edge being "about halfway" between the two nearest tick marks probably means that it could be 30.14 inches or 30.16 inches. So probably you should report that

$$W = 30.15 \, \text{inches} \pm 0.01 \, \text{inches}. \tag{0.14}$$

But, to continue the example, suppose you pick up the yard stick and hand it to your friend and ask her to measure the same quantity (the width of the same table). She plunks the yardstick down, carefully lines up one end of it with one edge of the table, and then looks to see where the other edge lines up with the tick marks. She finds that the other edge is about two thirds of the way between the "30.0 inch" and "30.1" inch marks and so, taking into account the uncertainty assciated with that interpolation, she reports that

$$W = 30.07 \, \text{inches} \pm .01 \, \text{inches}. \tag{0.15}$$

But now it should be obvious that your value of W contradicts hers! You were saying that W is between 30.14 inches and 30.16 inches, and she is now saying that W is between 30.06 inches and 30.08 inches, and there is no width value – no number of inches – that is simultaneously in both of those ranges.

The point is that probably both of you have *under-estimated* the uncertainty. Sure, you both took into account the uncertainty associated with interpolating between the smallest marks on the yardstick, but there are other additional sources of uncertainty. For example, there is a whole range of yardstick positions (relative to the table) for which you would say "the end of the yardstick is perfectly lined up with the left edge of the table". That is, you can only judge whether the edge of the yardstick is lined up with the edge of the table, with a certain finite accuracy. You *might* be able to do that with an accuracy that is about the same as the accuracy associated with the interpolation between the smallest tick marks on the yardstick. But it might also be less accurate if, for example, the edge of the table is rough or curved (so that it's not even clear exactly where "the edge" should be considered to be).

OK, so that's one additional source of uncertainty. And there are (potentially) many others. Maybe after you lined up the first end of the yardstick with the first edge of the table, you just left it sitting there as you walked around to the other side of the table, but you bumped the table slightly and jostled its position a little bit. Perhaps you weren't too careful about looking from exactly above, but were instead viewing the alignment between the edge of the table and the yardstick from some unknown angle to the side. Maybe the yardstick is made of wood, and expands/contracts slightly depending on the temperature or the humidity. And maybe a million other things that could be thought up.

The point is, there are lots of ways of accounting for the discrepancy between your initial measurement and your friend's. The thing to do in this situation is simply to revise your estimate of the uncertainty. Suppose you go on taking turns, measuring W again several times each, and you always find values that are between 30.0 inches and 30.2 inches. Then it would be reasonable to report that

$$W = 30.1 \text{ inches} \pm 0.1 \text{ inches.} \tag{0.16}$$

Why does this uncertainty exist? The lesson here is that there can be random fluctuations in various subtle and uncontrollable details about the measuring procedure itself. These fluctuations (which can be there even when the thing you are trying to measure is not itself fluctuating in any way) also prevent you from claiming any one precise value for the measured quantity; instead, you need to report a range of possible values, i.e., an uncertainty.

The practical upshot here is that measured values fluctuate due to various (known and unknown) sources of "random error". To learn the size of the uncertainty associated with these random fluctuations, the best procedure is simply to make a number of independent measurements to establish a range that the results seem to fall within. If one has just a few such measurements, it is reasonable to (say) take the largest value (from the set of individual measurement values) as A_{max} and the smallest value as A_{min}, and then (if needed) report a value for A in "standard form" by computing A_{best} and δA from Equations (0.10) and (0.11).

If one has a larger number of individual measurements, it is often appropriate to be a little bit more formal and utilize some statistical concepts. It is fairly intuitive that if one has N individual measurement values – A_1, A_2, ..., A_N – which differ due to random error, the *mean* is suitable to report as a best value:

$$A_{best} = \bar{A} = \frac{A_1 + A_2 + \cdots + A_N}{N} = \frac{1}{N} \sum_{i=1}^{N} A_i. \tag{0.17}$$

For the uncertainty δA we want some quantity that tells us the typical deviation of the individual measurement values from their mean. The "average deviation"

$$\bar{d} = \frac{1}{N} \sum_{i=1}^{N} d_i = \frac{1}{N} \sum_{i=1}^{N} \left(A_i - \bar{A} \right) \tag{0.18}$$

turns out to always be *zero*: the definition of the mean implies that there are (in an appropriate sense) just as many positive deviations as negative ones. So that doesn't tell us what we want.

The average of the absolute values of the deviations would work fine, but a more standard technique is to first *square* the deviations (so that one is working with a positive quantity for each data point), average those, and then take the square root at the end (to undo the squaring and give something with the right units). This is the so-called "standard deviation"

$$\sigma_A = \sqrt{\frac{1}{N} \sum_{i=1}^{N} \left(A_i - \bar{A}\right)^2}. \tag{0.19}$$

It is often suitable to use the standard deviation as an uncertainty

$$\delta A = \sigma_A \tag{0.20}$$

just on the grounds that "most" of the individual measured values (roughly 2/3 in normal situations) will lie between $\bar{A} - \sigma_A$ and $\bar{A} + \sigma_A$. If one prefers that, say, something more like 95% of the values should lie within δA of A_{best} one could instead take

$$\delta A = 2\,\sigma_A. \tag{0.21}$$

You should not lose too much sleep worrying about which formula for δA is right. At the end of the day, there is an ineliminable vagueness in the meaning of "uncertainty" – remember that the uncertainty is just a way of referring to the width of the range of values we are "reasonably sure" the true value of some quantity lies in. But what, exactly, does "reasonably sure" mean? There is no clean answer to that question, so in practice the thing to do is to do something that actually makes sense to you (as opposed to blindly following some recipe that you don't understand), explain clearly and explicitly to your audience what you are doing and why, and be ready to revise your uncertainty estimates as you learn more.

A few further notes are in order.

First, remember that any reported uncertainty (including the standard deviation of a set of repeated measurements) is a property not just of the quantity being measured, but also of the method used to measure it. For example, if you were to measure the width of the table using a giant pair of precision calipers, or a laser interferometer, you might end up with an uncertainty that was a thousandth or millionth of an inch, instead of a tenth of an inch. It's extremely important to get in the habit of estimating and reporting the uncertainties that your measurements do in fact have. That is, it's good to be open and honest with yourself and others about the accuracy of what you are doing. But it's also good to not lose sight of the possibility of revising your measurement methods to achieve better accuracy.

Second, there are actually two slightly different versions of the "standard deviation" formula (called the "sample" and "population" standard deviations). The one we gave above is easier to understand (it's just the square root of the average of the squared deviations of the individual values from the mean, i.e., it is the "root mean square" deviation). The other one is the same except that instead of adding up the squared deviations and, before taking the square root, dividing by N (to compute the average squared deviation), one instead divides by $N-1$. One way to understand the rationale for this alternative formula is as follows: if one only has a single data point, one really has no idea how much random error there might be, so it is kind of nice if σ comes out mathematically undefined (zero over zero) instead of coming out zero (which might fool you into thinking – ridiculously – that there is no uncertainty at all). That said, in practice, you should only ever bother to calculate the standard deviation if you have more than a few data points (certainly more than just one!) and as long as N is big, it doesn't really make much difference if you divide by N or instead $N-1$ inside the square root. So this is also not worth losing too much sleep over.

Third, in a situation where it is known that the individual values in a data set are fluctuating in a genuinely random way, and assuming that one is reporting the *mean* $\bar{A}$ of the individual values as A_{best}, it can be appropriate to report, as the uncertainty δA, not the standard deviation σ_A but instead the related but distinct quantity called the "standard deviation of the mean". This is defined as

$$\bar{\sigma}_A = \frac{\sigma_A}{\sqrt{N}}, \tag{0.22}$$

i.e., the standard deviation of the mean (SDOM) is just the regular standard deviation (SD) divided by the square root of the number of data points in your sample. The justification for this is a little too technical for us here, but it should make sense intuitively that, as you acquire more and more data, you improve the accuracy with which the quantity you are measuring is known, i.e., the uncertainty should decrease. (In general, the SD associated with a set of measurements performed in a certain way is roughly independent of how many measurements you make; but dividing by $\sqrt{N}$ produces something which will systematically decrease as N increases.)

However, if one is saying

$$\delta A = \bar{\sigma}_A = \frac{\sigma_A}{\sqrt{N}} \tag{0.23}$$

and N gets very very large, obviously the uncertainty δA will become very small. Indeed, it can often happen, in a situation where it is easy to collect lots of data points quickly, that if you use the SDOM as an uncertainty, your uncertainty will be *unrealistically* small, i.e., you will fool yourself into thinking you have achieved more accuracy than you in fact did. This is because, in practice, it is never true (or, at least, it is never *known*) that the individual values in a data set fluctuate in a perfectly random way. There is always, in addition to the random error, some "systematic error" – the subject to which we turn next.

§ 0.3 Systematic Error

A "random error" is, by definition, one which tends to average to zero across repeated measurements. Basically, on a given measurement attempt, the value is just as likely to come out a little too high, as a little too low.

But there are errors which always push the measured value in a certain direction, and hence do not tend to average to zero. These are called "systematic errors". To give a clear (but slightly ridiculous) example, suppose the yardsticks (that we were using to measure the width of the table in the earlier discussion) were misprinted at the factory: the tick mark at the very end of the stick is labeled "one inch" (instead of "zero"), the mark one inch from the end is labeled "two inches" (instead of "one inch"), etc. If we didn't notice this before, then all of our measurements of W would be *too big* by one whole inch. And no amount of averaging over many separate measurements would fix this.

You can see how, in a situation like this, it could be easy to really fool yourself about the accuracy of your measurements, and end up reporting that the true value "almost certainly" lies in some narrow range that in fact it is nowhere near. Of course, in this particular example, you might imagine that the factory misprint would be easy to notice and then correct for. And that's probably true. But unfortunately most systematic errors are less obvious. For example, suppose that the factory was very careful about printing correctly-labeled and accurately-spaced tick marks on the rulers, when they were made, but that the wood they used was freshly harvested and still somewhat moist. And suppose that in the following months, the wood dried out and contracted by 3% along its length. Well then every measurement made with the now dried and slightly-shrunken ruler will yield a value that is too big by about 3%.

This example is also slightly exaggerated and unrealistic, but systematic errors of this sort *are* always present and influencing measurement outcomes. There is unfortunately no way to simply eliminate them, and also no "magic formula" to tell you how big they are. So, in practice, what do you do? Here are several thoughts.

First, recognize that one obvious category of systematic errors (illustrated by the two examples above) involve imperfect *calibration* of measuring instruments. So the very first thing to say here is: don't forget to calibrate your instruments. This means, at very least, checking that it reads "zero" when it should read zero. And, whenever possible, it is good to check in addition that it reads some other, known, non-zero value when it should. A good rule of thumb to keep in mind here is that the ultimate, end-of-the-day accuracy of a measurement can never be better than the accuracy with which the measurement apparatus was calibrated.

A second general point is just that you should spend time understanding your experimental setup and equipment, thinking about what systematic errors might exist, and then working to either eliminate them or estimate their size and correct for them. For example, if you are using a stopwatch to measure the duration of some process, is there a reaction-time delay at the beginning? For example, your partner says "Go!" but it takes you half a second to push the "start" button. If you are able to hit "stop" at the correct time, the stopwatch readings might all be too short by about half a second. Perhaps you could eliminate the reaction-time systematic error by having the same person initiate the process (with one hand) and press "start" (with the other). Or if that's not feasible, maybe there's a way to correct for the systematic error – e.g., you could do a quick side experiment to *measure* your reaction time, and then add that amount of time to all the stopwatch readings. The point is, if you aren't being thoughtful and careful, more and bigger systematic errors will slip under the radar and impact your results.

A third point is that sometimes you can convert what might have been a systematic error, into a random error, and thereby get a handle on it using repeated measurements and statistics. Consider again the example of using a yardstick to measure the width of a table (and set aside for now the worries about the yardstick being miscalibrated). We discussed above how the alignment of the "zero" end of the yardstick with one edge of the table can affect the reading one extracts at the other edge of the table. If, for example, the "zero" end of the yardstick is hanging a tenth of an inch over the edge of the table, then whatever you read off at the far edge of the table will be too big by a tenth of an inch. If you just set the yardstick down a single time and then leave it in place, while, say, you and your friend take turns looking at the far edge from slightly different angles and reading values off, this imperfect alignment will introduce a systematic error. But if, instead, for each individual measurement, you physically lift the yardstick off the table, do the hokey pokey, and then set it back in place and try your best to line the "zero" end of the stick up with the edge of the table, before reading a value off at the other edge, the misalignment at that first edge is likely to be different each time. It may or may not average perfectly to zero over repeated trials, but at least now the imperfections of the "zero" alignment will have a statistical impact on the data set.[1]

When performing experiments, it can be very helpful to use a division-of-labor approach: one person releases the doo-dad, the other hits the "start button", etc. But the point here is that it can also be very helpful to explore what happens when you break out of this comfortable routine. Release the doo-dad with your *left* hand instead of your right. Sit on the *other* side of the table to hit the "start" button. Switch jobs with your partner. Re-do the calibration. Replace the string with a new one of the same length. And see to what extent the results change.

§ 0.4 Propagating Uncertainty

We often need to plug measured values into some mathematical formula. For example, suppose it is (somehow) known that the length L of a certain sheet of paper is precisely twice its width W. That is:

$$L = 2W. \tag{0.24}$$

---

[1]Incidentally, sometimes students dislike this suggestion on the grounds that it will make their uncertainty bigger, i.e., it will make their results less accurate, which is bad. But this attitude is completely backwards. The process described here might make the standard deviation, of your several individual measurement values, larger, but it does not diminish the accuracy of the measurement procedure at all. It simply makes the standard deviation better reflect the true accuracy of the measurement. The goal is not simply to *say* that the uncertainty – the width of the range you claim to be pretty sure the true value lies in – is small. Talk is cheap! The goal is instead to be *right* when you claim that the true value lies in some range!

If we have measured that

$$W = 11.3\,\text{cm} \pm 0.2\,\text{cm} \qquad (0.25)$$

it should strike you as reasonable that

$$L = 22.6\,\text{cm} \pm 0.4\,\text{cm}. \qquad (0.26)$$

After all, "$W = 11.3\,\text{cm} \pm 0.2\,\text{cm}$" *means* that W is (almost certainly) no smaller than 11.1 cm and no bigger than 11.5 cm. But then $L = 2W$ means that L is (almost certainly) no smaller than twice 11.1 cm, i.e., L is no smaller than 22.2 cm. And similarly, L is (almost certainly) no larger than twice 11.5 cm, i.e., L is no larger than 23.0 cm. But those two things together are just what Equation (0.26) *means*.

That may seem so simple and obvious that it is hardly worth explaining in such careful detail. But actually this example illustrates what could well be the one and only thing you really need to remember about propagating uncertainties: you can always figure out the uncertainty of some computed value by figuring out the biggest and smallest it could be (given the ranges of possible values for the inputs to the calculation).

Let's see how that principle plays out in a few simple cases. For example, suppose we measure A and B and then we need to compute

$$C = A + B. \qquad (0.27)$$

It is pretty obvious that the best value (i.e., the value in the middle of the range of possible values) of the sum of two quantities will just be the sum of their best values:

$$C_{best} = A_{best} + B_{best}. \qquad (0.28)$$

But what is the uncertainty in C? To find out, just consider the following: what is the *biggest* C could be? It is obvious, looking at Equation (0.27), that to make C as big as possible we should make both A and B as big as possible. That is,

$$C_{max} = A_{max} + B_{max} = (A_{best} + \delta A) + (B_{best} + \delta B). \qquad (0.29)$$

But then we know that the uncertainty in C is just the amount by which C_{max} exceeds C_{best}. So we can conclude that

$$\delta C = C_{max} - C_{best} = \delta A + \delta B. \qquad (0.30)$$

If we wanted, we could make that into a formal rule: the (absolute) uncertainty of a sum is just the sum of the (absolute) uncertainties of the things you are adding.

Here's another example. Suppose again that we measure A and B, but now we need to find their difference:

$$D = B - A. \qquad (0.31)$$

Again, it's obvious that the best value of D should be given by

$$D_{best} = B_{best} - A_{best}. \qquad (0.32)$$

But what is the uncertainty in D? Is it, perhaps, $\delta D = \delta B - \delta A$? Instead of guessing, let's use our technique. What is the biggest D could be? Looking at Equation (0.31), it is clear that to make D big as possible, we need to make B as big as possible but then make A (the thing we are subtracting from B) as *small* as possible. That is:

$$D_{max} = B_{max} - A_{min} = (B_{best} + \delta B) - (A_{best} - \delta A). \qquad (0.33)$$

So then apparently

$$\delta D = D_{max} - D_{best} = \delta B + \delta A. \tag{0.34}$$

Note that this is *not* the thing we guessed above, which (on further thought) wouldn't have made any sense. (The uncertainty on a difference should not be zero just because the two quantities we are subtracting have the same uncertainty!) We could again formalize this as a rule for propagating uncertainty through subtraction: the (absolute) uncertainty of a difference is the *sum* of the (absolute) uncertainties of the quantities you are subtracting.

Another commonly-encountered case is multiplication:

$$F = A \times B. \tag{0.35}$$

Obviously

$$F_{best} = A_{best} \times B_{best} \tag{0.36}$$

but what is δF? Well,

$$\begin{aligned} \delta F &= F_{max} - F_{best} \\ &= (A_{best} + \delta A) \times (B_{best} + \delta B) - A_{best} \times B_{best} \\ &= A_{best} \cdot \delta B + \delta A \cdot B_{best} + \delta A \cdot \delta B. \end{aligned} \tag{0.37}$$

We can make that look a little nicer by dividing both sides by F_{best} and then re-writing things in terms of the fractional uncertainties of each quantity. The result of this cleaning-up is

$$f_F = f_A + f_B + f_A \cdot f_B. \tag{0.38}$$

Usually it is the case that the fractional uncertainties are small (compared to one), so that the third term on the right – the product of the two fractional uncertainties – will be negligible compared to the first two terms. In that case, we can say that

$$f_F \approx f_A + f_B \tag{0.39}$$

which is arguably simple enough to formulate as a rule: the *fractional* uncertainty of a product is the sum of the *fractional* uncertainties of the factors.

We could go on like this, deriving separate rules for all kinds of different mathematical operations: addition, subtraction, multiplication, division, raising a quantity to a power, etc. Sometimes people find it very helpful to have all of these rules, and if that applies to you, I encourage you to derive them and use them. But I would encourage you, instead, to just use the "find the biggest possible value" technique that we've been using here to derive the rules. Using it prevents the need to memorize a bunch of different rules, and actually the technique is *better* than the rules: for example, Equation (0.39) is only approximate, whereas Equation (0.37) is exact.

Let me illustrate the technique I'm recommending one last time with a fully-concrete example which is likely to be more complicated than anything you'll actually encounter in the course. Suppose we have measured a time

$$t_1 = 3.7\,\text{s} \pm 0.3\,\text{s}, \tag{0.40}$$

another time

$$t_2 = 8.2\,\text{s} \pm 0.6\,\text{s}, \tag{0.41}$$

a mass

$$m = 43\,\text{kg} \pm 5\,\text{kg}, \tag{0.42}$$

and an angle

$$\theta = 87° \pm 6°. \tag{0.43}$$

And suppose for some reason (don't ask!) we need to calculate

$$Z = \frac{\sqrt{m}}{t_2^2 - t_1^2} \sin(\theta). \tag{0.44}$$

To begin with, we can calculate the best value of Z in the obvious way:

$$Z_{best} = \frac{\sqrt{43\,\text{kg}}}{(8.2\,\text{s})^2 - (3.7\,\text{s})^2} \sin(87°) = 0.122287\,\sqrt{\text{kg}}/\text{s}^2. \tag{0.45}$$

Now, to find the uncertainty in Z, we'll just compute Z_{max}. Looking at the structure of Equation (0.44), it's clear that to make Z as big as possible, we need to make m and $\sin(\theta)$ as big as possible, and we need to make the denominator as *small* as possible, which means making t_2 as small as possible and making t_1 as big as possible. Note that since the sine function reaches a local maximum at 90°, the maximum possible value of $\sin(\theta)$ – consistent with $\theta = 87° \pm 6°$ – is neither $\sin(81°)$ nor $\sin(93°)$, but is instead $\sin(90°) = 1$. (This sort of thing doesn't come up very often, but it's worth being aware that it might arise occasionally.)

Anyway, plugging everything in we have

$$Z_{max} = \frac{\sqrt{48\,\text{kg}}}{(7.6\,\text{s})^2 - (4.0\,\text{s})^2} \sin(90°) = 0.165905\,\sqrt{\text{kg}}/\text{s}^2 \tag{0.46}$$

which evidently means that

$$\delta Z = Z_{max} - Z_{best} = 0.043618\,\sqrt{\text{kg}}/\text{s}^2. \tag{0.47}$$

Usually that's all we would need to do, but let's press a little further just to raise a point that deserves to be discussed.

Note that by similar reasoning we could have computed

$$Z_{min} = \frac{\sqrt{38\,\text{kg}}}{(8.8\,\text{s})^2 - (3.4\,\text{s})^2} \sin(81°) = .092418\,\sqrt{\text{kg}}/\text{s}^2. \tag{0.48}$$

This then provides an alternative way to compute the uncertainty, namely

$$\delta Z = Z_{best} - Z_{min} = .029869\,\sqrt{\text{kg}}/\text{s}^2. \tag{0.49}$$

The point is that this comes out a little different from the first value we found for δZ. That is, Z_{max} is not quite the same amount higher than Z_{best}, as Z_{best} is higher than Z_{min}. That is, plugging the best values into a formula does not always give you the value at the middle of the range of possibilities for the result. The thing I kept saying above is "obvious" is not actually true! Sorry about that.

What should we *do* in the face of this fact? There are a couple of reasonable possibilities. One would be to forget the original Z_{best} that we calculated by just plugging in the best values of t_1, t_2, m, and θ, and instead use our Z_{max} and Z_{min} values to compute Z_{best} and δZ values using Equations (0.10) and (0.11). That is, instead of finding Z_{best} in the "obvious" (but not quite right!) way, and then finding Z_{max}, and subtracting to find δZ, this suggests that it might be better to always just compute Z_{max} and Z_{min}. This accurately defines the boundaries of the range that Z might lie in, and then we can always re-express this range in "standard form" with appropriately-calculated Z_{best} and δZ values.

That is in some sense the best version of the technique, but here's another pretty reasonable possibility. We could stick with the "obvious" (but not quite right!) Z_{best} value, and then report, as δZ, the *larger* of $Z_{max} - Z_{best}$ and $Z_{best} - Z_{min}$. That would have us reporting a range that, for sure, the true value of Z almost certainly lies in... it's just a range that's technically a little bigger than necessary. But that doesn't really do any harm. Of course, it would be slightly pointless to use this version of the technique, since it's more involved than the previously-described version. If you are going to calculate both Z_{max} and Z_{min} anyway, you might as well just do the previously-described thing, and not bother calculating also the "obvious" (but not quite right!) version of Z_{best}. But the point is, in practice, these two different versions hardly differ at all, and either one of them would be fine.

And there is a third thing that would also be just fine, namely, the thing we did first, where we just calculated Z_{best} in the "obvious" (but not quite right!) way, calculated Z_{max}, and then took δZ to be $Z_{max} - Z_{best}$. This, in practice, will differ only very slightly from the two previously-described versions of the technique, and so is usually just as good as either of them. And it has the benefit that it is a little easier: you get to use the "obvious" (but not quite correct!) version of Z_{best} which has the virtue of being very fast and easy to calculate: you just plug the best values of all the inputs into the formula. You still have to think a little bit to work out Z_{max} ("Hmm, to make Z as big as possible, should t_1 be as big as possible, or as small as possible, or something in between? What about θ?") but there'll only be this one calculation that requires a lot of thought.

So, to summarize, instead of deriving/learning/memorizing a bunch of separate rules for propagating uncertainties through various sorts of mathematical operations, use some version of the min/max/best technique that you are comfortable with. As to which exact version of the technique you should use, I simply don't care. If you do something that makes sense to you – and you explain what you are doing and why and show your work and do it correctly – we will all be happy.

§ 0.5 Rounding

The upshot of the example from the last section is that we could, at the end of the day, report the following:

$$Z = 0.122287\sqrt{\text{kg}}/\text{s}^2 \pm 0.043618\sqrt{kg}/\text{s}^2. \tag{0.50}$$

But... that should strike you as a lot of significant figures. Is it too many? Can we round the numbers off without losing any meaningful information?

Yes. To begin with, we have already said that the very concept of "uncertainty" – which is something like half the width of the range we are reasonably sure the true value lies in – is somewhat vague. What, for example, do we mean by "reasonably sure"? In some situations, this sort of vagueness could mean that the uncertainty is really only meaningfully defined to within a factor of 2, or perhaps even some bigger factor. In general, though, that's a little too extreme: there is often a meaningful difference bewteen "I'm pretty sure X is between 15 and 25" (i.e., $\delta X = 5$) and "I'm pretty sure X is between 10 and 30" (i.e., $\delta X = 10$).

But you'd be hard pressed to come up with a realistic situation where "I'm pretty sure X is between 15.0 and 25.0" (i.e., $\delta X = 5.0$) and "I'm pretty sure X is between 14.9 and 25.1" (i.e., $\delta X = 5.1$) were in any way meaningfully different assertions. That is, two uncertainties which differ by only a few percent should be regarded as literally meaning the same thing.

There's no particular sharp boundary between 100% (the difference between saying $\delta X = 5$ and saying $\delta X = 10$) and 2% (the difference between saying $\delta X = 5.0$ and saying $\delta X = 5.1$). But something of order 10% is kind of in the middle, so we can use that as a rough criterion: if two uncertainties differ by about 10% or less, they can be regarded as literally meaning the same thing, whereas if they differ by more than about 10%, they might mean different things.

This immediately suggests a convenient rule for rounding uncertainties: you can always safely round to two significant figures. Why? Because in rounding to two sig figs, you will always change the number by less than 10%. (Indeed, the most you could possibly change it by is about 5%.) And, if the number represents an uncertainty, we just agreed that this doesn't really change its meaning at all. So we might as well round it off. Indeed, in a lot of cases (namely, if the first significant digit in the uncertainty is "pretty big", like, say 7 or 8 or 9 as against 1 or 2) we could safely round to only one sig fig, and still be confident we weren't changing the number by more than 10%. So, by all means, round off the uncertainties to one or (at most) two sig figs.

What about the best value? The principle is the same: two ranges (that we might say some value probably lies in) should be considered as meaning the same thing if they overalap by something of order 90% (or more). So for the same reason it means the same thing to say either "X is between 15 and 25" or "X is between 14.9 and 25.1" (which was our basis for rounding the uncertainty to one or maybe two sig figs), we should also think of "X is between 15.1 and 25.1" as literally meaning the same thing as "X is between 15 and 25". That is, if we're saying $\delta X = 5$, there is really no point worrying about the difference between saying $X_{best} = 10.0$ and $X_{best} = 10.1$. The extra information in the tenths place just shifts the range over by a distance that is small compared to its width, so the two ranges almost completely overlap, and hence mean the same thing.

The rule that follows from this kind of consideration is: round the best value *at the same decimal place* where you rounded the uncertainty. So if $\delta X = 5$ (which has been rounded at the ones place) and some calculation spits out $X_{best} = 13.13732$, you should just round that to $X_{best} = 13$.

Note that rounding the best value at the same decimal place where you rounded the uncertainty *does not mean* the same thing as rounding the best value to the same number of sig figs (namely, one or two) to which you rounded the uncertainty. That would almost always be completely crazy. For example, suppose your fancy statistical calculations spit out

$$Y = 6{,}423{,}327.873 \pm 152.238. \tag{0.51}$$

Now, for sure, 6 sig figs in the uncertainty is excessive. We should round to 1 or 2 (probably 2 since the first significant digit, namely the "1" in the hundreds place, is "not big"), i.e., we should probably say

$$\delta Y = 150 \tag{0.52}$$

which, for the record, has been rounded at the tens place and now has two sig figs. Now, how should we round the best value? The correct thing is to round it also at the tens place, i.e., to say

$$Y_{best} = 6{,}423{,}330 \tag{0.53}$$

so that we are left with

$$Y = 6{,}423{,}330 \pm 150. \tag{0.54}$$

You should take a minute and convince yourself that the range of values described by Equation (0.51) and the range of values described by Equation (0.54) almost completely overlap, so they really do mean the same thing, and no harm has been done by rounding things off.

By contrast, if we instead rounded the best value to the same number of sig figs – namely, two – as we rounded the uncertainty, we would be left with

$$Y = 6{,}400{,}000 \pm 150 \tag{0.55}$$

which is a *completely different* range than the one described by Equation (0.51). If this isn't clear to you, draw it: draw a number line for Y, put some tick marks on it and label them, and then just shade in the ranges described by Equations (0.51), (0.54), and (0.55). You'll see that the range described by (0.55) doesn't overlap *at all* with the other two, and indeed, doesn't even come close to doing so.

§ 0.6 Comparing Values

It often happens that we want to know whether two measured values are the same or different. For example, if we change the mass of a pendulum from 200 grams to 400 grams, leaving everything else about it the same, does the oscillation period change?

We can answer this sort of question using a "t-test" as follows. Suppose the first value is

$$A = A_{best} \pm \delta A \qquad (0.56)$$

and the second is

$$B = B_{best} \pm \delta B. \qquad (0.57)$$

Then we can compute the quantity "t" which tells us about the extent to which the ranges for A and B overlap:

$$t = \frac{|A_{best} - B_{best}|}{\delta A + \delta B}. \qquad (0.58)$$

If $t = 1$, that means that the two ranges just bump up against each other, without overlapping, but also without any space in between. If $t < 1$, the ranges *do* overlap, and we can say that the two values are consistent with one another (which just means that there are at least some values that are within both ranges).

On the other hand, if $t > 1$, the ranges do not overlap at all, so A and B are not compatible.

In practice, though, we should not think of $t = 1$ as a sharp cutoff between two completely clear-cut situations, "Perfect agreement!" and "Contradiction!". After all, the uncertainties are only really meaningfully defined to a precision of something like 10%, i.e., the edges of the ranges corresponding to our A and B values are not themselves sharp. So what we should really say is that if t is smaller than about one, this suggests compatibility between the measured values of A and B, and if t is large compared to one, this suggests incompatibility between the measured values of A and B, and if t is "about 1" we are in a kind of in-between gray area where it is not really clear whether the values are compatible or not.

And actually it's a little unclear what to say even in the non-gray-area cases where t comes out smaller than 1 or substantially bigger than 1. For example, it could be that A and B are in fact completely different, but your experiments were not very precise (i.e., the uncertainties were large). So instead of saying that $t < 1$ indicates agreement, it is a little better to say that it indicates only the failure to detect a difference. Maybe the two quantities really are the same, or maybe you need to work a little harder (coming up with a more accurate measurement technique) to see the difference.

And similarly, a large t suggests, but does not conclusively establish, a difference: if one or both uncertainties were under-estimated (say, because some large systematic error went unrecognized), you might get $t = 100$ even though in fact A and B are exactly the same.

Still, despite the non-finality of the numerical t score, the t-test does at least give some helpful indication of whether the two quantities are likely different, or not.

§ 0.7 Goodness of Fit

Often we don't just have two values that we want to compare to each other, but instead we have a whole set of data that we want to compare to some theory or mathematical model.

For example, suppose we gather some data to test the hypothesis that the average air-speed velocity v of (African) swallows is proportional to their mass m. We go out and measure v and m for N randomly-chosen birds, and so have a collection of data points of the form $(m_i, v_i^{best}, \delta v_i)$ for $i = 1$ to N. (We imagine here, plausibly in my opinion, that the uncertainty on the masses is negligible, whereas the uncertainty on a given bird's "average air-speed velocity" – whatever that even means – is significant!)

The hypothesis can be represented by the equation

$$v = km \tag{0.59}$$

where k is a constant. The question is whether there is some value of k – or more precisely some *range* of values of k – for which this equation is consistent with the data.

For a particular value of k, the "theory" – Equation (0.59) – will "predict" what v should be for a given mass. For example, for the first mass value that we happen to have in our data set, m_1, there will be some theoretically predicted velocity, $v_1^{pred} = km_1$. (Note that if there is non-negligible uncertainty on the mass, this will imply a non-negligible uncertainty on v_1^{pred}, but for simplicity we set this possibility aside here.) The point is now that we can compare v_1^{pred} with the *measured* velocity (for that mass) with a t-test:

$$t_1 = \frac{\left| v_1^{best} - v_1^{pred} \right|}{\delta v_1}. \tag{0.60}$$

And of course we can do the same for all N of the data points. We can then *average* these individual t-scores, to get an overall sense of how well the theory (again, for some particular value of k) matches with the data:

$$G = \frac{1}{N} \sum_{i=1}^{N} t_i. \tag{0.61}$$

Note that what we are calling "G" (for "Goodness-of-fit") is closely related to the slightly-fancier (but basically equivalent) thing you would get by taking the square root of the so-called "χ^2 (chi-squared) per degree of freedom".

So far we have assumed a fixed value of the "fit parameter" k. But the t-scores for each data point, and hence their average, G, will depend on what value of k we pick. So the idea is to let k vary and to see whether or not there is some range of k values for which the fit is good, i.e., for which G is about 1 or smaller. If not – if there are no k values for which G is about 1 or smaller – then we conclude that the theory is incompatible with the data (subject, of course, to the same sorts of caveats that were discussed in the last section).

On the other hand, if there is some range of k values for which G is about 1 or smaller, then we can say that our data *is* compatible with the theory for a certain now-known range of k values. That is, we can learn the value (with uncertainty) of k that makes the theory fit the data.

All of this makes a little bit more sense graphically, so let's see how it plays out with some pictures. Here is a plot of some (made-up) data for the above example:

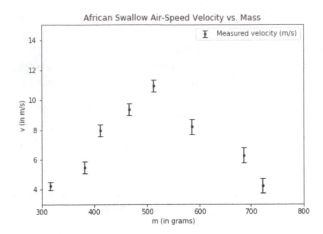

And here is a plot of the same data where we have included a graph of the theoretical formula $v = km$ for $k = 0.015$ m/s per gram, and also computed the G score for this particular value of k:

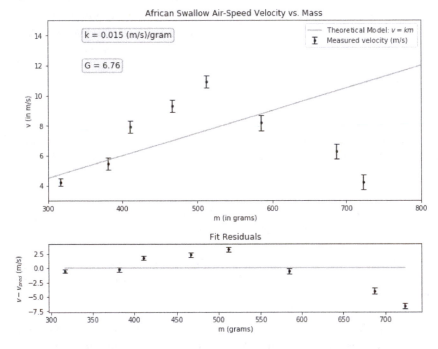

As you can see, the G score comes out between 6 and 7. Let's think through what this means. Each individual data point (in black) is a certain (vertical) distance from the theoretical formula (gray). The t-score for each individual data point is just this (vertical) distance, measured in "error bar size" steps. For example, for the first data point, it looks like the t-score is about 2 (because the vertical distance from the black dot to the gray line is about twice as big as the error bar on that data point). For the second data point, it looks like $t \approx 0.8$. For the third, it looks like $t \approx 3$. And so on.

Anyway, it is clear that the theoretical model (represented by the gray line) goes through or at least near *a few* of the data points, but it completely misses others. Just by eye, it seems like the t scores for the individual data points vary from about $t \approx 0.8$ (for that second data point whose error bar the gray line goes through) to about $t \approx 10$ (for the last data point). So it makes sense that the averge, the goodness-of-fit parameter G, comes out something like 6 or 7. And, of course, a G of 6 or 7 means that the data is really just not compatible with the theoretical model. The gray line does *not* do a good job of going through or near most of the data points!

So evidently the model $v = km$, for this particular value of k, is inconsistent with our data. And it is pretty clear from the graph that no other value of k (corresponding to a gray line with a different slope) will work any better. In fact, this particular value of k turns out to provide the *best fit* to the data (i.e., the smallest G score) that is possible for this model. So in this case we would conclude that the model $v = km$ is inconsistent with our data. (If we really trust our data, that means the model is wrong! But of course we should always consider the possibility that we have underestimated our uncertainties.)

Note the second graph, underneath the main one, that I included above. This is a plot of the so-called "residuals" for the fit. This just means that it is a plot of the *difference* between the measured and theoretically-predicted v values. It basically contains the same information as the main graph above, but sometimes the shape of the data overall makes it hard to see the detailed relationship between each data point and the model curve on the main graph, and plotting the residuals can give you a clearer visual perspective on how well the model fits the data.

OK, so that example illustrated the case where the data is just not compatible with the model. Let's illustrate the other case too. So, suppose that you now go out and measure the average air-speed velocity of *European* swallows, plot the data, and perform the same sort of curve-fit to the model $v = km$:

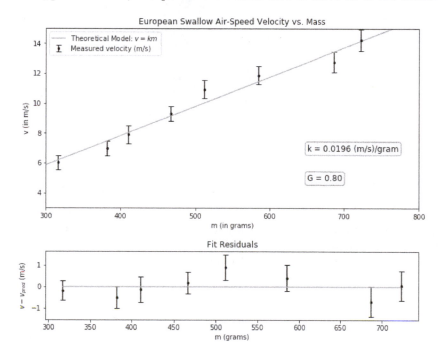

For this new data set, the best fit is now provided by $k = 0.0196$ (m/s)/gram, and the G score (of 0.8) for this value of k confirms what we can see in the graph: the fit is quite good!

(On the main graph, the fit being "quite good" is reflected in the fact that the gray line goes through, or at least almost goes through, almost all of the error-bar-ranges for the black data points. On the residuals graph, the fit being good is reflected in the fact that almost all of the residuals are compatible with zero, which is represented by the horizontal gray line on that graph, and there is no obvious structure to them: a few are a little above zero and a few are a little below zero, but things seem random in a nicely symmetrical way.)

It is legitimate to conclude at this point that the data is consistent with the model. We should not, however, conclude that

$$k = 0.0196 \, \text{m/s per gram} \tag{0.62}$$

because actually there is not just this one single k value that produces a good match to the data, but instead a whole *range* of k-values. We can find the width of this range by increasing k until the fit becomes "no longer good", as indicated, say, by G becoming bigger than 1 (or maybe 2 or something else in that ballpark).

For this particular data set, and taking (somewhat arbitrarily) $G < 2$ as my threshold for what counts as an "acceptably good fit", I find that k can be as large as 0.0217 (m/s)/gram:

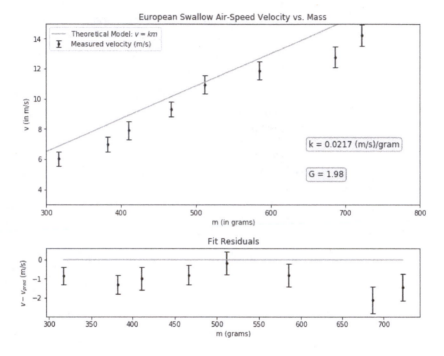

You can see that, for this value of k, the fit is only sort of borderline-decent: the model is at least *near* all of the data points, but it tends to miss all of them on the high-side. That is, the residuals are mostly reasonably close to zero, but they are almost all negative. That is what we should expect for the edge case.

Anyway, taking

$$k^{max} = .0217 \, \text{m/s per gram} \tag{0.63}$$

and

$$k^{best} = 0.0196 \, \text{m/s per gram} \tag{0.64}$$

we can conclude that, for European swallows, the model $v = km$ is compatibe with the data for

$$k = (0.0196 \pm 0.0021) \, \text{m/s per gram} \tag{0.65}$$

or, rounding things off as discussed earlier,

$$k = (0.020 \pm 0.002) \, \text{m/s per gram.} \tag{0.66}$$

§ 0.8 Summary

No experiment is infinitely accurate. So whenever you measure something – or somehow compute or infer something based on measured values – you are really constructing a case that the quantity in question lies in a certain finite *range* of possible values. Properly understanding and explaining the implications of your work thus requires knowing and reporting this range. Typically this is most effectively done by finding and reporting the "best value" (which identifies the center of the range) and the "uncertainty" (which is half of its width). There are many detailed techniques for calculating, estimating, propagating, and rounding uncertainties, but all of them are easier to understand and properly implement if one just remembers that an uncertainty always means (half) the width of the range in which we are reasonably certain the true value of the quantity in question lies.

Exercises:

E1. Use an actual meter/yard stick to measure the width of an actual table. You should, of course, think carefully about how to do this so you can confidently and accurately report both the best value and the absolute uncertainty of your result. Compare your result with another person or group who measured the width of an identical table and discuss and resolve any discrepancies.

E2. Show that the "average deviation" defined in Equation (0.18) really is identically zero as claimed in the text.

E3. Derive a formal rule for propagating uncertainty through division: if $C = A/B$, what is f_C (the fractional uncertainty of C) in terms of f_A and f_B? The exact result is not very pretty, but you can make it look nicer by using the binomial approximation: for $x \ll 1$, $(1+x)^n \approx 1 + nx$, so, for example, if f_B is small, then $1/(1 - f_B) \approx 1 + f_B$.

E4. By what percentage do you change $X = 85.4791$ if you round it to two sig figs? How about $X = 25.4791$? What is the biggest possible percentage change when rounding a number to two sig figs?

E5. Draw an appropriately-scaled number line and shade in the three ranges from Equations (0.51), (0.54), and (0.55). Also draw a zoomed-in sketch of the part of the number line that allows you to see in detail the extent of the overlap between the ranges in Equations (0.51) and (0.54).

Projects:

P1. Make a pendulum whose length L is exactly one meter with some bob-mass m. Measure, as carefully as you can, its oscillation period T when it swings with small amplitude. Compare your result, using a t-test, with the result of another person or group whose one-meter-length pendulum had a different bob-mass. Now systematically explore the dependence of the period T on the length L: measure T for (roughly) ten different L values between a few centimeters and a couple of meters. Plot your data and see if you can find a relationship of the form $L = kT^n$ (for constants k and n) that fits the data well.

Part I

Astronomy and Gravity

CHAPTER 1

Solar System Models

T HE first part of our tour through the historical development of physics will lead up to Newton's discovery of the laws of motion and universal gravitation. Newton arrived at his theory of gravity and his articulation of the laws of motion primarily by reflecting on the work of two astronomers from the previous generation, Galileo and Kepler , who, in turn, were early champions of the revolutionary idea that had been put forward by Nicolas Copernicus, who aimed to overthrow the geo-centric (Earth-centered) model of the solar system that had dominated since the time of the ancient Greeks. So we will begin at the beginning, with that ancient geo-centric worldview, articulated especially by the astronomer Ptolemy.

But before diving in, perhaps it is worth acknowledging that Ptolemy's model is famously and spectacularly *wrong* – so you might reasonably wonder why we would bother learning about it. Part of the answer is that, in this course, I want to help you develop a realistic, big picture appreciation for the development of scientific ideas, and it is just a fact that scientists are not always and automatically right about everything from the very beginning. Things have to be figured out, sometimes from a state of sheer ignorance and sometimes from a state of prior belief in some wrong or confused idea. This figuring-out process involves messy complications and controversies. It is very different from the "sterilized" version of science that is typically presented in textbooks. And getting accustomed to this, by studying some history, will make you a sharper, more sophisticated scientist who is better prepared to navigate the history of tomorrow, i.e., the scientific frontier of today.

We will also use our study of the ancient models of the solar system as an opportunity to develop some (21st century) modeling tools as well as some mathematical technology (having to do with angular kinematics) that will prove very practical in this course and beyond. So, let's jump in!

§ 1.1 Basic Observations and the Ancient Cosmology

The cosmological view of the Ancient Greeks, roughly sketched in Figure 1.1, had a lot of features that we, living in the 21st Century, are trained, from an early age, to regard as rather silly. They thought the Earth was at rest, in the center of the universe, with the stars in effect painted on the interior of a big spherical shell which rotated around us each day. They thought the Sun and the Moon and the 5 planets they knew about (Mercury, Venus, Mars, Jupiter, and Saturn – these are not shown in the Figure) orbited around the Earth in various complicated ways. They believed everything below the orbit of the Moon was made of just four terrestrial elements (earth, air, fire, and water), governed by a simple law of physics: everything just wanted to move to its "natural place" (as close to the center of the universe as possible) and then stay there. And they believed that the heavenly bodies (the Moon, Sun, 5 planets,

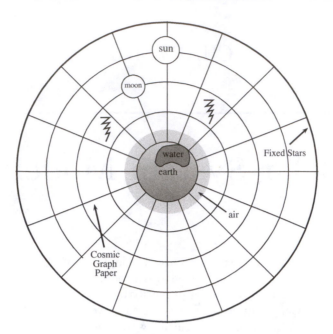

Figure 1.1: Simple sketch of Aristotle's universe. The matter (earth, water, air, fire, and aether) arranges itself relative to a fixed background space (the "cosmic graph paper") with the earth clumped up nearest the center, then water, then air, then fire, and finally (above the sphere of the Moon) the aether making up the Sun, stars, and other heavenly bodies (not shown).

and stars) were made of a fifth element ("aether") for which the "natural motion" involved rotating in perfect, never-ending circles.

To us, it sounds crazy. But if you set aside all the things you've been told, and just go by what you can directly observe, many of these ideas start to seem less absurd and can even begin to strike you as rather plausible. For example, rocks and other heavy things *do* seem perfectly happy to just sit on the ground, evidently as close to the center of the earth as they can get. If you intervene by lifting one up and then letting go, it will spontaneously return to its seemingly natural place, and then simply remain there until or unless you intervene again. Rest – stillness – really does seem like the natural state for inanimate objects (and even, ultimately, for us animate ones) in the terrestrial realm. As to the Earth itself being at rest: does it *feel* to you like it's moving? Of course not.

And things in the heavens do seem very different. The Sun rises and sets each day and moves across the sky in a big circular arc in between. (Notice that the very terminology of "rising" and "setting" implies that the Sun is moving while the Earth remains at rest.) The stars retain fixed positions in the nighttime sky (allowing us to identify constellations) and do indeed appear to constitute a giant shell that rotates around us: see Figure 1.2.

You might have heard that it was explorers (such as Christopher Columbus) in the 15th century who first hypothesized that the Earth was spherical rather than flat. This is not true. As reflected in Figure 1.1, the ancient Greeks knew that the world was round. They based this conclusion on perfectly reasonable evidence, e.g., that the masts of departing ships remain visible even after the body of the ship has dropped below the horizon (and similarly, inland mountains are visible, from a ship approaching land, before the coastline); new stars become visible as one travels south and the angle of the Sun (and various stars) above the horizon also varies with (what we now call) latitude in just the way you'd expect if the Earth were spherical; and finally, during a lunar eclipse the Earth's shadow (cast onto the face of the moon) is

Figure 1.2: A long-exposure photograph, taken at night with the camera facing north, will reveal circular "star trails" as shown here in artist's conception. It is as if the stars are dots of light, painted on the interior of an enormous hollow rigid sphere which rotates around the Earth once per day, about an axis that goes through the Earth's north and south poles. The fixed dot in the image, about which the other stars appear to move, is Polaris, i.e., "the north star". (Google "star trails" to see some real, and quite beautiful, images of this sort.)

visibly *circular*.

The Greeks had also determined the *size* of the spherical Earth quite accurately, and had even measured (less accurately, but still using ingeneious methods that in principle are perfectly valid) the distances to the Moon and Sun. (You can explore these methods in this week's Exercises.)

In short, the ancient Greeks were far from stupid, and their geo-centric (Earth-centered) picture of the universe – despite turning out to be wrong – was not absurd or crazy. We should respect it as a plausible first attempt at making sense of the large-scale structure of the world and our place in it.

§ 1.2 The Sun

Let's study, in a little more detail, the way the Sun appears to move. The first thing to say is that the daily rising and setting of the Sun is closely related to the motion of the stars, which also rotate daily. On any particular day, if you look east just before sunrise and see a certain constellation, you will see that same constellation to the west just after sunset. That is, it appears that the Sun basically just maintains a fixed position on the shell of fixed stars as that shell rotates around us.

But if you watched carefully over several days and weeks, you would notice that actually the Sun "slides" or "drifts" gradually, to the east, relative to the stars. In particular, each day the sun rises and sets about four minutes *later* than the day before, relative to the stars: if, on a given morning, you are looking to the east just before sunrise and notice a certain group of stars that is just visible on the horizon before the sunlight makes it impossible to see them, then, on the following morning, you could see that same group of stars rising slightly above the horizon for four minutes before the sunrise rendered them invisible.

This same fact could also be put in the following way: the stars appear to rotate around us not once per day, exactly (a day being a period of time defined in terms of the motion of the Sun), but rather once every 23 hours 56 minutes. So on two successive days (on, which, for simplicity, sunrise occurs at the

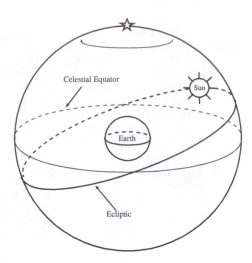

Figure 1.3: The two-sphere model of the universe, including the Sun and the ecliptic (the Sun's path through the fixed stars). Note that the ecliptic and the celestial equator can be thought of as circles – specifically, intersections of the sphere of fixed stars with a plane passing through the center of the earth. The plane that cuts through the celestial equator and the plane that cuts through the ecliptic make approximately a 23.5° angle with one another. Thus, during the course of the year, the Sun is as much as 23.5° north of the celestial equator, and as much as 23.5° south of the celestial equator. It reaches these two extremes on the Summer and Winter Solstices, respectively, and goes back and forth between them in between, crossing the equator on the spring and fall equinoxes. The star at the top – the "North star" – corresponds to the center point of the circular star trails in Figure 1.2. For us today this is the star called "Polaris", although due to a subtle long-term variation called the "precession of the equinoxes" – which, amazingly, the Ancient Greek astronomers already knew about – it was a different star in their time and will be different again in the future.

same time) the shell of fixed stars will have rotated all the way around once *plus a little bit* during the time – one day exactly – between successive sunrises.

It turns out that this gradual motion of the Sun – with respect to the background of fixed stars – is also cyclic, with a period that is (by definition) one year. That is, we define the "year" to be the amount of time it takes the Sun to gradually wander its way, through the 12 constellations of the Zodiac, and return to its initial location. The exact path that the Sun takes through the stars, over the course of a year, is called the "ecliptic" and it can be helpfully visualized as a circle on the shell of fixed stars. See Figure 1.3.

It is important that the ecliptic is tilted, by an angle of about 23.5°, relative to the celestial equator (which is just the points on the shell of fixed stars that are directly above points on the Earth's equator, i.e., all of the points on the shell of fixed stars that are 90° away from the north star). This tilt of the ecliptic relative to the equator gives rise to the seasons.

On the days we call the vernal (spring) and autumnal (fall) equinoxes, the Sun is at the places along the ecliptic where the ecliptic and the celestial equator intersect. On these days, the Sun appears to move around a circular path that lies just above the equator. So the sunrise occurs directly to the east (see Fig. 1.4) and the sunset occurs (exactly 12 hours later) directly to the west. If one lives on the equator, the Sun will be direcly overhead ("straight up") at noon. More generally, the noontime elevation of the Sun above the horizon to the south will be the complement of the latitude at one's location. So for example, if one is at 30 degrees North latitude (say, in the southern USA), at noon on the equinox the Sun will be

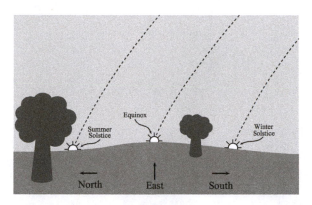

Figure 1.4: On the spring and autumn equinoxes, when the Sun is on a point on the ecliptic that is also on the celestial equator, the Sun rises above the horizon in the morning directly to the east. On the summer solstice, when the Sun is at the point on the ecliptic that is maximally north of the celestial equator, the Sun rises considerably north of due east. And on the winter solstice, when the Sun is at the point on the ecliptic that is maximally south of the celestial equator, the Sun rises considerably south of due east.

60 degrees above the horizon to the south (i.e., 30 degrees down, toward the south, from straight up).

During the half of the year between the vernal and autumnal equinox – i.e., during the summer – the Sun is moving along the part of the ecliptic that is *north* of the celestial equator. This deviation reaches a maximum of 23.5° on the summer solstice. During the summer, the Sun rises north of due east, sets north of due west, and is up for more than 12 hours. It is also more directly overhead during the daytime. For example, at 30 degrees North latitude, the Sun is a full 83.5° above the horizon to the south (i.e., a mere 6.5° down from straight up) at noon on the summer solstice.

During the other half of the year, the Sun is *south* of the celestial equator (a full 23.5° south on the winter solstice), so the time between sunrise and sunset is shorter, and the Sun's arc across the sky is lower: sunrise occurs *south* of east, the elevation of the Sun at noon is *smaller* than the complement of one's latitude, and sunset occurs *south* of west.

Some of what we've just been saying applies only for middle northern latitudes. You should try to stretch your thinking by figuring out how to use the two-sphere model to understand how the seasons work if one is, for example, north of the Arctic circle (66.5° north latitude), or south of the Tropic of Cancer (23.5° north latitude), or in the Southern hemisphere.

We've been discussing how the Sun moves relative to the rotating shell of fixed stars, over the course of a year, along a particular circular path called the ecliptic. Let's now get a little more quantitative about its motion. We can describe the position of the Sun along the ecliptic by giving its angle, $\phi(t)$, east of some arbitrarily-chosen reference location (such as its location on the vernal equinox). As a rough approximation, then, we could say that

$$\phi(t) = \omega t \tag{1.1}$$

where ω – the ecliptic angular velocity – is 2π radians/year, or just a little less than one degree per day.

But more careful observation of the Sun's motion reveals that its ecliptic angular velocity

$$\omega = \frac{d\phi}{dt} \tag{1.2}$$

is not exactly constant. During half of the year the ecliptic angular velocity is a little bit higher than 2π radians per year, and during the other half it is a little bit lower. See Figure 1.5.

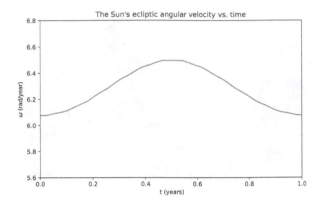

Figure 1.5: Over the course of a year, the Sun's ecliptic angular velocity varies above and below its average value of 2π radians/year.

Explaining exactly how the Sun moves along the ecliptic (including its not-quite-constant rate of progression) will be a major focus of the modeling efforts you'll learn about later in this reading and then work with in class. But before turning to that, let us quickly survey the qualitative motions of the Moon and 5 Planets.

§ 1.3 The Moon and the 5 Planets

In all but a few details, the motion of the Moon is exactly like the motion of the Sun. The Moon shares the (rough) daily motion of the Sun and stars, and also shares with the Sun a more subtle motion with respect to the stars: it, too, moves (approximately) along the path through the stars called the ecliptic. The only difference is that the Moon moves eastward along the ecliptic *faster* than the Sun. Where the Sun takes a whole year to complete its circuit around the ecliptic, the Moon takes only about a month (27.3 days to be precise).

A few more words are in order about the parenthetical "approximately" in that last paragraph. The ecliptic is *defined* as the path of the Sun across the background of fixed stars, so the Sun's yearly journey through the stars lies *precisely* along the ecliptic, by definition. The Moon, on the other hand is always seen *near* the ecliptic, but it deviates during its monthly cycles by a couple of degrees on either side. We could think of the angle "ϕ" introduced previously as the "ecliptic longitude", and describe deviations to the north or south of the ecliptic in terms of another angle, $\psi(t)$, called the "ecliptic latitude". Then we would say that while the Sun's ecliptic latitude is always zero, so its motion relative to the stars is perfectly captured by its ecliptic longitude $\phi(t)$, the Moon's motion is more complicated, and involves non-trivial $\phi(t)$ and $\psi(t)$. In general, the Moon's motion is complicated enough that we will set it aside and not study it in detail. But what we've just said for the Moon applies to the 5 planets as well. So just be aware that when we study the planets, we'll be making the simplifying assumption that the small but nonzero latitudinal deviations from the ecliptic are negligible, and that we can basically understand most of what's important by focusing on the ecliptic longitude $\phi(t)$.

The Moon exhibits one other unique feature, too: phases. During its monthly circuit around the ecliptic, the Moon alternates between "full" (when the entire circular disk of the Moon is illuminated) and "new" (when the entire circular disk is dark). These phases are readily explainable by the assumption that the Moon's light is not intrinsic, but is rather reflected light from the Sun. The Moon presents as full when it is just opposite the Sun in the sky, such that, from here on earth, it is precisely the bright side of the Moon (the side facing the Sun) that is visible. New Moon occurs when the Moon is very close to the Sun in the sky, meaning that the side of the Moon that is illuminated faces away from earth, with only the dark side being "visible" from here. And so forth for all of the intermediate (crescent, half, gibbous) phases. Note that this explanation requires that the Moon be closer to the Earth than the Sun. The ancient Greeks, incidentally, already understood all of this; see again the relative positioning of the Sun and Moon in Figure 1.1.

It is worth mentioning here another occurrence involving the Moon: eclipses. There are two types. A lunar eclipse happens when the Moon passes through the shadow cast by the earth and thus appears dark for a short period of time right around full Moon. (Do you see why a lunar eclipse can only happen at full Moon? It doesn't happen *every* full Moon because the Moon's path isn't *exactly* along the ecliptic – it is rather within a couple of degrees of the ecliptic, but this small deviation is enough that most of the time it doesn't pass directly through the earth's shadow.) The other type of eclipse is a solar eclipse. This occurs when the Moon gets right between the earth and the Sun, so that the view of some or all of the Sun is blocked. And, again, this doesn't happen *every* time there is a "new" Moon, because the Moon's path is only roughly along (within a couple of degrees of) the ecliptic.

The Moon and Sun have several things in common as against the stars. First, unlike the stars, they are not *fixed* in their positions relative to (other) stars. Rather, they move (more or less) slowly through the stars, along the ecliptic. And second, the Sun and Moon just look different than stars: stars look like little points of light, while the Sun and Moon both present a large disc.

Careful observation of the heavens, however, reveals several additional objects which *look* like stars (in the sense of the second point just mentioned) but which have the first point in common with the Sun and Moon. That is, these objects look like little points of light (though they are typically as bright as some of the brightest stars), yet their positions are not fixed. Like the Sun and Moon, they *wander*. There are five such "planets" (from the Greek word for "wanderer") that are visible to the naked eye and hence were known about by the Greeks: Mercury, Venus, Mars, Jupiter, and Saturn. Actually, not surprisingly, the Greeks tended to think of all *seven* of the wandering objects we've talked about as "planets."

Not only do these additional five planets, like the Sun and Moon, wander – they wander in much the same way. Each of them (in addition, of course, to sharing the daily rotation of the stars) moves with a roughly-steady eastward drift (approximately!) along the ecliptic. Mercury and Venus each move around the ecliptic in (on average) one year, just like the Sun. The other three planets take longer: about two years for Mars, about twelve years for Jupiter, and about thirty years for Saturn.

The Greeks basically just assumed that the correlation between distance-from-earth and ecliptic-period which held for the Moon and Sun, continued to hold for the other planets as well. So they inferred that the seven planets had distances from earth in the following ascending order:

- Moon

- Mercury, Venus, Sun (order ambiguous!)

- Mars

- Jupiter

- Saturn

Note that Mercury, Venus, and the Sun cannot be placed unambiguously on this list because they all take, on average, exactly one year to go around the ecliptic.

To understand why it is necessary to say "on average" we must clarify a further important detail about the observed motion of the planets. Whereas the Sun and Moon *always* move eastward along the ecliptic,

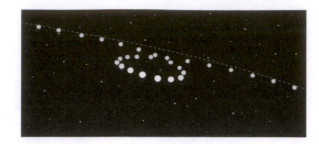

Figure 1.6: The motion of Mars, relative to the background stars, during the several-month period surrounding a retrograde episode. Here Mars starts at the right and initially moves to the left (eastward) along the ecliptic (indicated here by the dashed curve). But the eastward motion eventually slows and stops and Mars moves for a couple of months to the west (this is the so-called "retrograde motion") before again turning around and resuming its normal motion to the east along the ecliptic. Note that the (always fairly small) deviation of the planet from the ecliptic (that is, its ecliptic latitude which we largely ignore in the discussion in the text) is greatest, and also that the planet becomes noticeably brighter, during the retrograde episode.

the five planets only do this most of the time. They also occasionally stop their eastward drift, move for some time *to the west*, and then stop and return to their normal eastward motion. See Figure 1.6 for an illustration.

This bizarre behavior is referred to as "retrograde" (backward) motion. Each planet retrogrades at regular, periodic intervals, but the time between retrograde episodes – the so-called "synodic period" – varies from planet to planet. On average, Saturn does it every 378 days, Jupiter every 398 days, Mars every 779 days, Venus every 584 days, and Mercury every 116 days. So there is no obvious correlation here between the (distance) order of the planets and their frequencies of retrograding.

Another curious feature is that the planets are not uniformly bright. A given planet (say, Mars) is sometimes brighter and sometimes dimmer than its average brightness, and (curiously) the planets Mars, Jupiter, and Saturn achieve their maximum brightness just as they retrograde.

It is a little harder to determine the brightnesses of Venus and Mercury, since both planets are always near the Sun in the sky. Think of this in terms of their motion along the ecliptic. Most of the time, Venus moves eastward along the ecliptic at a rate just a little faster than the rate of the Sun. But then, when its ecliptic longitude ϕ gets about 45° ahead of the Sun's, Venus reverses direction and moves for a time *westward* along the ecliptic (i.e., with a *negative* ecliptic angular velocity), until it is about 45° behind the Sun, at which point it resumes its eastward motion. It's important that the motion is "centered" on the Sun: the "normal" eastward motion of Venus along the ecliptic always has it catching up to and then overtaking the Sun, and then it passes it again in the other direction as it retrogrades, only to start over again (584 days later). Mercury does basically the same thing, only it goes back and forth faster, and doesn't get as far away from the Sun on either side.

This somewhat complicated motion of the planets relative to the stars is indicated in Figure 1.7, which shows a graph of the ecliptic longitude of Mars over the course of several years. Note in particular the relatively-steady increase of the ecliptic coordinate (ϕ_{ecliptic}) with time – corresponding to the relatively-steady eastward drift of the planet along the ecliptic – punctuated, every couple of years in the case of Mars, by a retrograde episode, in which the planet moves, for a short time, "backwards" (i.e., westward) along the ecliptic.

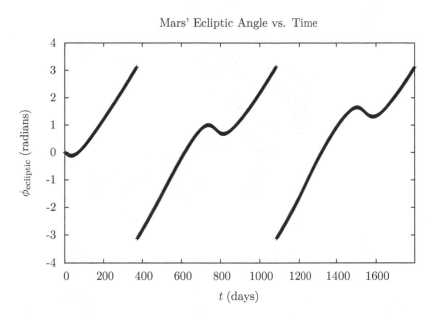

Figure 1.7: Graph showing how the angular position of Mars along the ecliptic changes with time. The general, steady increase in $\phi_{ecliptic}$ over time corresponds to the general, steady eastward drift of the planet along the ecliptic. The funny "kinks" in the graph appear when Mars retrogrades – i.e., when it temporarily halts its eastward motion and instead moves westward along the ecliptic. Note that we have selected $\phi_{ecliptic} = 0$ and $t = 0$ to correspond to the middle of one of Mars' retrograding episodes; this makes it easy to see in the graph that Mars retrogrades again after roughly 779 days, and then again after another 779 days or so. Note that, by convention, $\phi_{ecliptic}$ lies between $-\pi$ and π. So nothing particularly dramatic is actually happening at the times when the graph wraps around: an ecliptic angle of π and an ecliptic angle of $-\pi$ actually refer to Mars being at precisely the same location relative to the stars.

§ 1.4 Ptolemy's Models

Claudius Ptolemy (85 - 165 AD), who lived and worked in Alexandria, Egypt, was the most important of the ancient astronomers, partly because he systematized and cataloged many of the things that had been done by earlier Greek astronomers (such as Eudoxus, Aristarchus, Eratosthenes, and Hipparchus). But Ptolemy also helped develop and improve the kinds of observations we have been summarizing, and – most importantly for us here – he systematically developed a detailed model to account for the observed motions of the planets (including the Sun and Moon).

The Ptolemaic theory basically starts with the two-sphere model described previously, and incorporates the Sun, Moon, and the other planets in roughly the way we've already suggested. Thus, according to Ptolemy's theory, the earth is at rest at the center, with a big rotating sphere of fixed stars on the outside. The Sun, Moon, and planets are placed in the region between the earth and the stars, in the order we've already mentioned. To begin with, each of these seven planetary objects is pulled around (some way or other, either mechanically or just mathematically) by the rotating sphere of stars. This accounts for the shared gross daily motion of all the heavenly bodies. The Sun, Moon, and 5 planets are then each given an additional motion relative to the stars. These additional motions involve an eastward drift along the ecliptic – but, as we have seen, a simple, perfectly uniform circular motion along the ecliptic would not match the observed behavior of any of these bodies. Something more complicated was needed.

Figure 1.8: Claudius Ptolemy (c. AD 85 - c. 165)

Ptolemy's major innovation, then, was a series of mathematically detailed schemes to account for the observed motions of these heavenly bodies, while maintaining at least some kind of plausible allegiance to the idea that the heavenly bodies' motions were, in some sense, the perfectly circular type appropriate to objects made of the 5th element, aether.

Here is Ptolemy's own statement of the guiding principle of his work:

> "it is first necessary to assume in general that the motions of the planets in the direction contrary to the movement of the heavens are all regular and circular by nature, like the movement of the universe in the other direction.... But the cause of this irregular appearance can be accounted for by [several] simple hypotheses. For if their movement is considered with respect to a circle in the plane of the ecliptic concentric with the cosmos so that our eye is the center, then it is necessary to suppose that they make their regular movements either along circles not concentric with the cosmos, or along concentric cricles; not with these simply, but with other circles borne up on them called epicycles. For according to either hypothesis it will appear possible for the planets seemingly to pass, in equal periods of time, through unequal arcs of the ecliptic circle which is concentric with the cosmos."

To summarize, the idea is to account for the observed motions of the planets – including especially the fact that they *don't* just move uniformly along the ecliptic circle relative to the stars – by compounding or otherwise fiddling with circular motions. Ptolemy here mentions several devices for achieving this, which we now consider in more detail.

1.4.1   The Eccentric

The angular velocity

$$\omega = \frac{d\phi}{dt} \tag{1.3}$$

of the Sun along the ecliptic is *roughly* constant, with a value of 2π radians per year, or about one degree per day. But it is not *exactly* constant: it cycles back and forth around this average value over the course of the year, as shown in Figure 1.5. Ptolemy introduced two devices, called the "eccentric" and the "equant", either of which could account for the observed deviations from uniformity for the Sun.

The first device involves having the Sun move in a circular orbit with uniform angular velocity relative

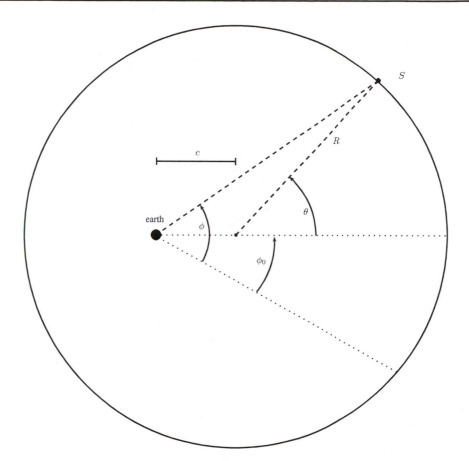

Figure 1.9: An eccentric circle. The point S (which represents the Sun) moves around the circle with uniformly-increasing θ. But since the center of the circle doesn't coincide with the earth, the observed angle of S relative to the (background) stars will be φ, which will increase, over the course of the year, in a not-quite-uniform way.

to the center of the circle – but displacing the center of that circle somewhat from the Earth, i.e., letting the circle be "eccentric". See Figure 1.9.

It can be read off the Figure that the relationship between the angles θ and ϕ is as follows:

$$\tan(\phi - \phi_0) = \frac{R\sin(\theta)}{R\cos(\theta) + c}. \tag{1.4}$$

With the angle θ increasing uniformly in time, i.e.,

$$\theta = \omega_0 t + \theta_0 \tag{1.5}$$

(where $\omega_0 = 2\pi$ radians/year), we then have that

$$\phi(t) = \phi_0 + \arctan\left(\frac{\sin(\omega_0 t + \theta_0)}{\cos(\omega_0 t + \theta_0) + c/R}\right). \tag{1.6}$$

In class we will work through the mathematical details and see how well we can fit the sort of data for $\phi(t)$ that the Greeks might have had, with this model.

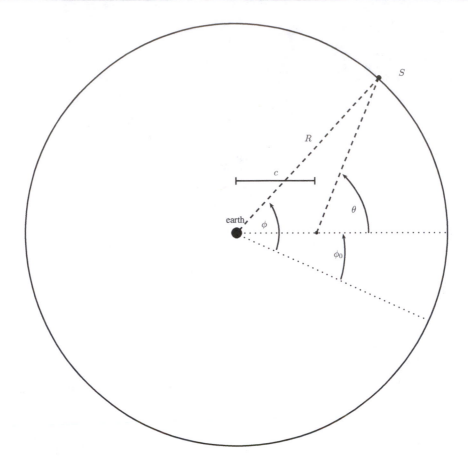

Figure 1.10: The equant: the point S (again representing the Sun) moves around a circle centered at the Earth. But it moves with a uniformly-increasing θ (its angular coordinate as measured from the so-called "equant" point, displaced by distance c from Earth), which makes the ecliptic coordinate φ, as observed from Earth, change in a non-uniform way.

1.4.2   The Equant

An alternative mathematical scheme for accounting for the not-quite-uniform motion of the Sun around the ecliptic was called the "equant". In this scheme, the Sun moves around a circle which *is* centered at the Earth. But its angular velocity is uniform with respect to a different, so-called "equant" point, that is displaced by some distance from the center of the circle. This makes the ecliptic angle as seen from Earth – ϕ in Figure 1.10 – increase in a non-uniform way over the course of the year.

The mathematics here turns out to be a little more complex. In place of Equation (1.4), the relationship between θ and ϕ here turns out to be

$$\tan(\theta) = \frac{R\sin(\phi - \phi_0)}{R\cos(\phi - \phi_0) - c} \tag{1.7}$$

which you can see is a little harder to solve for ϕ. It is possible, though! If you use $\sin^2(\phi - \phi_0) + \cos^2(\phi - \phi_0) = 1$ to eliminate either $\sin(\phi - \phi_0)$ or $\cos(\phi - \phi_0)$, the resulting equation turns out to be quadratic in the remaining trig function of $\phi - \phi_0$. The nicest way I could find of writing the result involves making both substitutions and solving two quadratic equations to develop formulas for both $\sin(\phi - \phi_0)$ and

$\cos(\phi - \phi_0)$, and then dividing these to give:

$$\tan(\phi - \phi_0) = \tan(\theta) \frac{\sqrt{1 + \left(1 - \frac{c^2}{R^2}\right)\tan^2(\theta)} \mp \frac{c}{R}}{\sqrt{1 + \left(1 - \frac{c^2}{R^2}\right)\tan^2(\theta)} \pm \frac{c}{R}\tan^2(\theta)} \tag{1.8}$$

so that

$$\phi(t) = \phi_0 + \arctan\left[\tan(\theta)\frac{\sqrt{1 + \left(1 - \frac{c^2}{R^2}\right)\tan^2(\theta)} \mp \frac{c}{R}}{\sqrt{1 + \left(1 - \frac{c^2}{R^2}\right)\tan^2(\theta)} \pm \frac{c}{R}\tan^2(\theta)}\right] \tag{1.9}$$

with, again, $\theta = \omega_0 t + \theta_0$.

The non-uniform motion (for the Sun around the ecliptic) produced by this equant model is qualitatively similar to, but not precisely the same as, that produced by the eccentric model. We will see in class that, taking into account the accuracy with which the Sun's position was known to the ancient Greeks, either model suffices to account for the observed motion of the Sun, relative to the fixed stars, over the course of the year.

In Ptolemy's terminology, the Sun had only a single "anomaly" – that is, its motion along the ecliptic, relative to the background sphere of fixed stars, deviated from uniformity in a way that could be accounted for with just one of these corrective devices.

We describe both constructions here mostly because both were needed, and used, in Ptolemy's models for the motions of the 5 planets. Each of the 5 planets had *three* anomalies and hence required three corrective devices: eccentric circles, equant points, and the third Ptolemaic construction to which we turn next.

1.4.3   The Epicycle

The most important of Ptolemy's corrective devices (which had actually been first introduced by an earlier astronomer, Apollonius) was a clever scheme for accounting for the occasional retrograde motions of the 5 planets. The basic idea is to *compound* two circular motions for each planet. The first would account for the average easterly drift along the ecliptic, while the second would account for the occasional retrograde motion. The way it works is sketched in Figure 1.11. It should be clear (looking at the figure) how this compounding of two circular motions (one circular motion relative to another point which is itself undergoing circular motion) can give rise to precisely the sort of behavior observed for the planets. In particular, by adjusting the relative sizes of the two circles and the two speeds involved for each planet, one can match pretty well the observed motions.

To maintain (at least some kind of) allegiance to the idea that things in the heavens move with perfectly uniform circular motion (even as we are attempting to model the precise ways in which they don't!), let us assume that the deferent point moves with constant angular velocity ω_d around the deferent circle, while the planet moves with constant angular velocity ω_e around the epicycle. To keep the math as simple as possible, suppose we agree to call "$t = 0$" the middle of a retrograde episode, and suppose we call the planet's position along the ecliptic at that moment $\phi = 0$. Then one can see, from contemplation of Figure 1.12, that the angle θ_d (which is the angular position of the deferent point) will be given by

$$\theta_d = \omega_d t \tag{1.10}$$

and the angle θ_e (which is the angular position of the planet on the epicycle) will be given by

$$\theta_e = \pi + \omega_e t. \tag{1.11}$$

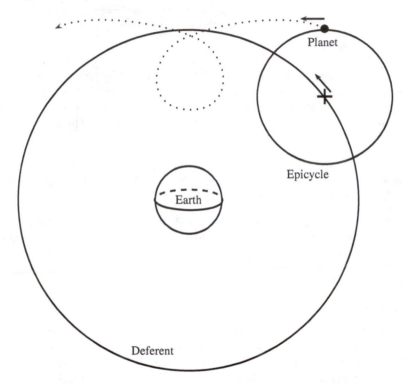

Figure 1.11: Sketch of the basic deferent-epicycle combination in Ptolemy's theory. The deferent point, marked "+", moves uniformly around the deferent circle while the planet moves uniformly around the epicycle (which is centered at the + and pulled around the deferent as it moves). Both circles – the deferent and the epicycle – lie (approximately) in the plane of the ecliptic. This compounding of two circular motions gives rise to a trajectory like that sketched in the dotted line. As seen from the earth, the motion is generally counter-clockwise (which here means eastward along the ecliptic), but the occasional retrograde motion is also accounted for. Note too that the theory automatically explains the observed correlation between retrograde motion and brightness: the planet retrogrades when it is closest to the earth, which accounts for its increased apparent brightness.

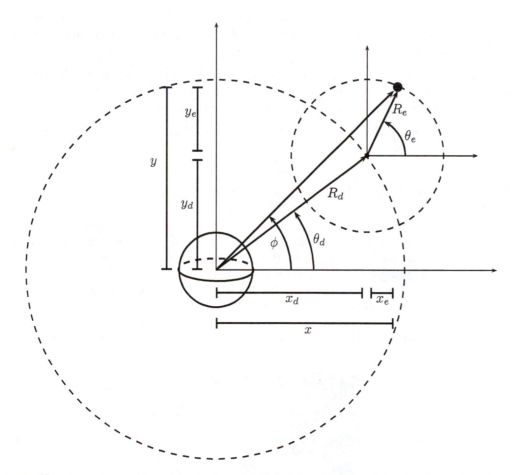

Figure 1.12: *How the observable angle ϕ of a planet on the ecliptic relates to the radii (R_d and R_e) of the deferent and epicycle, and the angles θ_d and θ_e.*

So then the (tangent of the) planet's ecliptic angle coordinate, at time t, will be given by

$$
\begin{aligned}
\tan(\phi) &= \frac{y_d + y_e}{x_d + x_e} \\
&= \frac{R_d \sin(\theta_d) + R_e \sin(\theta_e)}{R_d \cos(\theta_d) + R_e \cos(\theta_e)} \\
&= \frac{(R_d/R_e) \sin(\omega_d t) + \sin(\omega_e t + \pi)}{(R_d/R_e) \cos(\omega_d t) + \cos(\omega_e t + \pi)}.
\end{aligned}
\tag{1.12}
$$

Note that only the ratio, R_d/R_e, of the sizes of the two circles (the deferent and the epicycle) appears in the formula, so there are really just three parameters – (R_d/R_e), ω_d, and ω_e – to adjust to try to account for the type of behavior shown in Figure 1.7.

We will see in class that we can do pretty well, fitting data for the planets' $\phi(t)$ to the epicycle model. The results will not be perfect, though, which is why Ptolemy in fact found it necessary to combine all three devices – the epicycle, the equant, and the eccentric – for each of the 5 planets.

Fast-forwarding many centuries, let us next turn our attention to the alternative model of the solar system that was put forward by Nicolas Copernicus.

§ 1.5 Copernicus' Models

Our use of the word "revolution" to mean some historically-important event or discovery which "changes everything" is based on Copernicus' 1543 book, *On the Revolution of the Heavenly Spheres*, and its enormous cultural, historical, and scientific impact. But the book is actually surprisingly un-revolutionary. Although Copernicus is arguing for a radical reconception of the place of earth (and hence mankind) in the universe, he accepts and implements almost all of the cosmological and astronomical premises of the Greeks: for example, that the universe is spherical and bounded by a sphere of fixed stars, that the proper motion of heavenly bodies is eternal circular motion, and that the small irregularities or anomalies in the motion of the heavenly bodies should be accounted for with Ptolemaic correctives such as eccentrics and

Figure 1.13: Nicolas Copernicus (1473-1543)

epicycles. Indeed, one of Copernicus' main arguments for his heliocentric system was a seemingly very marginal point of detail: it allowed him to do without the particular Ptolemaic device called the equant, which Copernicus regarded as an abhorrent departure from the basic axiom of uniform circular motion. (To me, it has always been puzzling to single out the equant here; all three devices seem to contradict the axiom of uniform circular motion to about the same extent.) Copernicus was, in short, a surprisingly conservative revolutionary.

But nevertheless, Copernicus' work did begin a revolution that ultimately culminnated in Newton's theory of universal gravitation. So let us examine it.

Copernicus' first big claim is that the (apparent) daily westward rotation of the entire heavens is best understood (instead) as a daily *eastward* rotation of the earth:

> "Although there are so many authorities for saying that the Earth rests in the centre of the world that people think the contrary supposition inopinable and even ridiculous; if however we consider the thing attentively, we will see that the question has not yet been decided and accordingly is by no means to be scorned. For every apparent change of place occurs on account of the movement either of the thing seen or of the spectator, or on account of the necessarily unequal movement of both. For no movement is perceptible relatively to things moved equally in the same directions – I mean relatively to the thing seen and the spectator. Now it is from the Earth that the celestial circuit is beheld and presented to our sight. Therefore, if some movement should belong to the Earth it will appear, in the parts of the universe which are outside, as the same movement but in the opposite direction, as though the things outside were passing over. And the daily revolution in especial is such a movement. For the daily revolution appears to carry the whole universe along, with the exception of the Earth and the things around it. And if you admit that the heavens possess none of this movement, but that the Earth turns from west to east, you will find – if you make a serious examination – that as regards the apparent rising and setting of the Sun, moon, and stars the case is so."

Part of the argument for the rotating earth is that it gives a *simpler* explanation of the apparent motion of the heavens. What observation supports is merely the claim that the entire extra-terrestrial universe (the stars, the planets, the Sun, and the moon) moves in a certain way *relative to the earth*. So why have all those other objects move, when the motion of just one object – the earth – will equally well account for the observations?

Interestingly, Copernicus also rests his argument in favor of the rotating Earth on the fact that the Earth is a sphere, and circular or rotational motion is (he claims) natural and proper for a spherical object:

> "the movement of the celestial bodies is circular. For the motion of a sphere is to turn in a circle; by this very act expressing its form, in the most simple body, where beginning and end cannot be discovered or distinguished from one another, while it moves through the same parts in itself."

Thus – since "the Earth is held together between its two poles and terminates in a spherical surface" –

> "Why therefore should we hesitate any longer to grant to it the movement which accords naturally with its form, rather than put the whole world in a commotion – the world whose limits we do not and cannot know? And why not admit that the appearance of daily revolution belongs to the heavens but the reality belongs to the Earth?"

If, on either of these arguments, you are willing to accept that the earth *rotates*, it is then only a little more of the same to accept that the yearly (apparent) motion of the Sun around the ecliptic, is in fact due to the yearly orbit of the earth around the (stationary) Sun. Thus, according to Copernicus, it is the

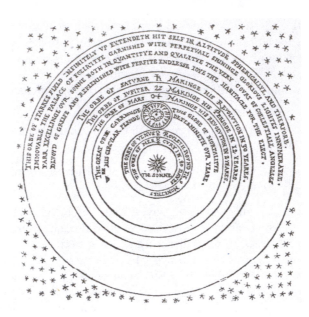

Figure 1.14: A famous depiction of Copernicus' model of the universe, from a 1576 text by Thomas Digges which included the first English translation of much of Copernicus' De Revolutionibus. The Sun lies at the center, surrounded in turn by the orbits of Mercury, Venus, Earth (with its moon), Mars, Jupiter, and Saturn. Note that Digges (unlike Copernicus) depicts the solar system as embedded in an expanse of stars, rather than a fixed shell. This is a natural extension from what Copernicus proposed since, unlike Ptolemy, Copernicus doesn't require the fixed stars to move. There is hence no particular reason they should be attached together (on a spherical shell) rather than spread out through space.

Sun which is at rest at the center of the universe. The earth is then just another planet, orbiting around the Sun in a more-or-less circular trajectory. The basic scheme is sketched in Figure 1.14.

Let's try to understand in detail how the apparent motion of the Sun and planets is accounted for in Copernicus' helio-centric theory. Figure 1.15 shows how the annual eastward movement of the Sun around the ecliptic can be understood in terms of the posited annual orbit of the Earth around a fixed Sun. For Copernicus, the ecliptic – whose original definition is just the set of points that the Sun occupies on the map of fixed stars – can be understood in terms of the Earth's orbit. That orbit lies in a *plane* which intersects the Sun, and so the set of possible apparent positions of the Sun against the background of fixed stars is simply the circle on which the Earth's orbit plane intersects the sphere of fixed stars.

What about the progression of the seasons? The observational fact that the ecliptic is tilted with respect to the celestial equator is explained by Copernicus as follows: the axis about which the earth's daily rotation occurs, is not perpendicular to the plane of its yearly orbit around the Sun. The rotation axis does not point to the pole of the ecliptic, but is instead tilted by 23.5° and hence points toward the celestial pole (currently near the star Polaris, the "north star"). The earth's rotation axis remains (more or less) fixed in space (pointing to Polaris) as the earth orbits the Sun, as illustrated in Figure 1.16.

As already mentioned and sketched in Figure 1.14, the five planets (Mercury, Venus, Mars, Jupiter, and Saturn) will – like Earth – make roughly circular orbits around the Sun. For Mercury and Venus, whose orbits are *interior* to the Earth's orbit, it is pretty clear how their motion will be perceived, from Earth, as a kind of back-and-forth motion along the ecliptic that is always centered at the Sun. In particular, the retrograde motion of the two "inferior" planets will be explained, in Copernicus' theory, by the motion of the planets when they are on the near side of the Sun (as seen from Earth). But what about the three

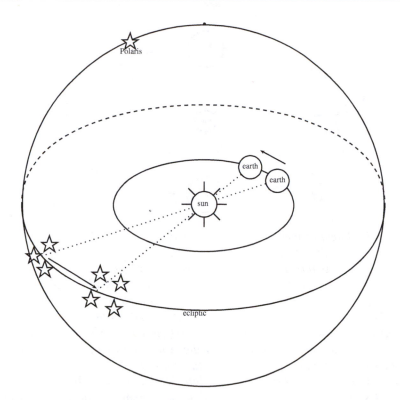

Figure 1.15: Motion of the Earth around the stationary Sun results in an apparent motion of the Sun with respect to the fixed stars. Note that in the Copernican system the ecliptic is understood in terms of the plane in which the earth's orbit lies. It is because we (on earth) always view the Sun from some point on that orbit, that the Sun always appears at some point on the ecliptic. That is, the ecliptic (thought of as a path along the sphere of fixed stars) is the intersection of the plane of the earth's orbit with the sphere of fixed stars.

"superior" planets: Mars, Jupiter, and Saturn? How is their occasional retrograde motion accounted for by Copernicus?

In Ptolemy's theory, the motion of these planets was analyzed in terms of two conjoined circular motion: a "big" circular motion (of the deferent point around earth), and a "small" circular motion (of the planet on an epicycle around the deferent point). The key to understanding Copernicus' explanation for the apparent motion of these planets is to grasp that, at the level of mathematical description, the account is precisely the same! Here too the motion of (say) Mars *with respect to Earth* can be analyzed in terms of the conjunction of two circular motions – namely, the motion of Mars around the Sun (which corresponds in Ptolemy's theory to the "big" circular motion, the deferent), and the motion of Earth around the Sun (which corresponds in Ptolemy's theory to the "small" circular motion, the epicycle). That is: Copernicus replaces the epicycle for each planet (which Ptolemy introduced precisely to account for the retrograde motion) with the motion of the Earth around the Sun. Figure 1.17 is an attempt to sketch the mathematical argument that the two schemes must make the same observational predictions for the motion of Mars (and Jupiter and Saturn) with respect to Earth.

Nevertheless, it is somehow harder to see intuitively how retrograde motion arises in the Copernican system, than in the Ptolemaic system. Figure 1.18 sketches the way to understand this, and the caption explains how the explanation can be understood to apply to both the inferior and superior planets.

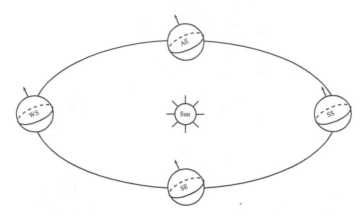

Figure 1.16: The earth is shown at four points in its yearly orbit around the Sun. At the summer solstice (SS) the 23.5° tilt of the earth's rotation axis down from the pole of the ecliptic is directly toward the Sun. On that day, the Sun will therefore be directly overhead at noon for an observer on the Tropic of Cancer (23.5° North Latitude). For observers in more northerly latitudes, the Sun will be higher in the sky at noon than at any other day during the year. Six months later, on the winter solstice (WS), the earth's rotation axis is tilted away *from the Sun. (That, at least, is how we describe it in the Northern Hemisphere!) On that day, an observer on the Arctic Circle (66.5° North Latitude) won't see the Sun rise above the horizon at all. Observers in moderate northern latitudes will see the Sun lower in the sky at noon than on any other day during the year. Also pictured are the Autumn and Spring equinoxes (AE and SE), when the plane defined by the Earth's equator intersects the Sun. On these days all observers will see the Sun rising and setting precisely to the east and west, respectively, and the time between sunrise and sunset will be precisely half a day.*

Now we can finally see the major sense in which Copernicus' model of the solar system is *simpler* than Ptolemy's. Ptolemy's theory required a completely independent construction (deferent and epicycle) for each planet. But it turns out that the same number – the angular velocity of the Sun – shows up again and again in these separate constructions. Each planet is somehow or other "infected" with the motion of the Sun. For the two inferior planets, their *deferents* rotate at the same rate as the Sun, while for the three superior planets, their *epicycles* rotate at the same rate as the Sun. There is no reason for this in Ptolemy's system. It is just a *coincidence* that emerges when one fits the model to real data.

Copernicus removes and explains the coincidence by letting the same one circle – the Earth's orbit around the Sun – do all the jobs that were done by the coincidentally-identical circles in Ptolemy's theory. That is, Copernicus replaces five of Ptolemy's circles (the deferents of Mercury and Venus, and the epicycles of Mars, Jupiter, and Saturn) with just one circle (the Earth's orbit around the Sun).

But we have glossed over something important. So far we have only spoken of one property of Ptolemy's circles: their speeds or angular velocities. But circles have another important property, too: *size*! And recall that, in Ptolemy's system, only the *relative* sizes of the deferent and epicycle was determined by the data – i.e., only the *ratio* R_d/R_e could be fit to observation. But in Copernicus' model, one or the other of these two radii (depending on which planet one is talking about) actually refers to the size of the Earth's orbit around the Sun. So unlike Ptolemy's model, Copernicus' model allowed the sizes of the planets' orbits, relative to the Earth-Sun distance, to be determined.

Copernicus quite properly advertises this aspect of his theory as one of its major virtues. By in effect reducing the number of free parameters in the theory, the whole system becomes much more tightly integrated, logically speaking, such that nothing can be adjusted without affecting the rest:

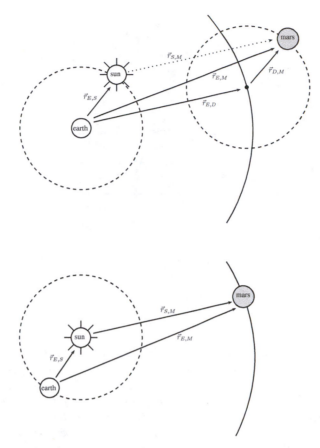

Figure 1.17: *The top and bottom parts show (respectively) how the position of a planet (Mars) is analyzed in Ptolemy's theory and Copernicus' theory. For Ptolemy, the position of Mars with respect to Earth is given by the sum of two vectors (each of which maintains a constant magnitude but changes its direction uniformly in time):* $\vec{r}_{E,M} = \vec{r}_{E,D} + \vec{r}_{D,M}$. *It is the coincidence noted in the text that the vector representing the position of Mars on its epicycle –* $\vec{r}_{D,M}$ *– is just the same as the vector representing the position of the Sun relative to the Earth:* $\vec{r}_{E,S}$. *According to Ptolemy, the vector labeled* $\vec{r}_{S,M}$ *is not directly relevant to the motion of Mars – it is only by coincidence equal to his* $\vec{r}_{E,D}$. *By instead referencing the position of Mars directly to the Sun, Copernicus removes the coincidence and accounts for the position of Mars with respect to Earth as shown in the bottom part of the figure:* $\vec{r}_{E,M} = \vec{r}_{E,S} + \vec{r}_{S,M}$. *Recall that, in Ptolemy's system, the absolute sizes of the deferent and epicycle could not be determined – only the ratio of their radii could be fixed by observational data. But with Ptolemy's deferent corresponding, in the Copernican system, to the planet's orbit around the Sun and Ptolemy's epicycle corresponding, in the Copernican system, to the Earth's orbit around the Sun, the same observational data fixes the ratio of those two orbital radii for Copernicus. Which means, once the absolute distance between the Earth and Sun has been measured, the absolute size of the planet's orbit is determined as well. The top part of the figure, as drawn, may obscure this important point, because it shows the epicycle for Mars and the Sun's orbit around Earth as having the same size. In Ptolemy's system, there is no reason these circles would need to have the same size. We have drawn them with the same size here only to make the correspondence between the two models as clear as possible.*

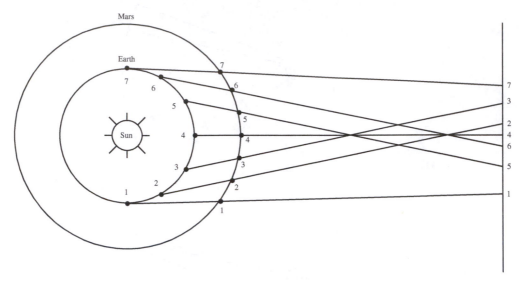

Figure 1.18: How retrograde motion arises in the Copernican system: the positions of Earth and Mars are shown at seven times over the course of half a year, as are the apparent positions of Mars against the backdrop of fixed stars. Between 1 and 2, Mars appears to move relatively quickly (to the east) relative to the stars. Between 2 and 3 it continues to move eastward, but more slowly. At 3, though, it reverses its motion and retrogresses (i.e., moves to the west) until 5, at which point it again reverses and continues its normal (easterly) motion relative to the stars. The same process also explains how the other two superior planets (Jupiter and Saturn) come to retrogress. Note that the distance to the fixed stars is (contrary to the figure) really supposed to be much *larger than the size of Earth's orbit. So really one can perceive the retrograde motion merely from the way the angle of the "lines of sight" evolve over time: from 1 to 3 the line of sight rotates counter-clockwise, which corresponds to an apparent motion of Mars to the east. Then from 3 to 5 the line of sight rotates clockwise, corresponding to an apparent westerly Martian motion. Then from 5 to 7 the line of sight rotates again counter-clockwise. This way of understanding the origin of retrograde motion is helpful, because it allows one to immediately infer, for example, that if one lived on Mars and was charting the apparent position of the Earth relative to the (unshown) fixed stars (to the left), one would also observe retrograde motion during this same period. (Clockwise rotation of the line of sight corresponds to westerly apparent motion, whether one thinks of the line of sight extending to the right or to the left in the figure!) And one can then re-label Mars in the figure as "Earth" and re-label Earth in the figure as (say) "Venus" and hence understand (from the same figure) how the inferior planets come to retrogress. The general rule for retrograde motion can be formulated this way: a given planet will undergo retrograde motion as seen from earth when the relatively-inferior planet (which is Earth if we're talking about Mars, Jupiter, or Saturn, and is the planet itself if we're talking about Mercury or Venus) overtakes or "passes under" the relatively-superior planet.*

"I found after much and long observation, that if the motions of the other planets were added to the motions of the earth, ... not only did the apparent behavior of the others follow from this, but the system so connects the orders and sizes of the planets and their orbits, and of the whole heaven, that no single feature can be altered without confusion among the other parts and in all the Universe. For this reason, therefore, ... have I followed this system."

As we will see next week, the fact that Copernicus' theory allows the sizes of the planetary orbits to be determined, yields great additional fruit in the hands of Johannes Kepler, one of Copernicus' two great followers.

Let us briefly discuss one more of the observational facts discovered by the Greeks and how it is explained in Copernicus' system: the precession of the equinoxes. In Ptolemy's theory, this was explained by attributing, in addition to the daily rotation, a second (more subtle) motion to the sphere of fixed stars: a sort of "wobble" by which the pole of the ecliptic moved (very slowly) about the celestial pole such that, over the course of thousands of years, the celestial pole would migrate around a circle (centered at the pole of the ecliptic). In Ptolemy's theory this motion, just like the daily rotation, was "inherited" by all of the other objects in the universe *except* the Earth which was, of course, fixed at the center. So the same argument we began with is clearly going to motivate Copernicus to instead attribute this motion to the Earth: the rotational axis of the earth does not stay *precisely* fixed in direction as the earth orbits the Sun; rather it turns by a little less than a degree each century such that (refering again to Figure 1.16) after some 13,000 orbits around the Sun, the rotation axis will be tilted 23.5° to the *right*. The star Polaris will then be a whole 47° away from the north celestial pole, and it will be the middle of winter when the Earth is in the part of its orbit (to the right in the Figure) that now corresponds to summer. We will discuss the physical cause of this precession again in a few weeks.

§ 1.6 Comparing the Two Models

We can summarize Copernicus' model, as explained so far, as follows. Copernicus attributes three motions to the earth: a daily rotation, a yearly orbit, and a slow precession of the rotation axis (to account for the precession of the equinoxes). This allows him to get rid of the daily rotation of the stars, planets, Sun, and moon; the motion of the Sun relative to the stars; and the long-period wobble. He also gets rid of the planetary epicycles, and, in so doing, determines the absolute sizes of all the planets' orbits. At this level of description, the theory is really elegant and extremely useful. But there are a couple of serious caveats.

First, there were a number of reasons that Copernicus' model was hard to accept. For example, we noted at the very beginning that it does not *feel* like the Earth is moving. Not only is this undeniably true, but people had come up with a number of rather convincing arguments purporting to prove that everyday observable things would appear very different than they in fact do appear, if the Earth were moving in the way required by Copernicus' model. We will talk about these more next week, but just to give the flavor, Copernicus' model implies that (even leaving aside the even-faster motion associated with Earth's yearly orbit) the surface of the earth is moving at (what we would now describe as) hundreds of meters per second to the east. But then, if you throw a ball straight up in the air, shouldn't it land hundreds of meters to the west? Shouldn't birds starve to death because, upon seeing a worm in the ground below and leaping from a branch to fly down, the worm will be carried hundreds of meters to the east before the bird arrives? Shouldn't there always be an extremely powerful westerly wind, as the earth is rotating to the east under the atmosphere? You get the idea.

You might be able to appreciate how these sorts of objections tacitly rely on assumptions having to do with the Aristotelian physics of "natural place" and "natural motion". Those underlying assumptions have to be identified, challenged, and ultimately rejected, to clear a path for the acceptance of Copernicus' model. But Copernicus himself didn't really do that, and his answers to these sorts of objections were therefore not terribly convincing. Basically he just said that, apparently, the air and birds and rocks share in the various motions that he claimed for the Earth:

> "what would we say about the clouds and the other things floating in the air or falling or rising up, except that not only the Earth and the watery element with which it is conjoined are moved in this way but also no small part of the air and whatever other things have a similar kinship with the Earth? .... Hence the air which is nearest to the Earth and the things floating in it will appear tranquil..."

But Copernicus couldn't really give any convincing physical explanation for *why* this should be the case, and it seemed to contradict assumptions people had about how the relevant physics worked.

A second objection to Copernicus' model had to do with a phenomenon called "parallax". This refers to a difference in the apparent position or size of some object due to different perspectives of the observer. You'll have the opportunity to work through how the Greeks used this phenomenon to measure the distance to the Moon. A much simpler example is provided by holding one's finger at arm's length, and looking at its position (relative to the background objects on the far side of the room) as one alternates between looking with one's left eye (with the right eye closed) and looking with one's right eye (with the left eye closed). The finger appears to jump back and forth between two positions relative to the backdrop, not because the finger itself moves, but because one's perspective changes.

According to Copernicus, our perspective on the heavens changes *a lot*, as the Earth moves around a very large circular orbit over the course of the year. Even if the stars were painted on a large spherical shell, our distance from a given part of that shell would vary over the course of the year. And so, for example, the measureable angle between two nearby stars should be expected to vary over the course of the year. And if the stars are not painted on a spherical shell, but are instead distributed throughout three-dimensional space, as suggested in Figure 1.14, we would expect to see even more complicated variations in the stars' apparent positions, with respect to one another, over the course of the year. But no such parallax effects had been observed in Copernicus' time, so this also made people understandably skeptical of Copernicus' model.

Copernicus' answer to this variety of skepticism was simply to assert that the distances to the stars was much larger than previously believed:

> "...the dimensions of the world are so vast that though the distance from the Sun to the earth appears very large compared with the size of the orbs of some planets, yet compared with the dimensions of the sphere of fixed stars, it is as nothing."

Of course it turns out to be true that the distance from the Earth to the Sun, though vast, is tiny compared to the distance to even the nearest stars. But you can see why it didn't go very far in convincing many of his contemporaries. Copernicus was asking them to believe that the universe was incomprehensibly more vast than they had thought. Not only were we no longer at the center, but our whole terrestrial world was a mere tiny insignifcant speck.

The third big problem with Copernicus' model is that the simple helio-centric theory as presented so far fails to match the observational data. To be specific: it matches the data just as well as – or probably one should say just as *poorly* as – a Ptolemaic theory which has *just* a single deferent-epicycle construction for each planet, and simple uniform circular motion for the Sun. But precisely because this theory didn't account for the details of the observed motion of the Sun, Moon, and 5 planets, Copernicus – just like Ptolemy – had to introduce a number of suspicious devices (such as eccentric circles and additional epicycles). For Ptolemy, these devices seemed somehow marginal or minor: what's the big deal about making the orbit of Mars, for example, slightly eccentric, when it already has a big epicycle? But precisely because the basic version of Copernicus' theory is so elegant and coherent compared to Ptolemy's, the needed eccentrics and epicycles stand out rather dramatically as ugly blemishes.

Let us discuss in a little more detail a couple of these blemishes.

One has to do with the fact that, in Copernicus' model, the Earth's orbit around the Sun is an *eccentric* circle. That is, the Sun was *not* located at the center of the Earth's circular orbit – the point shown in Figure 1.19 that Copernicus called the "Mean Sun". Furthermore, the (eccentric) orbits of the other planets were referred to this empty point in space – the "Mean Sun" – rather than to the actual location of the Sun. The system, therefore, was not really "helio-centric" (Sun-centered) at all, which meant there was a kind of tension between the fully detailed version of the theory, and its supposed overall philosophy, which Copernicus described like this:

> "In the center of all rests the Sun. For who would place this lamp of a very beautiful temple in another or better place than this wherefrom it can illuminate everything at the same time? As a matter of fact, not unhappily do some call it the lantern; others, the mind and still others, the pilot of the world... And so the Sun, as if resting on a kingly throne, governs the family of stars which wheel around."

Another blemish has to do with the fact that, in the centuries between Ptolemy's time and Copernicus' time, the eccentricity associated with the Sun-Earth system had changed. As Copernicus puts it:

> "the distance of the Sun from the centre of the orbital circle... has now become approximately 1/31st [of the radius of the Earth's orbital circle], though to Ptolemy it seemed to be 1/24th. And [the aphelion, the Earth's farthest point from the Sun], which was at that time 24.5° to the west of the summer solstice, is now 6 2/3° to the east of it."

That is, both the magnitude and direction of the eccentricity of the Earth's orbit had changed.

How did Copernicus account for this in his model? He let the center of the Earth's orbit – the Mean Sun – move around a small circle centered at the point labeled "Mean Sun" in Figure 1.19. This is a really small irregularity which only manifests itself over long periods of time. We will cycle back and discuss its causes in a few weeks. It is mentioned here only to provide a further sense of the ways that Copernicus needed to mar the initial elegance of his theory, with a slew of Ptolemy-style epicycles and other devices, in order to achieve the same sort of match to the observational data that Ptolemy's theory had enjoyed.

So you can start to see how everything is not quite as rosy for Copernicus' model as it might have appeared at first. Comparing just the crudest versions of the Ptolemaic and Copernican models, it seems that Copernicus is able to explain the observations in a massively simpler way: the independent deferent-epicycle constructions for the 5 planets (a total of 10 circles, each with an associated size and angular velocity that is picked to match observation) are replaced with just 6 circles (for the 5 original planets plus the Earth). There are thus about *half* as many "free parameters" in the crude Copernican system as there are in the crude Ptolemaic system, and the strange "infection" of each planet's motion with the motion of the Sun (the thing that is just a sheer coincidence for the Ptolemaic system) is explained in a beautifully compelling way.

But when we instead compare the fully-detailed versions of each system – including the eccentricities of all the circles, the small tilts in their orientations with respect to the ecliptic (needed to account for the small variations in ecliptic latitude that we've been ignoring), etc. – the two systems are really quite similar, in terms of overall complexity. The main virtue Copernicus claimed for his system was that it did without the equant, but that is not very convincing since it's really just not clear why the equant is any more problematic than the other corrective devices.

In short: it is by no means clear that Copernicus' model represents a significant improvement over Ptolemy's, at least in so far as a scientific understanding of the motion of the planets is concerned. It's

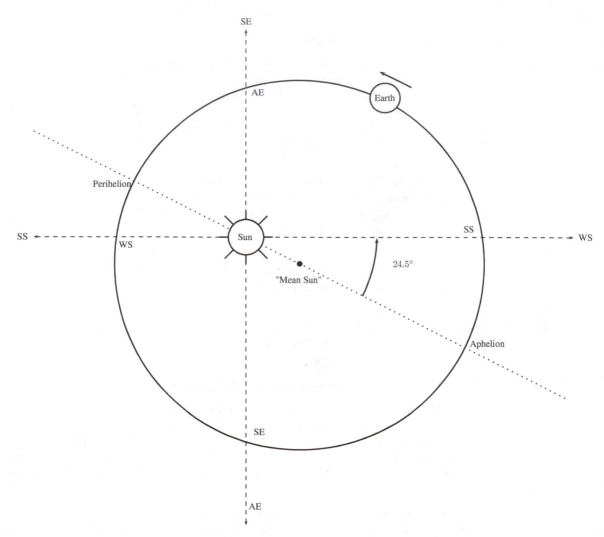

Figure 1.19: The Earth's eccentric orbit around the Sun. The horizontal and vertical lines point to the apparent position of the Sun against the distant fixed stars, as seen from earth on the solstices and equinoxes. The same symbol, SS, labels both the position of the Earth at the Summer Solstice and the apparent position of the Sun against the fixed stars at the Summer Solstice; likewise for the Autumn Equinox (AE), Winter Solstice (WS), and Spring Equinox (SE). According to Ptolemy, the Earth-Sun distance is greatest when the Sun is about 24.5° west of the Summer Solstice point. In Copernicus' model, this means the Earth reaches "aphelion" (the point in its eccentric orbit where it is farthest from the Sun) when it is 24.5° shy (as measured from the Sun) of the point on its orbit labelled SS. The center of the Earth's circular orbit is labelled "Mean Sun" (which is Copernicus' somewhat odd terminology). The observational data requires an eccentricity of about one part in twenty, meaning that the "Mean Sun" is displaced from the Actual Sun by about 1/20 the radius of the Earth's orbit. (Note that Copernicus refers to the actual location of the Sun not, as I just have, as the "Actual Sun" but rather as the "Apparent Sun" – because that is where it appears to be! You can see that it's slightly embarrassing for him that the Sun is displaced from the center of the Earth's orbit – i.e., that the system really isn't heliocentric!)

surely a reasonable hypothesis worth pursuing, but the motion of the planets don't exactly contain any knock-down conclusive proof that Copernicus is right and the Greeks were wrong. Add to this the fact that Copernicus' theory runs up against the various objections to putting the earth in motion, and you can start to see why Copernicus' ideas took some additional time to gain wide acceptance.

Next week we'll see how two astronomers from the next generation – Galileo and Kepler – provided several different varieties of new evidence that eventually turned the tide in favor of Copernicus' model.

Questions:

Q1. What other observations can you think of that are consistent with – and might have been taken as evidence for – the Greek cosmology sketched in Figure 1.1? Can you think of any observations that are definitely contrary to this picture?

Q2. Orient yourself spatially, i.e., figure out which way is north. Now indicate with a sweep of the arm the path that the Sun will take across the sky today. Now indicate (again with your arm) how this path changes over the course of the year.

Q3. Describe how the daily trajectory of the Sun across the sky would vary over the course of the year if you were in Ecuador (which, of course, is on the equator). How about at the north pole? What is the significance of the Tropic of Cancer and the Arctic Circle (in the northern hemisphere, and correspondingly the Tropic of Capricorn and Antarctic Circle in the southern)? What are the latitudes of these "special" points, and why are those numbers significant?

Q4. The Moon can only exhibit the full range of observed phases (new, crescent, half, gibbous, full) if the Moon is closer to the earth than the Sun. Explain why. For example, suppose to the contrary that the Moon and Sun both orbited earth in circles, but the Sun's orbit was smaller/closer than the Moon's. What range of lunar phases would then be observed?

Q5. The Moon completes its monthly cycle around the ecliptic in (on average) 27.3 days. But the time between successive full Moons is longer – 29.5 days on average. Explain why these two periods are different, according to both Ptolemy's model and Copernicus'.

Q6. You should understand how the deferent-epicycle device sketched in Figure 1.11 can account for the retrograde motion of a planet. But look more carefully at Figure 1.6. What aspects of the actual motion of Mars shown in Figure 1.7 *cannot* be accounted for by the deferent-epicycle device as shown in Figure 1.11? How could this be fixed in Ptolemy's theory?

Q7. Can observers at different locations on earth disagree about whether a lunar eclipse is "total" (i.e., whether all of the Moon enters the earth's shadow cone)? Can observers at different locations disagree about whether a solar eclipse is total (i.e., whether the Sun is completely covered by the Moon)?

Exercises:

E1. Eratosthenes measured the size of the Earth as follows. On the summer solstice, the Sun is approximately directly overhead at noon in a certain city in Southern Egypt. But on the same day, 500 ± 30 miles north in Alexandria, a vertical pole of height h casts a shadow of length s at noon. So the angle θ in the Figure can be determined from $\sin(\theta) = s/h$. Suppose Eratosthenes found that $\theta = 7.2° \pm 0.3°$. What is the radius of the Earth?

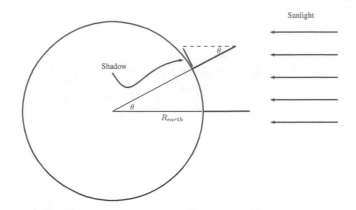

E2. The ancient Greek astronomer Hipparchus determined the distance to the Moon, using the concept of "parallax", in the following way. On a day that there was a solar eclipse, the eclipse was "total" as seen from a certain location in modern Turkey. But as seen from $B = 450 \pm 50$ miles away, in Alexandria, Egypt, the eclipse was only "partial" – and, in particular, the Moon was about 1/5 of the Sun's half-degree angular diameter away from covering the entire Sun. That is, the parallax angle θ in the Figure was about a tenth of a degree. Suppose we say $\theta = 0.10° \pm 0.02°$. What is the distance to the Moon? It is convenient to convert your answer into Earth radii. (Note that the Figure is not at all to scale: the distance to the Moon is much greater than the radius of the Earth, and then the distance to the Sun is much greater than the distance to the Moon. This is relevant to understanding how the observable angular mis-alignment of the Moon and Sun is equivalent to the angle θ in the Figure!)

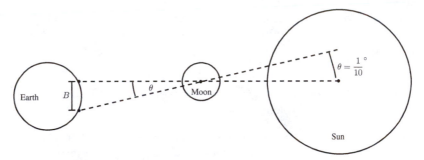

E3. The Greek astronomer Aristarchus tried to determine the distance to the Sun in the following way. When the Moon is precisely half-full, the Sun, Moon, and Earth evidently form a right triangle, as shown in the Figure. And so, knowing already the distance to the Moon, one can determine the distance to the Sun by measuring the angle θ between the Sun and Moon, and using $\cos(\theta) = D_{moon}/D_{sun}$. Aristarchus reported that $\theta = 87°$ (without reporting an uncertainty). If we assume an uncertainty of $\pm 3°$, so $\theta = 87° \pm 3°$, what can we say about the distance to the Sun?

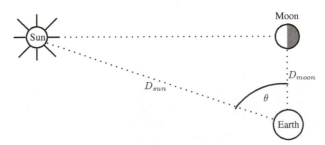

E4. Measure the length L of the shadow cast, around noon, by a vertical stick of height h. You should then be able to determine, using trigonometry, the angle θ shown in the Figure. By looking up the Sun's "declination" D (this is just the angle by which the Sun is north of the celestial equator) on the day you made the measurements, you should be able to determine ϕ, which is the co-latitude (i.e., the complement of the latitude) at your location.

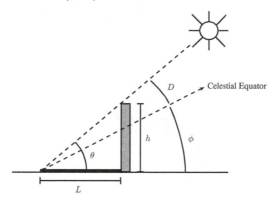

Projects:

P1. You will be given some data for the angular position (along the ecliptic) of the Sun, over the course of a year. How well can the data be fitted with the eccentric circle construction described in the text? What is the eccentricity c/R?

P2. You will be given some data for the angular position (along the ecliptic) of a planet, over the course of several years. Perform a curve-fit to find the values of ω_d, ω_e, and R_d/R_e – from Equation 1.12 – that provide a good fit to the data. Use the data developed by the class to calculate the sizes (in units of AU, i.e., the distance from the Earth to the Sun) of the orbits of all the planets and also their orbital periods. (Use the Copernican idea that the size of one of the two circles in each Ptolemaic deferent-epicycle construction is 1.0 AU. You can also find the orbital period for each planet from the appropriate ω in the Ptolemaic construction.)

P3. According to Copernicus, retrograde motion (for a given superior planet) occurs whenever the Earth passes between that planet and the Sun. (For inferior planets, retrograde motion occurs when the planet passes between Earth and the Sun.) The rate at which these passes occur depends on the *difference* between the orbital angular velocities of the two planets involved. Use this fact to develop a formula for the time T_r between retrograde episodes (i.e., the synodic period) for a given planet, in terms of that planet's orbital period T_p and the orbital period of Earth, T_e (which is just one year). Use the T_p values determined in the previous project, P2, to calculate the synodic period for each of the 5 (non-Earth) planets and compare to the values reported in the text: $T_r = 378$ days for Saturn, 398 days for Jupiter, 779 days for Mars, 584 days for Venus, and 116 days for Mercury.

CHAPTER 2

Evidence for (and updates to) the Copernican Model

THIS week we will study the evidence that three great followers of Copernicus – Galileo Galilei, Tycho Brahe, and Johannes Kepler – brought to bear on the structure of the solar system during the late 1500s and early 1600.

§ 2.1 Galileo on Kinematics, Inertia, and Relativity

Galileo Galilei made a number of important discoveries about how objects move, simply by studying them more carefully than anyone had bothered to do previously. Many of his conclusions will be familiar to you already from previous physics courses. Here we will just sketch those aspects of his work which helped strengthen the case for a Copernican universe.

Galileo's first major contribution to the Copernican revolution had its origins in a perhaps unlikely-seeming place: his careful experiments with familiar terrestrial objects such as pendulums, balls rolling down ramps, and projectiles. His crucial discovery in this area was that free vertical motion is motion

Figure 2.1: Galileo Galilei (1564-1642)

with constant downward acceleration and that free horizontal motion is motion with constant velocity. Furthermore, the motion of an object like a thrown rock – which moves both horizontally and vertically – can be analyzed into its separate *non-interacting* horizontal and vertical components. In particular, the downward acceleration experienced by a ball after you throw it sideways in no way influences or arises at the expense of the horizontal motion you imparted to it when you threw it. Rather, the horizontal motion just continues uniformly and indefinitely (to the extent that air resistance can be neglected).

This is the origin of the concept "inertia" which represents a rejection of several of the Ancient Greek ideas about motion. Inertia refers to the fact that, in the absense of a resistive force like air drag, an object will simply retain its velocity. As it would be later clarified in the first and second laws of Newton, an applied force is needed to explain *changes* in velocity, but not velocity itself. That is: uniform motion at constant speed – such as the horizontal component of the motion of a projectile – is just as "natural" as rest.

Galileo's understanding of the principle of inertia was, however, flawed in two important ways. First, Galileo's inertia really only applied to the horizontal aspect of motion, since he conceptualized "gravity" not as an external force acting on things like balls, but as a kind of inherent tendency for them to accelerate downward. (This is a last remnant of the ancient Aristotelian notion of "proper place".) Thus, for Galileo, the "natural" motion of terrestrial objects like balls is (simultaneously) constant downward acceleration and constant horizontal velocity. The second flaw in Galileo's idea of inertia was a kind of equivocation on the word "horizontal." A horizontal line extended far enough will increase in altitude and eventually stray far from the Earth. Does horizontal inertia imply that a freely moving object will follow such a trajectory? No, said Galileo. For example, if you imagine rolling a ball across the planar surface of a giant table, it seems clear that it won't just keep going forever. To do so, it would have to be going increasingly *uphill* and would hence slow down and turn around. But if the table *curved* so as to maintain a constant height with respect to the surface of the earth, then the ball *would* just keep rolling forever (assuming again an absense of friction). Hence, concludes Galileo, the "natural," constant speed, horizontal aspect of motion is really *circular*, not rectilinear. It's just that, since the relevant circle is about as big as the circumference of the Earth, we don't notice it in the flight of thrown balls.

Despite these minor flaws, Galileo's experimentally-rooted concept of inertia was a profound advance. But what does it have to do with Copernicus? Recall from last week the following argument against the possibility of a moving Earth, as phrased here by Galileo:

> "...the strongest reason of all [for rejecting the idea of a moving earth] is adduced that of heavy bodies, which, falling down from on high, go by a straight and vertical line to the surface of the earth. This is considered an irrefutable argument for the earth being motionless. For if it made the diurnal [i.e., daily] rotation, a tower from whose top a rock was let fall, being carried by the whirling of the earth, would travel many hundreds of yards to the east in the time the rock would consume in its fall, and the rock ought to strike the earth that distance away from the base of the tower. .... This argument is fortified with the experiment of a projectile sent a very great distance upward; this might be a ball shot from a cannon aimed perpendicular to the horizon. In its flight and return this consumes so much time that in our latitude the cannon and we would be carried together many miles eastward by the earth, so that the ball, falling, could never come back near the gun, but would fall as far to the west as the earth had run on ahead."

Galileo rebuts the argument by pointing out that the rocks and balls in these kinds of situations would, during their flights, *maintain* whatever horizontal speed they had *initially* by virtue of the earth's eastward rotation – and would hence land right at the base of the tower or cannon, just as is in fact observed in this kind of situation.

Of course, Copernicus had already claimed the same thing – that, somehow, projectiles (and clouds, birds, etc.) partake of the uniform circular motion that (he argued) was proper and natural for the Earth. So what has Galileo added? Only a rigorous experimental proof that this is in fact how objects really move!

It is clarifying to understand how earlier thinkers had wrongly understood the visibly curved trajectory of projectiles. They had the idea that the horizontal aspect of the motion was "violent" and "unnatural" (i.e., something artifically and externally imposed by your hand or whatever) *and therefore fleeting*. The idea was that after the ball left your hand, the vertically downward motion (toward the ball's natural or proper place) increased, while the unnatural horizontal component of the motion died out, causing the trajectory to curve downward. This had the implication that, should the ball stay in the air long enough, it should eventually be found to be moving *straight down*. And so things like birds or arrows shot from bows – things which do stay in the air for a reasonably long period of time – should therefore, at least by the end of their motions, be moving straight up and down. Which would mean they would be observed to race to the west at a thousand feet per second if the earth were rotating!

Galileo gives a number of vivid examples which help explain how his concepts of motion (to use an ironic turn of phrase) remove the ground from under this sort of objection. For instance, Galileo mentions "another experiment, which is to drop a lead ball from the top of the mast of a boat at rest, noting the place where it hits, which is close to the foot of the mast." But, according to his opponents, "if the same ball is dropped from the same place when the boat is moving, it will strike at that distance from the foot of the mast which the boat will have run during the time of fall of the lead, and for no other reason than that the natural movement of the ball when set free is in a straight line toward the center of the earth." That is, an observer watching from the shore will allegedly see the ball fall straight down, whether the boat is moving out from underneath it or not. And so, if the boat *is* moving, the ball will hit the deck some distance behind the base of the mast.

According to Galileo, however, this is just factually, observably wrong. According to an observer watching from the shore, the ball will retain its horizontal speed as it falls, and will hence trace out a curved (parabolic) trajectory. But since its horizontal speed is *maintained* and just matches that of the ship, the stone still manages to strike the ship's deck just at the base of the mast – just exactly as it would do if the ship weren't moving:

> "anyone who does it will find that the experiment shows exactly the opposite of what is written; that is, it will show that the stone always falls in the same place on the ship, whether the ship is standing still or moving with any speed you please. Therefore, the same cause holding good on the earth as on the ship, nothing can be inferred about the earth's motion or rest from the stone falling always perpendicularly to the foot of the tower."

More generally, if you are locked inside a windowless compartment (such as the hold of a ship), there is no experiment you can do that would distinguish whether you are at rest or moving with uniform velocity. In the 20th century, this principle has become a cornerstone of Einstein's Special Theory of Relativity. But it was formulated quite beautifully three centuries earlier by Galileo:

> "For a final indication of the nullity of the experiments brought forth, this seems to me the place to show you a way to test them all very easily. Shut yourself up with some friend in the main cabin below decks on some large ship, and have with you there some flies, butterflies, and other small flying animals. Have a large bowl of water with some fish in it; hang up a bottle that empties drop by drop into a narrow-mouthed vessel beneath it. With the ship standing still, observe carefully how the little animals fly with equal speed to all sides of the cabin. The fish swim indifferently in all directions; the drops fall into the vessel beneath; and, in throwing something to your friend, you need throw it no more strongly in one direction than another, the distances being equal; jumping with your feet together, you pass equal spaces in every direction. When you have observed all these things carefully (though there is no doubt that when the ship is standing still everything must happen in this way), have the

ship proceed with any speed you like, so long as the motion is uniform and not fluctuating this way and that. You will discover not the least change in all the effects named, nor could you tell from any of them whether the ship was moving or standing still. In jumping, you will pass on the floor the same spaces as before, nor will you make larger jumps toward the stern than toward the prow even though the ship is moving quite rapidly, despite the fact that during the time that you are in the air the floor under you will be going in a direction opposite to your jump. In throwing something to your companion, you will need no more force to get it to him whether he is in the direction of the bow or the stern, with yourself situated opposite. The droplets will fall as before into the vessel beneath without dropping toward the stern, although while the drops are in the air the ship runs many spans. The fish in their water will swim toward the front of their bowl with no more effort than toward the back, and will go with equal ease to bait placed anywhere around the edges of the bowl. Finally the butterflies and flies will continue their flights indifferently toward every side, nor will it ever happen that they are concentrated toward the stern, as if tired out from keeping up with the course of the ship, from which they will have been separated during long intervals by keeping themselves in the air. And if smoke is made by burning some incense, it will be seen going up in the form of a little cloud, remaining still and moving no more toward one side than the other. The cause of all these correspondences of effects is the fact that the ship's motion is common to all the things contained in it, and to the air also. That is why I said you should be below decks; for if this took place above in the open air, which would not follow the course of the ship, more or less noticeable differences would be seen in some of the effects noted..."

The argument – as it bears on the Copernican revolution – is that one can replace the hold of the ship with the whole Earth, and the still water outside the ship with space surrounding the Sun, and everything would stay the same. That is, contrary to the assumption of Copernicus' opponents, we *wouldn't* have noticed it if the Earth were rotating on its axis once per day and also orbiting the Sun once per year. Notice that this represents the complete rejection of the "cosmic graph paper" dynamics sketched at the beginning of Chapter 1.

§ 2.2 Tycho

Tycho Brahe was a Danish astronomer who spent his life working heroically to improve the quantity and quality of observations of the positions of the stars and planets. Early in his career, he observed a new star – a "nova" (which we now understand to be an exploding star that grows incredibly bright for a period of weeks or months) – that had important implications for the debates about the nature of the universe:

> "Last year [1572], in the month of November, on the eleventh day of that month, in the evening, after sunset, when, according to my habit, I was contemplating the stars in a clear sky, I noticed that a new and unusual star, surpassing the other stars in brilliancy, was shining almost directly above my head; and since I had, almost from boyhood, known all the stars of the heavens perfectly (there is no great difficulty in attaining that knowledge), it was quite evident to me that there had never before been any star in that place in the sky, even the smallest, to say nothing of a star so conspicuously bright as this. I was so astonished at this sight that I was not ashamed to doubt the trustworthiness of my own eyes."

But the heavens were thought to be a realm of unchanging timeless perfection. As Tycho reported, "all philosophers agree ... that in the ethereal region of the celestial world no change, in the way either of generation or corruption, takes place." And yet here was this new star.

Others suggested that maybe the nova was an atmospheric, rather than a celestial, phenomenon. To

Figure 2.2: Tycho Brahe (1546-1601), depicted (larger than life) managing operations at his great mural quadrant.

test this idea, Tycho carefully measured the angle between the nova and other nearby stars to determine whether the nova would display parallax. But he found none: the nova remained perfectly still with respect to the rest of the stars and was therefore farther away than the Moon. (Recall that the Moon does display an easily measureable parallax, which the Greeks had used to determine, with surprising accuracy, the distance to the Moon.)

This appearance of a new star in the heavens was a first crack in the received view of the celestial realm as incorruptible and unchanging. But it was nothing compared to the flood of new evidence that would be revealed when Galileo pointed the newly-invented telescope to the heavens in the early 1600s.

Before turning to that, however, let us quicky summarize Tycho's main contribution to astronomy, which was mapping out the positions of the stars and tracking the positions of the Sun, Moon, and planets, with unprecedented care. He invented, implemented, revised, and improved novel methods and instruments for observing the stars and measuring the positions of planets as well as comets. He used his impressively accurate techniques to search for stellar parallax (small annual variations in the relative positions of stars due to the motion of the Earth around the Sun) and – finding none – showed that the distance to the stars must be significantly larger even that Copernicus had implied, if Copernicus' model was correct. Indeed, the implied distance to the stars – several million Earth radii – was so large that Tycho himself could not accept that the Earth moved in the way Copernicus had described. This is one of the reasons that Tycho put forward his own model of the solar system, a kind of hybrid of the Ptolemaic and Copernican systems in which the Earth is at rest at the center, with the Sun orbiting it, but with the other planets orbiting the (moving) Sun.

But Tycho's importance lies not in this theoretical model, but simply in the quality and precision of the astronomical data that he collected and eventually bequeathed to his assistant, Johannes Kepler. Just to give one indication of the seriousness with which Tycho took the collection of data, note that the Danish King funded the construction of a dedicated observatory, Uraniborg, on the island of Hven. But later, Tycho had a second, independent observatory (Stjerneborg, the "star castle") built, so that measurements could be taken independently at the two sites and then compared, to better understand the uncertainties involved. It is hard to say which is more important: the fact that Brahe's methods resulted in observations which were accurate to approximately one arc minute (1/60 of a degree, about a factor of ten better than any of his predecessors); or the fact that he genuinely *knew* that the observations were this accurate.

§ 2.3 Kepler

Johannes Kepler, roughly a contemporary of Galileo, made significant contributions to the growing understanding of the solar system which are in many ways complementary to Galileo's. In particular, where Galileo's discoveries are largely observational in nature, Kepler's are more technical and theoretical. But like any good theory, Kepler's ideas – at least the ones that play an important role in our story here – were grounded in observational data, especially the data of Tycho Brahe.

Near the end of his life, Brahe hired the young Johannes Kepler as an assistant, and Kepler succeeded Brahe in the post of Imperial Mathematician after Brahe's death. Kepler's major innovation was the discovery of three mathematical laws describing the precise nature of the planets' motions around the Sun. We will mainly focus on these three laws, after first discussing some preliminary points that played important roles in Kepler's thinking.

Kepler was a supporter of the Copernican system before he discovered the three laws of planetary motion for which he is now remembered. This is partly a result of a sort of neo-Platonic mystical Sun-worship that represents another side of his personality than the side we will focus on here. For example, Kepler claims that "the Sun is the first cause of the movement of the planets and the first mover of the universe, even by reason of its own body." What is this claim for the causal primacy of the Sun based on? "[T]hese arguments are drawn from the dignity of the Sun and that of the place, and from the Sun's office of vivification and illumination in the world." Or as he put it more poetically elsewhere:

Figure 2.3: Johannes Kepler (1571-1630)

"[The Sun] is a fountain of light, rich in fruitful heat, most fair, limpid, and pure to the sight, the source of vision, portrayer of all colors, though himself empty of color, called king of the planets for his motion, heart of the world for his power, its eye for his beauty, and which alone we should judge worthy of the Most High God, should he be pleased with a material domicile and choose a place in which to dwell with the blessed angles.... For if the Germans elect him as Caesar who has most power in the whole empire, who would hesitate to confer the votes of the celestial motions on him who already has been administering all other movements and changes by the benefit of the light which is entirely his possession? ... [Hence] by the highest right we return to the Sun, who alone appears, by virtue of his dignity and power, suited for this motive duty and worthy to become the home of God himself, not to say the first mover."

Another example of the somewhat mystical side of Kepler is the idea he had, early in his career, for explaining the number of planets and the relative sizes of their orbits. On the Copernican model, there were, of course, six planets: Mercury, Venus, Earth, Mars, Jupiter, and Saturn. And it had been known, since the time of the ancient Greeks, that there were precisely five so-called "Platonic solids". These are closed three-dimensional figures in which each face is an identical shape. For example, the cube (with six square faces) is one, the tetrahedron (with four faces shaped like equilateral triangles) is another, the dodecahedron (with twelve pentagonal faces) is another, and so on.

Kepler's idea was to imagine a kind of Russian nesting doll construction in which each of the 5 Platonic solids was alternated with a sphere (whose size corresponded to the orbit of one of the planets). In particular: the sphere of Mercury was surrounded by an octahedron, which was surrounded by the sphere of Venus, which was surrounded by an icosahedron, which was surrounded by the sphere of Earth, which was surrounded by a dodecahedron, which was surrounded by the sphere of Mars, which was surrounded by a tetrahedron, which was surrounded by the sphere of Jupiter, which was surrounded by a cube, which was surrounded by the sphere of Saturn. With each layer being made to just touch the next layer in, a particular sequence of sizes for the six spheres was implied, and this roughly corresponded to the relative sizes of the planetary orbits according to Copernicus' model.

But the match wasn't perfect, and anyway, there were 120 different orders in which the 5 Platonic solids could be put, each of which would correspond to some particular sequence of orbital radii for the planets. So it is hardly surprising that (at least) one of these happened to match the actual sequence reasonably well. Nevertheless, Kepler argued that this particular sequence was necessitated by various reasons (that we would now regard as mystical and numerological), and was convinced that he had uncovered a profound truth about the architecture of the heavens.

But Kepler had a more scientifically-respectable side, too. For example, early in his career, he discovered that if the orbital planes of the various planets (which remember are all tilted slightly with respect to the ecliptic) intersect one another – not at the center of the Earth's orbit (Copernicus' "Mean Sun") – but at the actual location of the Sun, the complicated and somewhat suspicious variations in ecliptic *latitude* (suspicious because these variations, for the other planets, seemed to be mysteriously tied to the motion of the Earth) were greatly simplified. This was a strong early indication that the motions of the planets were really relative to, and almost certainly in some sense caused by, the Sun:

"Accordingly because the Sun is the node common to all the systems: therefore.... For the planets the Sun is a fixed mark, which all their revolutions regard."

As to the nature of the Sun's causal influence on the planets, Kepler convinced himself (after reading the influential 1600 book of William Gilbert, which we will encounter again in a few weeks) that the Sun controlled the planets through a *magnetic magnetism* influence:

"I am much occupied with the investigation of the physical causes. My aim in this is to show that the celestial machine is to be likened not to a divine organism but rather to a clockwork..., in so far as nearly all the manifold movements are carried out by means of a single, quite simple magnetic force, as in the case of a clockwork all motions [are caused] by a simple weight."

Kepler's idea was roughly that the Sun exerted a long-range magnetic force on the planets, which pushed them around in their orbits.

This was more than a merely qualitative idea: Kepler argued that, since the only job of this force emanating from the Sun was to move the planets in their orbits, and since the planets lie essentially in the plane of the ecliptic, the force would spread out in this two-dimensional plane – its strength therefore decreasing in proportion to the inverse of the distance from the Sun:

$$F \sim \frac{1}{r}. \tag{2.1}$$

Kepler also cited the rotation of the Sun – which he argued for independently, but which was also established by Galileo's telescopic observations that we will discuss shortly – as evidence for the idea that a magnetic force emanating from the Sun swept the planets around in their orbits.

The crucial upshot of this (essentially wrong) idea, was an early formulation of what would later become Kepler's second law of planetary motion. Kepler argued as follows: since the Solar force (allegedly) falls off with distance from the Sun as $1/r$, and since the force exerted on a body is (allegedy) proportional to its speed, the speed of a planet in its eccentric circular orbit around the Sun should vary in inverse proportion to its distance from the Sun:

$$v \sim \frac{1}{r}. \tag{2.2}$$

According to Equation 2.2 a planet should be moving fastest when it is closest to the Sun (perihelion) and slowest when it is farthest away (aphelion), with the above equation giving a precise description of the variation in the speed. Qualitatively, this same behavior can be produced using a circular orbit with both an eccentric and equant (and with the eccentric and equant points located on opposite sides of the circle's center). But the quantitative details of an eccentric circular orbit will be slightly different if governed by Equation 2.2 than if it were governed by the eccentric-equant construction.

Anyway, these were the sorts of ideas bouncing around in Kepler's mind when he began his careful study of the solar system by trying to understand the orbit of Mars, a problem that had been assigned to him by Brahe. Adopting first the methods of his predecessors, Kepler attempted to design an eccentric circular orbit for the planet (with, of course, the Sun at the eccentric point) and with the planet's orbital speed governed by Equation 2.2. This sounds straightforward, but is actually incredibly complex because we have no data for the position of Mars with respect to the Sun. Instead, we have data for the position of both Mars and the Sun with respect to Earth. Of course, if we knew both the angular positions *and* *distances* of both of these bodies, it would be a simple matter to compute their positions in space, and hence the position of one relative to the other by (vector) subtraction. But while it is easy to measure the angular position of the Sun or a planet (against the background of fixed stars), there was no means available to Kepler to determine with any accuracy or absoluteness the *distances* to these objects. The Greeks had been wrong by a factor of about 20 in determining the distance from the Earth to the Sun, and this error had not yet been corrected in the 17th century!

Of course, the *relative* sizes of Earth's and Mars' orbits around the Sun were determinable by Copernicus – but only *approximately*, and Kepler's goal was no mere approximate treatment. He required extremely precise models in order to match the extreme precision of Brahe's data. And so Kepler embarked on an almost unthinkable, decade-long "battle with Mars" in which he explored, largely through trial and error, various assumptions for the precise (relative) sizes for the circular orbits of Earth and Mars, and the various associated Ptolemaic/Copernican fixes: making the circles eccentric to various degrees, introducing equants, and utilizing corrective epicycles. At some point, through a combination of ingenious methods and ferocious tenacity, he convinced himself that it could not be done: circular orbits simply could not be made consistent with Brahe's data.

It is important to appreciate that it was *close*. Kepler's best circular models reproduced the observed positions of Mars with an accuracy of better than a tenth of a degree (i.e., a tenth of 60 arc minutes, or 6 arc minutes), and would therefore have been considered perfectly adequate by any of his predecessors. But Brahe's data was accurate to a single arc minute, and so discrepancies of seven or eight arc minutes simply could not be tolerated:

> "Since the divine goodness has given to us in Tycho Brahe a most careful observer, from whose observations the error of 8' is shown in this calculation... it is right that we should with gratitude recognize and make use of this gift of God .... For if I could have treated 8' of longitude as negligible I should have already corrected sufficiently the hypothesis.... But as they could not be neglected, these 8' alone have led the way toward the complete reformation of astronomy, and have been made the subject-matter of a great part of this work."

2.3.1   The first law

The "complete reformation of astronomy" begins with Kepler's first law of planetary motion, stumbled upon finally toward the end of his battle with Mars: planets move in elliptical orbits with the Sun at one focus.

There are two significant aspects to the first law. Most obviously, it asserts that the trajectory of a planet is not a circle at all, but is rather an elongated shape, an ellipse. With this identification, Kepler finally overthrows the ancient axiom that the motion of heavenly bodies must be circular. But equally important in the first law is that a special, defining point of the elliptical orbit – its "focus" – coincides with the location of the Sun. For this mathematically embodies Kepler's intuition that the Sun is somehow or other dynamically responsible for the motion of the planets. Let us step back and discuss the mathematics of ellipses before moving on to the other two laws.

An ellipse is most simply defined as the planar figure whose points have constant summed distances from two fixed points (the focus points). That is, for each point on the ellipse, its distance from one focus point *plus* its distance from the other focus point, is a constant – the same sum one would get for any other point on the ellipse:

$$d_1 + d_2 = \text{constant} \tag{2.3}$$

where d_1 and d_2 are as shown in Figure 2.4. Thus, one can draw an ellipse on paper by pinning the two ends of a piece of string at two points (the foci), and then moving a pencil such that the string is kept taut on both sides.

The equation satisfied by the points on the ellipse is

$$\frac{x^2}{a^2} + \frac{y^2}{b^2} = 1 \tag{2.4}$$

which is like the equation for a circle ($x^2 + y^2 = R^2$) but with different "radii" along the x and y directions.

The eccentricity of the ellipse is a quantitative measure of its departure from circularity. The eccentricity can either be written as the ratio of the center-focus distance (s) to the semi-major axis (a), or in terms of the ratio of semi-minor-axis (b) and semi-major-axis (a):

$$\epsilon = \frac{s}{a} = \sqrt{1 - \frac{b^2}{a^2}}. \tag{2.5}$$

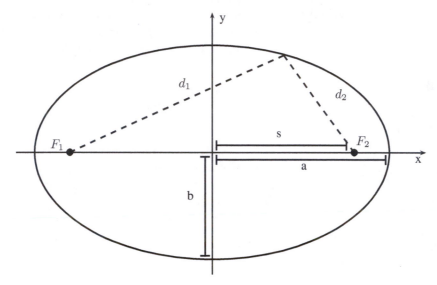

Figure 2.4: *An example of an ellipse. We have here chosen the center of the ellipse as the origin of the x, y coordinate system. The two focus points are labeled F_1 and F_2. The two dotted-line distances (from a point on the ellipse to the two foci, respectively) add to the same constant value for any point on the ellipse. Here the "major axis" of the ellipse coincides with the x-axis, and the ellipse's semi-major-axis is labeled a. Its semi-minor-axis is labeled b. We have also labeled s, the distance from the center out to one of the foci along the major axis. The eccentricity is defined as $\epsilon = s/a = \sqrt{1 - b^2/a^2}$. Here, with $a = 5$, $b = 3$, and $s = 4$, we have $\epsilon = 4/5$ – a much larger eccentricity than the ellipses traced out by any of the planets.*

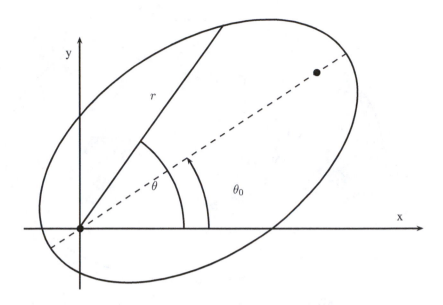

Figure 2.5: Another ellipse, now with the origin of coordinates at one of the foci, and with the major-axis rotated by θ_0 relative to the x-axis. r and θ are then related, for points on the ellipse, as in Equation 2.7.

Finally, by placing the origin of one's coordinate system at one of the focus points instead of at the center, it can be shown that the ellipse satisfies a simple equation in polar coordinates:

$$r(\theta) = \frac{a(1 - \epsilon^2)}{1 - \epsilon \cos(\theta)} \tag{2.6}$$

where r and θ are as shown in Figure 2.5. This expression still assumes that the major-axis of the ellipse is along the x-axis, but polar coordinates make it easy to write the more general expression for an ellipse of eccentricity ϵ and semi-major-axis a whose major-axis makes an angle θ_0 with the x-axis:

$$r(\theta) = \frac{a(1 - \epsilon^2)}{1 - \epsilon \cos(\theta - \theta_0)}. \tag{2.7}$$

You'll be able to work with and prove some of these mathematical relations in the Exercises this week.

2.3.2 The second law

Kepler's second law, as it is accepted today, is a slightly modified version of his $v \sim 1/r$ rule for how a planet's speed varies as its distance from the Sun varies. The rule is formulated in a somewhat unfamiliar way: the planet is said to "sweep out equal areas in equal times." What this means is shown in Figure 2.6: as the planet moves around the Sun, the areas swept out by its coordinate vector from the Sun (e.g., the three distorted pizza-slice shapes shown) will be equal, for parts of the trajectory completed in equal times. This implies that the planet moves *faster* when it is closer to the Sun, and *slower* when it is farther from the Sun, as is evident in the Figure: the actual *distance* covered by the planet during a fixed time interval increases as the planet's distance from the Sun decreases, as is clearly necessary to keep the areas swept out in equal times equal. Qualitatively, then, Kepler's second law is the same as his original idea that the planet's speed is inversely proportional to its distance from the Sun. But as we will explore further in the Projects, the two formulations are not precisely equivalent.

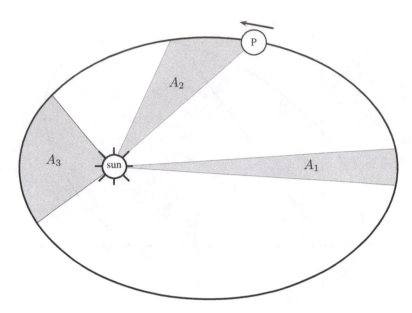

Figure 2.6: According to Kepler's second law, the areas "swept out" by a planet in equal times (e.g., the areas A_1, A_2, and A_3 shown) will be equal. This requires that the planet move a smaller distance in a given time when it is farther from the Sun, and to move a greater distance in that same time as it gets closer to the Sun – i.e., the planet's orbital speed is large when its distance from the Sun is small, and vice versa.

It is important to appreciate that Kepler's first two laws, which describe the motion of planets around the Sun, both make central reference to the position of the actual Sun: the orbits are ellipses *with the Sun at one focus*, and the speed varies such that equal areas are swept out *from the position of the Sun* in equal times. The two laws are therefore more than just descriptions of the planets' motions. They are descriptions of the planets' motions *about the Sun* – i.e., they contain strong empirical evidence in support of the Copernican concept of the Sun as the central orchestrator of the planets' motions. If the Sun weren't somehow controlling the planets from the center, but were (as in Ptolemy's theory) just another heavenly body orbiting around the Earth, it would be a bizarre and unthinkable coincidence that the motions of the planets would take on such a simple and elegant mathematical form when referenced to the position of the Sun.

Kepler's first two laws characterize the orbits of individual planets, giving precise rules for the shapes of their trajectories and the speeds with which the planets trace them out. But so far there is nothing that relates the various planets together: each planet's elliptical trajectory will have a unique size, eccentricity, and orientation; and each planet will "sweep out area" at a (uniform) rate that is different from all the other planets.

But, as we discussed earlier, Kepler believed deeply that there must be some hidden unity behind the apparently-unrelated orbits of the different planets. For example, he had "discovered" early in his career that the relative sizes of the planets orbits could be "explained" by what we now consider the bizarre numerological construction involving the 5 Platonic solids.

2.3.3   The third law

But much later in his career, about a decade after discovering that orbits obeying his first two laws allowed for perfect matches with Brahe's planetary data, Kepler discovered what is now known as his third law. It states that the cube of a planet's orbital radius, divided by the square of its orbital period, is a constant – the *same* constant for *all* of the planets. That is:

$$\frac{R^3}{T^2} = \text{constant.} \tag{2.8}$$

It is hard to describe the motivation that led Kepler to this discovery, since, despite arriving at a relation that would be crucial to Newton's formulation of his theory of gravitation, Kepler's quest was tinged with a sort of mysticism that is now properly regarded as unscientific. It is also difficult to overstate the kind of rapture this discovery triggered in Kepler:

> "...after I had by unceasing toil through a long period of time, using the observations of Brahe, discovered the true distances of the orbits, at last, at last, the true relation ... overcame by storm the shadows of my mind, with such fullness of agreement between my seventeen years' labor on the observations of Brahe and this present study of mine that I at first believed that I was dreaming..."

One question that may already have occured to you is: what is meant by the "orbital radius" of a planet (in Equation 2.8) since the orbits of the planets are not (according to Kepler's first law) circles? In Kepler's own formulation of the law, the R in Equation 2.8 stood for the *average* radius of the planet, though even that is not unambiguous: does this mean the average distance from the Sun of all the points on the orbit, or the distance from the Sun averaged over *time*? These will not be quite the same since (as per Kepler's second law) the planet moves *faster* through the points on its orbit that have smaller radii, and vice versa.

This is not a huge concern, since the actual orbits of the planets are not very eccentric, so the two different senses of "average" will produce almost the same R. We mention it here mostly because it is an interesting feature of the historical development of these ideas that Newton, after using Kepler's laws to arrive at his theory of gravitation, realized – by deduction from his theory – that the correct statement of Kepler's third law would have the R in Equation 2.8 being the *semi-major-axis*, i.e., the simple average of the distances of closest approach and farthest departure from the Sun.

This leaves us in a good position to understand how Isaac Newton arrived at his legendary theory of gravitation, the topic of Chapter 3. But before turning to that, let's return to Galileo and survey some additional evidence he found in support of the Copernican model of the solar system.

§ 2.4 Galileo's Observations with the Telescope

In addition to his work involving the concept of inertia, Galileo contributed significantly to the Copernican revolution with a series of observations of heavenly bodies:

> "... A report reached my ears that a certain Fleming had constructed a spyglass by means of which visible objects, though very distant from the eye of the observer, were distinctly seen as if nearby. Of this truly remarkable effect several experiences were related, to which some persons gave credence while others denied them. A few days later the report was confirmed to me .... which caused me to apply myself wholeheartedly to inquire into the means by which I might arrive at the invention of a similar instrument. This I did shortly afterwards, my basis being the theory of refraction. First I prepared a tube of lead, at the ends of which I fitted two glass lenses, both plane on one side while on the other side one was spherically convex and the other concave. Then placing my eye near the concave lens I perceived objects satisfactorily large and near, for they appeared three times closer and nine times larger than when seen with the naked eye. Next I constructed another one, more accurate, which represented objects

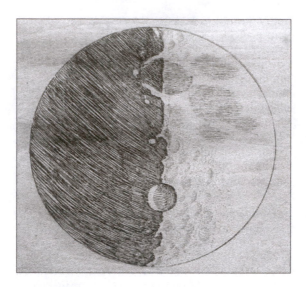

Figure 2.7: One of Galileo's hand-drawn sketches of the Moon, as he saw it through his telescope starting in 1609. Notice the sunlit peaks to the left of the lunar night/day boundary (the "terminator").

as enlarged more than sixty times. Finally, sparing neither labor nor expense, I succeeded in constructing for myself so excellent an instrument that objects seen by means of it appeared nearly one thousand times larger and over thirty times closer than when regarded with our natural vision."

What did Galileo see when he became the first person to examine the heavens through a telescope? Lots of things. And all of them supported, in one way or another, the Copernican worldview:

- *New stars.* As soon as he pointed the telescope to the sky, Galileo was "overwhelmed by the vast quantity of stars" – "...more than five hundred new stars distributed among the old ones within limits of one or two degrees of arc." The Milky Way was revealed as a vast tract of individual stars, which blended and blurred together when seen with the naked eye. And some stars were revealed by the telescope to be *double* – two stars so close together that they could not be individually discriminated with the naked eye. None of this provided any sort of direct confirmation of the Copernican theory. But it was, like Tycho's nova, indirect evidence that the ancient dogmas were based on shamefully incomplete or downright erroneous information, and were therefore to be doubted.

- *The Moon.* When Galileo turned his telescope toward the moon, he saw

  "that the surface of the moon is not smooth, uniform, and precisely spherical as a great number of philosophers believe it (and the other heavenly bodies) to be, but is uneven, rough, and full of cavities and prominences, being not unlike the face of the earth, relieved by chains of mountains and deep valleys."

He went on to describe the lunar sunrise, as seen from afar:

  "[N]ot only are the boundaries of shadow and light in the moon seen to be uneven and wavey, but still more astonishingly many bright points appear within the darkened portion of the moon, completely divided and separated from the illuminated part and at a considerable distance from it. After a time these gradually increase in size and brightness, and an hour or two later they become joined with the rest of the lighted part which has now increased in size. Meanwhile more and more peaks shoot up as if sprouting now here, now there, lighting up within the shadowed portion; these become larger, and finally they too are united with that same luminous surface which extends further. And on the earth, before the rising of the Sun, are not the highest peaks of the mountains illuminated by the Sun's rays while the plains remain in shadow? Does not the light go on spreading while the larger central parts of these mounts are becoming illuminated? And when the Sun has finally risen, does not the illumination of plains and hills finally become one? But on the moon the variety of elevations and depressions appears to surpass in every way the roughness of the terrestrial surface..."

One of Galileo's sketches of the beautifully-textured lunar surface is shown in Figure 2.7. You are also encouraged to look at some more modern images online or, if you have the opportunity and especially if you have not done so before, to look at the Moon through a telescope with your own eyes.

- *Sunspots.* Galileo also used his telescope to observe the Sun. He saw that it too, like the Moon, displayed an imperfect, pock-marked surface. And the marks gradually drifted across the surface, from west to east, over the course of about two weeks – implying that the Sun *rotated*, just as Copernicus required the Earth to do. (See Figure 2.8 for one of Galileo's early sketches. And you are again encouraged to google "sunspots" and look at some modern images.)

Orthodox thinkers tried to insist that the Sunspots were some other celestial or atmospheric phenomenon, merely passing in front of the Sun (as opposed to being blemishes inherent to it). Galileo's careful observations of the spots' motion, however, revealed that their apparent motion increased and then slowed again as they crossed the visible face of the Sun, just as they should be expected to do if they are surface features of a rotating body:

"I have finally concluded, and believe I can demonstrate necessarily, that [the sunspots] are contiguous to the surface of the solar body, where they are continually generated and

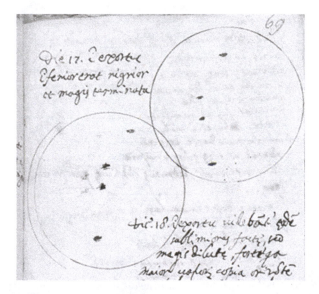

Figure 2.8: Sketches of several sunspots that Galileo observed, on subsequent days, in about 1612. Note the drift of the spots, to the right in the image, from the first day (in the upper right) to the second (in the lower left), demonstrating that the Sun rotates to the east, just like the Earth.

dissolved, just like clouds around the earth, and are carried around by the Sun itself, which turns on itself in a lunar month with a revolution similar [in direction] to those others of the planets, that is, from west to east around the poles of the ecliptic; which news I think will be the funeral, or rather the extremity and Last Judgment of pseudophilosophy, of which signs were already seen in the stars, in the moon, and in the Sun."

- *Jupiter's moons.*

    "There remains the matter which in my opinion deserves to be considered the most important of all – the disclosure of four *planets* never seen from the creation of the world up to our own time, together with the occasion of my having discovered and studied them, their arrangements, and the observations made of their movements and alterations during the past two months. I invite all astronomers to apply themselves to examine them and determine their periodic times, something which has so far been quite impossible to complete, owing to the shortness of the time."

Subsequent observations revealed that the new planets followed Jupiter "in both its retrograde and direct movements in a constant manner" and were, he reasoned, therefore in orbit around Jupiter, just as (in the Copernican system) the Moon orbits around the Earth. Subsequent observations revealed that the four moons had periods of revolution about Jupiter of, respectively: 1 day 18.5

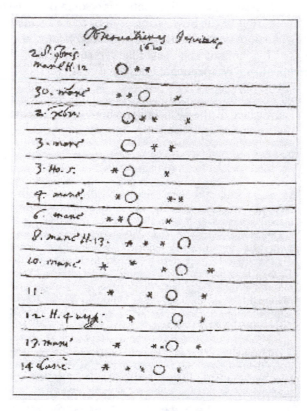

Figure 2.9: Galileo tracked the positions, over time, of what he described as several new planets (or, as we would now call them, moons) orbiting Jupiter. Each row in the table depicts the arrangement of Jupiter and its visible moons on a different night, with the circle in each row representing the planet Jupiter itself and the little stars (which oscillate, over time, from one side of the planet to the other) the moons.

hours; 3 days 13.3 hours; 7 days 4 hours; and 16 days 18 hours.

Here is Galileo's description of the significance of this discovery:

> "Here we have a fine and elegant argument for quieting the doubts of those who, while accepting with tranquil mind the revolutions of the planets about the Sun in the Copernican system, are mightily disturbed to have the moon alone revolve about the earth and accompany it in an annual rotation about the Sun. Some have believed that this structure of the universe should be rejected as impossible. But now we have not just one planet rotating about another while both run through a great orit around the Sun; our own eyes show us four stars which wander around Jupiter as does the moon around the earth, while all together trace out a grand revolution about the Sun in the space of twelve years."

- *Phases of Venus.* Perhaps the most clear-cut and dramatic new evidence against the Ptolemaic worldview was the observation that, over the course of several months, the planet Venus displays a complete sequence of phases (just like the Earth's Moon), from "new" to slender crescent, to half, to gibbous, and finally "full." Recall that in the Ptolemaic system it was ambiguous whether Venus was closer or further to Earth than the Sun. But whichever it was, it must occupy that position for all time. Hence according to the Ptolemaic system, Venus should either display phases (roughly) on the "full" side of "half", or on the "new" side of "half" – but certainly not both. Yet it *does* display both, which proves that it must orbit

  > "around the Sun, as do also Mercury and all the other planets – something indeed believed by the Pythagoreans, Coeprnicus, Kepler, and myself, but not sensibly proved as it now is by Venus...."

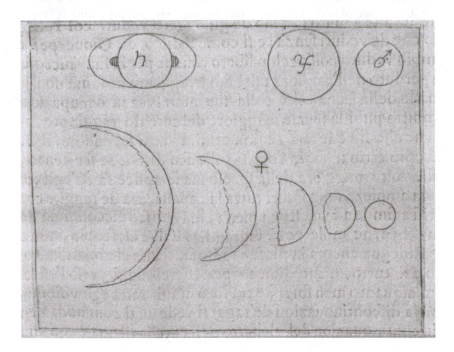

Figure 2.10: Galileo's sketches of the planets Saturn, Jupiter, Mars, and Venus as they appeared through his telescope. Note in particular the complete array of phases exhibited by Venus over the course of several months.

It is clear that Galileo's telescopic observations shift the weight of evidence rather dramatically in favor of Copernicus.

The Copernican model of the solar system shatters the ancient barrier between the terrestrial and celestial realms, implying as it does that the Earth is just another planet. Galileo's application of principles extracted from terrestrial experiments to celestial phenomena represented a first step down a road that would lead Newton to a complete unification of Heaven and Earth. Galileo was also one of the first consistent champions of the idea, the basis of all modern science and everything which rests on it, that knowledge is to be gained by careful observation and experimentation, not by passive contemplation of sacred texts.

For example, in his *Dialogue Concerning the Two Chief World Systems*, in the context of the discussion (already quoted above) about the experiment with the ball falling from the mast of a moving ship, Galileo has the naive Simplicio say that, although he has never actually performed the experiment, "I certainly believe that the authorities who adduced it had carefully observed it. Besides, the cause of the difference is so exactly known that there is no room for doubt." To which Salviati, Galileo's own mouthpiece in the dialogue, replies: "You yourself are sufficient evidence that those authorities may have offered it without having performed it, for you take it as certain without having done it, and commit yourself to the good faith of their dictum. Similarly it not only may be, but must be that they did the same thing too – I mean, put their faith in their predecessors, right on back without ever arriving at anyone who had performed it." (But then Salviati admits he hasn't performed it either! Nevertheless, Galileo had clearly done enough relevant experiments to generalize, and his assurance of what would happen in this particular case turns out to have been entirely warranted.)

Given all of this, it is not surprising that Galileo was considered dangerous and heretical – and hence severely restricted and persecuted – by the religious authorities of his time. For doing so much to move knowledge forward in the face of such resistance, Galileo deserves our profoundest respect and gratitude.

Questions:

Q1. Explain, in your own words, how Galileo's empirical findings about (for example) projectile motion, answer the objections (having to do with cannon balls shot straight up, etc.) to Copernicus' model.

Q2. What, precisely, is the advantage of measuring something twice (using completely independent procedures and equipment), as Tycho arranged to do for all of his astronomical observations by building a second observatory?

Q3. Kepler's third law states that the cube of the orbital radius is proportional to the square of the orbital period: $R^3 \sim T^2$. But the *speed* of a planet in its orbit is equal to the circumference of its orbit divided by the period: $v \sim R/T$. And so Kepler's third law can be rewritten as a statement about how the speed of the planetary orbits varies with their increasing distance from the Sun: $v^2 \sim 1/R$ so that $v \sim 1/\sqrt{R}$. But this is completely different from the $v \sim 1/r$ speed-distance rule that Kepler hypothesized early in his career, and then corrected in his second law of motion. Does this mean the second and third laws contradict each other? Explain why there is no contradiction, by clarifying what the speed of a given planet at a given moment is being compared to, in the second and third laws.

Q4. Summarize in your own words the way(s) in which Galileo's telescopic observations of the surface of the Moon, Sunspots, Jupiter's moons, and the phases of Venus constitute evidence in favor of the Copernican model of the solar system.

Exercises:

E1. Devise your own method to measure, as accurately as possible, the angle between two fixed distant landmarks (e.g., two trees across campus that are easily visible through a window from the classroom). Compare your result to the results of other groups who worked independently and (probably) used entirely different methods.

E2. Use the data for the radii of the planets' orbits and their periodic times, that the class found in P2 of Chapter 1, to check Kepler's third law.

E3. Show that the semi-major axis a, the semi-minor axis b, and the distance from the center to one of the foci s are related by: $a^2 = b^2 + s^2$. (Hint: consider the definition of the ellipse, $d_1 + d_2 = $ constant, for a point on the major axis and a point on the minor axis.)

E4. Show that the Cartesian form of the equation of the ellipse, $\dfrac{x^2}{a^2} + \dfrac{y^2}{b^2} = 1$, is equivalent to the other definition discussed in the reading: $d_1 + d_2 = $ constant.

E5. Show that the Cartesian form of the equation of the ellipse, $\dfrac{x^2}{a^2} + \dfrac{y^2}{b^2} = 1$, is equivalent to the polar form: $r(\theta) = \dfrac{a(1 - \epsilon^2)}{1 - \epsilon \cos(\theta)}$. (Remember that the two formulas use different coordinate systems, though: the Cartesian expression has the center of the ellipse at $x = 0$, $y = 0$, while the polar expression measures r from one of the foci.)

Projects:

P1. Reproduce Galileo's experiment in which a pendulum (with variable length) is used to time the freefall of a ball dropped from various heights.

P2. Kepler used an ingenious trick to convert the directly observable angular positions of the Sun and Mars into fuller knowledge of those bodies' positions in space, i.e., x and y coordinates. The trick involves recognizing that, whatever its shape exactly, Mars' orbit around the Sun is supposed to be *periodic*, with a period of 686.9 days. Thus, every 686.9 days, Mars will be at exactly the same place. Since this is a little less than two years, Earth will be at a different point in its orbit every 686.9 days. And so one can observe, from Earth, the angular positions of the Sun and Mars every 686.9 days, and use this information to "triangulate" one's exact position with respect to each of the two bodies for each of these observations, and hence discover the structure of Earth's orbit around the Sun by plotting its position at a number of these different times.

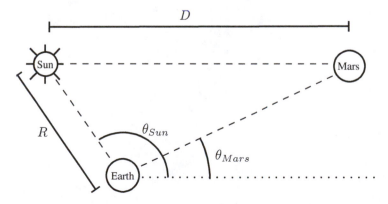

You will be given some data for the ecliptic angles of the Sun (θ_{Sun}) and Mars (θ_{Mars}) for a number of these points. Use the (fixed) distance D between the Sun and Mars as a distance unit. Then determine, for each data point, the distance R shown in the Figure. (Use the law of sines!) Then, again for each data point, use simple trigonometry to get the x and y coordinates of the Earth relative to the Sun. Make a graph of these x, y positions and you can see the trajectory of the Earth around the Sun emerge. Can the shape of Mars' orbit be revealed by a similar method, appropriately reversed?

P3. You will be given some data for the ecliptic angles of – *and distances to* – both the Sun and Mars over a period of several years. (This is unrealistic in that the distance data was not so easily available to Kepler! But we don't have 10 years, and this will give us the flavor of what Kepler did.) Convert the Mars data into Cartesian coordinates, and plot the trajectory of Mars, relative to Earth, over the given period. Now perform the appropriate vector subtraction to generate the x- and y-coordinates of Mars with respect to the Sun, and plot that as well. Marvel at how beautiful and simple the trajectory becomes when we reference the position to the Sun instead of the Earth! Show explicitly that the positions of Mars with respect to the Sun are consistent with one of our mathematical definitions of the ellipse. Now use your position data to generate data for the x- and y-components of Mars' *velocity* over time. Compute $|\vec{r}||\vec{v}|$. Does it stay constant in time? You should find that it is not exactly constant; note that this is equivalent to Kepler's preliminary speed law, $v \sim 1/r$, being not quite right. But maybe there is some other simple relationship between $\vec{r}$ and $\vec{v}$? Try computing the dot product $\vec{r} \cdot \vec{v} = xv_x + yv_y$. Is this constant in time? How about the cross-product $\vec{r} \times \vec{v}$, whose z-component can be written $xv_y - yv_x$? Reflect on the relationship with Kepler's second law as discussed in the reading.

P4. You will be given some data for the apparent positions along the ecliptic of Jupiter and its four moons, over the course of a couple of months. Make appropriate graphs to determine the periods of the orbits of the four moons, and also the relative *sizes* of their orbits. Can you find any mathematical relationship between the sizes and periods of the orbits, like Kepler found for the sizes and periods of the planets' orbits around the Sun?

CHAPTER 3

Gravitational Forces
and the Laws of Motion

Isaac Newton presented his theory of universal gravitation in one master stroke: his 1687 *Mathematical Principles of Natural Philosophy* or *Principia* for short. The book develops side-by-side not only his theory of gravitation, but also the whole edifice of Newtonian mechanics that you are already pretty familiar with from earlier physics classes. This includes the first consistent statement of rectilinear inertia (the first law of motion), $\vec{F} = m\vec{a}$ (the second law of motion, though Newton formulated this in the perhaps less-familiar form $\vec{F} = d\vec{p}/dt$), the principle of action and reaction (the third law of motion), the concept of momentum and its conservation, etc.

We will spend three whole weeks studying Newtonian mechanics. Our content focus will be on understanding how Newton arrived at the idea of universal gravitation and how his theory of gravity accounts for Kepler's laws of planetary motion (and some other astronomical phenomena that we have encountered already). Instead of approaching these topics in the way that Newton himself did, however, we will largely rely on modern technology – computer simulations in particular. So learning how to solve $F = ma$ numerically, using a computer, will be a big skills-focus of the week. We will also get a start developing the skill of "setting up integrals", something that we will continue in the coming weeks when we consider multi-particle systems and rotational motion.

Figure 3.1: Isaac Newton (1643-1727)

§ 3.1 The Concept of Gravity

Newton may have come up with a novel and deeper understanding of gravity, but the idea of "gravity" – as such – was not new to him. Earlier thinkers had understood that many objects naturally fall when impediments are removed, and had attributed this to "gravity." But as we have seen they tended to think of this as a kind of intrinsic "heaviness" that expressed the object's "desire" to reach its "natural place" at the center of the universe or Earth. Newton was the first to grasp fully that this "heaviness" is not innate to heavy objects, but is rather the expression of an *external force* exerted on them *by the Earth*. This recognition required a rejection of the ancient doctrine of "natural place," a clear formulation of the principle of rectilinear inertia, and its consistent application to both objects near the surface of the earth and heavenly bodies.

In addition to reconceiving "gravity" as an external force (rather than an innate tendency), Newton's conception of gravity was also novel in extending the applicability of the concept beyond terrestrial objects. There are two aspects to this extension. First, Newton's theory involves the extension of the Earth's gravitational influence far beyond the surface of the Earth – to the Moon and beyond. And second, Newton attributes this same sort of long-range gravitational attraction to bodies other than the Earth: the Moon, the Sun, the planets, comets, etc.

To further illuminate what was novel about Newton's proposal, let us see specifically how Newton's theory contrasts to the ideas of his predecessors.

Copernicus for example, had speculated that gravity existed not just for the Earth, but for other heavenly bodies as well, and believed that this could account for all these bodies' apparent sphericality:

> "I myself think that gravity or heaviness is nothing except a certain natural appetancy implanted in the parts by the divine providence of the universal Artisan, in order that they should unite with one another in their oneness and wholeness and come together in the form of a globe. It is believable that this affect is present in the sun, moon, and the other bright planets and that through its efficacy they remain in the spherical figure in which they are visible, though they nevertheless accomplish their circular movements in many different ways."

Thus, for Copernicus, the natural place of Earth-stuff is as close as possible to other Earth-stuff; the natural place of Moon-stuff is as close as possible to other Moon-stuff; and so on.

But Copernicus had no concept of a "gravity" that extended over long distances and allowed separate heavenly bodies to attract one another, and hence completely missed the role of gravity in causing not only the (spherical) shape of the planets, but the (roughly circular) shape of their *orbits*. Rather, he just accepted their (he thought) circular motions as "natural" and hence in need of no causal explanation via external forces applied by other bodies.

Some important steps towards Newton's theory of gravity were taken by Kepler, who remember believed strongly that the motion of the planets was governed by forces exerted by the Sun. Kepler had specifically posited (under the title *anima motrix*) that the rotating Sun sent "spokes" out into the plane of the ecliptic, with the rotating spokes exerting the tangential force supposedly needed to keep the planets moving in roughly circular orbits. He then speculated that the would-be circular orbits were distorted into ellipses by an alternating *magnetic* attraction and repulsion, as the magnetic Earth alternately presented its North and South poles toward the Sun during its yearly orbit (and, presumably, similarly for the other planets). Despite being essentially erroneous, Kepler's ideas were important because they represented the first suggestion that the planets' orbits were genuinely caused by some kind of force or influence exerted on them by the Sun.

Another great pre-Newtonian thinker, the philosopher and armchair scientist Rene Descartes, had influentially speculated that the motions of the planets could be explained by a great cosmic whirlpool (made of some unseen fluid aether) which carried the planets in their orbits around the Sun. Newton devoted considerable space in the *Principia* (whose title is essentially the same as Descartes' own earlier book on similar topics) to proving that the whirlpool theory is untenable, since it makes predictions which contradict one another and also observational data about the motions of planets, moons, and comets. In a summary statement, Newton writes:

> "I have tried to investigate the properties of vortices in order to test whether the celestial phenomena could be explained in any way by vortices. For it is a phenomenon that the periodic times of the secondary planets [i.e., moons] that revolve about Jupiter are as the 3/2 powers of the distances from the center of Jupiter; and the same rule applies to the planets that revolve about the sun. Moreover, these rules apply to both the primary and the secondary planets very exactly, as far as astronomical observations have shown up to now. And thus if those planets are carried along by vortices revolving about Jupiter and the sun, the vortices will also have to revolve according to the same law. But the periodic times of the parts of a vortex turned out [in an earlier, fluid-mechanics analysis of vortex motion] to be in the squared ratio of the distances from the center of motion, and that ratio cannot be decreased and reduced to the 3/2 power, unless either the matter of the vortex is the more fluid the further it is form the center, or the resistance arising from a deficiency in the slipperiness of the parts of the fluid ... is increased in a greater ratio than the ratio in which the velocity is increased. Yet neither of these seems reasonable. .... It is therefore up to philosophers to see how that phenomenon of the 3/2 power can be explained by vortices."

One of the crucial similarities between Kepler's and Descartes' ideas is their failure to grasp that the curved trajectories of the planets require and imply a *centripetal* force, as opposed to a *tangential* force or an underlying tangential motion with which the planets were swept along. This key idea from Newtonian mechanics – that uniform circular motion involves a force directed not along the direction of motion, but toward the center of the circle – represents the final renunciation of the old Greek ideas about motion, e.g., "natural place" and the affiliated idea that forces produce velocity, rather than acceleration.

With all of this as historical background, let us turn to the developments that served as proximate causes for Newton's full gravitational theory, as presented in the *Principia*.

§ 3.2 Evidence for an Inverse Square Force

Kepler's accurate descriptions of the trajectories of the planets around the Sun had raised the obvious next question: what causes the planets to move this way? Kepler's own speculations about magnetic forces were largely arbitrary, qualitative, and unconvincing. But by the 1660s, several thinkers were hot on the trail. In that decade, Newton and Christian Huygens independently discovered the "law of circular motion" – now understood as the claim that uniform motion at speed v around a circle of radius R involves a *centripetal acceleration* whose magnitude is given by

$$a_c = \frac{v^2}{R}. \tag{3.1}$$

This acceleration would evidently be caused by some associated (and also centripetal) force.

In the 1660s, however, both Newton and Huygens suffered from some (perhaps familiar) confusion between centripetal and centrifugal forces. Perhaps thinking about what one experiences when one whirls a rock on a string in circles about one's head, they conceptualized v^2/R not as describing the inward acceleration (or the associated inward force), but rather as quantifying the exact amount that a circularly-moving object "wants" to recede from the center. In effect, they understood this formula as describing the outward pull that one would feel from the whirling rock.

Thus armed with this correct, but somewhat misunderstood, formula, several thinkers hit on the idea of using it to analyze the motions of the planets – in particular, how the different planets' centrifugal "endeavors" vary with their distance from the Sun. The computation is readily made by combining the law of circular acceleration with Kepler's third law. This latter, recall, states that the planets' orbital periods vary as the 3/2 power of their orbital radii, i.e., the *squared* orbital periods of the planets are proportional to their *cubed* orbital radii:

$$T^2 \sim R^3 \tag{3.2}$$

where here we approximate the planetary orbits as *circles* and hence take R as the *radius* of the orbits.

Since the speed of a planet in its orbit is proportional to its orbital radius divided by the period

$$v = \frac{2\pi R}{T} \sim \frac{R}{T} \tag{3.3}$$

Kepler's third law can be re-written as

$$\frac{v^2}{R} \sim \frac{1}{R^2}. \tag{3.4}$$

But the left hand side is immediately recognizable as (proportional to) the "centrifugal force" of a given planet. So Kepler's third law implies that these forces fall off as the *inverse square* of the planets' distance from the Sun.

Some argument like this had convinced several thinkers that an "inverse square law" force might be involved in the motion of the planets around the Sun. But, lacking Newtonian mechanics (and in particular a correct concept of centripetal force/acceleration), nobody had been able to really do anything with this pregnant hypothesis.

Newton had been working on just this problem when, in 1684, his friend the astronomer Edmond Halley (yes, of the comet) approached him with a challenge. Halley had heard that Robert Hooke (of the spring force law) and some other scientists were trying to work out the precise trajectory of a planet moving under a central inverse square force. The idea was that, if it could be proved mathematically that the resulting trajectory was an ellipse, this would provide strong support for the idea that the Sun exerted an inverse square force on the planets. Newton claimed that he had solved this problem earlier, but was (he claimed) unable to locate his notes. Whether Newton had actually solved the problem by 1684 is, however, immaterial. In either case, the news from Halley that others were working on ideas similar to his own was perceived as a threat, and caused Newton – prodded and supported by Halley – to redouble his efforts at producing and/or writing up and publishing his work. The result, after several years of furious work, was the *Principia*.

The proof that an inverse square force produces elliptical orbits (or, really, more generally, conic sections – circles, ellipses, parabolae, or hyperbolae) is mathematically quite difficult, requiring the solution of a complicated differential equation. Instead of tackling this problem with analytical techniques (which are modern adaptations of the techniques used by – and really one should say developed by – Newton), we will develop computer simulations of the planetary orbits, training the computer to solve $\vec{F} = m\vec{a}$ numerically and then comparing the resulting planetary orbits to Kepler's laws.

But despite being a central piece of evidence for the theory, Newton's proof that an inverse square force produces trajectories that match those actually followed by the planets is only one part of his theory. For this proof alone would only suggest that some force with this character is operative in the solar system; it would in no way suggest that the force is *gravitational*, i.e., the same sort of force that pulls heavy objects down toward the earth.

We develop this aspect of Newton's argument in the next section.

§ 3.3 The Apple and the Moon

The story of "Newton's apple" is legendary (though in most respects probably apocryphal). He was (supposedly) sitting under an apple tree when a falling apple – which by some accounts hit him on the head – inspired him to conceive of Universal Gravitation. What was going on in his mind that allowed him to make this connection?

First of all, he wasn't thinking about apples but about the Moon. The moon orbits the Earth in a roughly circular orbit whose radius is about 60 Earth radii, and with a period of some 27.3 days. The centripetal acceleration of an object moving in uniform circular motion is given by

$$a_c = \frac{v^2}{R} \tag{3.5}$$

where v is its speed and R the radius of its orbit. Since the speed is related to the period through $v = 2\pi R/T$, this can be re-written

$$a_c = \frac{4\pi^2 R}{T^2}. \tag{3.6}$$

Plugging in the known values for the radius and period of the Moon's orbit gives an acceleration for the moon

$$a_{moon} = 20,340 \, \text{km/day}^2 = 0.0027 \, \text{m/s}^2. \tag{3.7}$$

The moon's acceleration is thus very small compared to the gravitational acceleration of heavy objects near the surface of the earth (such as falling apples), the familiar

$$g = 9.8 \, \text{m/s}^2. \tag{3.8}$$

Is there any relationship between these two very different accelerations?

The simplest way to compare the two numerical acceleration values is to compute their ratio:

$$\frac{a_{moon}}{g} = \frac{.0027 \, \text{m/s}^2}{9.81 \, \text{m/s}^2} = .000275 \approx \frac{1}{3600} = \frac{1}{(60)^2} \tag{3.9}$$

which Newton would immediately have recognized as significant, because 60 is precisely the ratio of the *distances* of the moon and apple, respectively, from the center of the earth – the moon being at a distance of 60 Earth radii, and the apple being at *one* Earth radius!

The fact that the accelerations of the apple and the Moon are in the same ratio as the inverse square of those objects' distances from the Earth's center, obviously suggests a connection to the inverse square force that Newton and others had already been contemplating as somehow relelvant to the motion of the planets.

Moreover, the numerical connection between the accelerations of the apple and the Moon may also have helped trigger the needed clarification in Newton's understanding of the direction of the force involved in circular motion. The apple's acceleration (and hence the gravitational force acting on it) was obviously *downward*, i.e., toward the center of the Earth. Perhaps contemplation of the falling apple helped Newton realize that, in its roughly-circular orbit around the Earth, the Moon is *also* accelerating – *falling* we could even say – toward the center of the Earth.

But if the force on the apple and the force on the Moon are both directed toward the center of the Earth, and there is this numerical connection between them, perhaps those two forces are of the same type – i.e., perhaps the force on the Moon is *also* "gravitational". As Newton would explain the thinking later,

"Therefore, since both forces ... are directed toward the center of the earth and are similar to each other ... they will ... have the same cause. And therefore that force by which the moon is kept in its orbit is the very one that we generally call gravity."

This idea implies that the Earth's gravitational influence extends far beyond its surface. But the numerical connection between the accelerations of the apple and the Moon implies that the strength of the Earth's gravitational influence decreases as the inverse square of the distance – just like the force that the Sun seems to exert on the planets. So, again, on the assumption that things which appear to be the same are in fact the same – i.e., on the assumption that Nature has chosen the simplest possible mechanism consistent with our observations – the forces exerted by the Sun on the planets (and, indeed, the forces exerted by Jupiter and Saturn on their moons) are also gravitational in nature.

Of course, all of this would require more careful and rigorous analysis. But you can see how the simple comparison of the Moon's and apple's accelerations suggests both aspects of Newton's novel idea: that the Earth's gravity extends (with an inverse-square-law influence) to the Moon and beyond, and that it is similar inverse-square-law *gravitational* forces which govern the motion of the heavens.

§ 3.4 Further Evidence for Universal Gravitation

So far we have discussed Newton's arguments that (a) the forces governing the motion of the planets around the Sun (as well as the motion of moons around Jupiter and Saturn) must fall off in strength as the inverse square of the distance from the center, to account for the period-radius relationship that was identified in Kepler's third law, and (b) the Earth's gravitational influence extends (at least) to the Moon and reduces its strength in proportion to the inverse square of the distance.

Newton also showed (as we will explore this week with our computer simulations) that an inverse square law force – and only an inverse square law force – will produce elliptical orbits with the Sun at one focus. And (as we will explore in detail in the coming weeks) Newton also showed that Kepler's second law required that the force governing the motion of a planet must be directed toward the Sun, just as gravitational forces on objects near the Earth are evidently directed toward the Earth.

The conclusion Newton was thus leaning toward was not merely that the forces on the planets and moons were gravitational in nature, but that gravitation was a *universal* phenomenon: *all* bodies exert inverse-square gravitational forces on all other bodies. In order to move this claim from a probability to a certainty, Newton carefully worked out the mathematical implications of this idea and compared them to observational data. He found stunning confirmations everywhere he looked.

For example, according to the hypothesis of universal gravitation, not only should the Moon be attracted to the Earth by a gravitational force, but the Earth should be in turn gravitated toward the Moon. Indeed, according to Newtonian mechanics, the Moon does not exactly orbit around the Earth, so much as the Earth and Moon jointly orbit around their mutual center of mass (a concept we will study and develop in detail next week). This should produce small monthly deviations of the Earth's orbit from its Keplerian ellipse. Such deviations were in fact confirmed by detailed astronomical observation.

More interestingly, the Moon's gravitational tug on the Earth falls off with distance. So the material on the side of the Earth facing the Moon is pulled toward the Moon slightly more strongly than average (and hence tends to move toward the Moon, relative to the center of the Earth). Conversely, the material on the far side of the Earth is pulled toward the Moon slightly less strongly than average (and hence tends to move away from the Moon, relative to the center of the Earth). This effect causes ocean waters to rise up in altitude on these opposite sides of the Earth. And since the Earth rotates on its axis daily, a given observer will notice the seas rising and falling twice per day. Newton's theory of universal gravitation thus produced the first correct explanation of the phenomenon of tides!

(The Sun also produces a tidal effect on the Earth, though it is smaller because the Sun is so much further away than the Moon. But an important aspect of Newton's theory is that the Solar and Lunar tides can work either in concert or at cross-purposes. Thus, around full moon and new moon, one expects higher-than-average high tides and lower-than-average low tides; at half moon, on the other hand, the Moon tends to make the seas rise just where the Sun is making them fall, and vice versa, so the high tides are not very high and the low tides are not very low. This is exactly how the tides do in fact behave.)

Newton's theory also predicted that the planets should exert gravitational influences on one another. Since Jupiter and Saturn are the heaviest planets and their orbits are adjacent to one another, Newton suggested that astronomers of his time look for small perturbations in the orbits of these two planets around the time Jupiter passes "under" Saturn. (The inverse square force will be greatest around this time.) Such perturbations were eventually observed, as were similar perturbations on the other planets, including Earth. You can explore this phenomenon – deviations from Kepler's laws caused by inter-planetary gravitational forces – in more detail in the Projects.

Newton's theory also predicted that objects could move in parabola- or hyperbola-shaped trajectories about the Sun. Careful analysis of the motion of certain comets eventually revealed that they moved in precisely these ways, with other comets moving in (extremely eccentric) elliptical orbits. Halley's comet, for example, has an elliptical orbit which brings it into the inner solar system with a period of approximately 75 years. Its orbit is very eccentric, coming all the way in past the orbit of Earth to a distance from the Sun of about half an AU, and then returning again to about 35 AU, far beyond the orbit of Saturn. Notably, the comet's small deviations from a perfectly elliptical orbit and a perfectly periodic motion can be precisely explained by the gravitational forces exerted on the comet by the planets. Careful analysis of the comet's previous trajectories including calculations of the perturbing effects of Jupiter and Saturn (based on Newton's theory) allowed scientists to predict – within a matter of weeks – the subsequent returns of Halley's comet. The accuracy of these predictions was hailed as a major piece of evidence in support of Newton's ideas.

There are several other pieces of evidence as well, some of which we may encounter in the coming weeks. To summarize, though, not only does Newton's theory account, in precise mathematical detail, for the motions of the moons and planets as these were described by Kepler; it also predicts and explains a number of small deviations from Keplerian orbits which are produced by the relatively small gravitational forces exerted by the moons and planets on each other, as well as accounting naturally for certain previously unexplained processes such as the tides. As Newton summarizes:

> "All the planets are heavy toward one another.... And hence Jupiter and Saturn near conjunction, by attracting each other, sensibly perturb each other's motions, the sun perturbs the lunar motions, and the sun and moon perturb our sea..."

Gravity, according to Newton's theory, is a universal phenomenon in which every massive body attracts – is heavy toward – every other massive body. His theory predicts and explains these effects with precise, mathematical rigor. One could not reasonably ask for a more conclusive array of evidence in support of a theory.

§ 3.5 The Precise Form of the Gravitational Force

Having surveyed the evidence for Newton's theory and discussed one quantitative feature of the force law (that the force varies as the inverse square of the distance between the two gravitating bodies), let us here develop a more precise statement of the basic equation describing the gravitational force. We will here consider the gravitational force between two point masses, and postpone until the final section a discussion of the gravitational forces produced by an extended object (composed of many individually-gravitating point masses).

To begin with, recall the crucial fact identified by Galileo: in the absense of appreciable air resistance (i.e., when the gravitational force is the only one acting), projectiles move with a constant downward acceleration *independent of their mass*. For example, a dropped baseball and a dropped bowling ball will both, despite their different masses, accelerate toward the ground with the same $(9.8\,\mathrm{m/s^2})$ acceleration.

But according to Newton's second law, $F = ma$, the acceleration should be given by

$$a = \frac{F}{m}. \tag{3.10}$$

It should thus be clear that the acceleration can only be independent of the mass m, if the force F is itself *proportional* to m.

In support of this idea – that the gravitational force exerted *on* a certain body is proportional to its mass – Newton also cites the identical rates at which Jupiter and its moons are accelerated toward the Sun by its gravitational attraction:

> "Further, that the weights of Jupiter and its satellites toward the sun are proportional to the quantities of their matter [i.e., their masses] is evident from the extremely regular motion of the satellites..."

That is, despite the very different masses of Jupiter and its several moons, all of these objects accelerate toward the Sun at the same rate, allowing the Jovian system to maintain its coherent, stable, periodic evolution. So evidently their weights toward the Sun – that is, the gravitational forces exerted on them by the Sun – must be proportional to their masses (or what Newton called their "quantities of matter").

Newton also undertook his own precise experimental test of this principle, by comparing the periods of pendulums made of different substances:

> "Others have long since observed that the falling of all heavy bodies toward the earth ... takes place in equal times, and it is possible to discern that equality of the times, to a very high degree of accuracy, by using pendulums. I have tested this with gold, silver, lead, glass, sand, common salt, wood, water, and wheat. I got two wooden boxes, round and equal. I filled one of them with wood, and I suspended the same weight of gold (as exactly as I could) in the center of oscillation of the other. The boxes, hanging by equal eleven-foot cords, made pendulums exactly like each other with respect to their weight, shape, and air resistance. Then, when placed close to each other [and set into vibration], they kept swiging back and forth together with equal oscillations for a very long time. Accordingly, the amount of matter in [i.e., the mass of] the gold ... was to the amount of matter in the wood as the action of the motive force upon all the gold to the action of the motive force upon all the wood – that is, as the weight of one to the weight of the other."

So there is very strong evidence that the weight of one object toward another – i.e., the gravitational force exerted on the one object by the other – is indeed proportional to the mass of the object the force is exerted on.

But Newton's *third* law implies that the gravitational force exerted by A on B must be equal (in magnitude, but of course the direction is opposite) to the gravitational force exerted by B on A. The idea we were just discussing implies that the first of these forces (the force exerted by A on B) must be proportional to the mass of B, and also that the second of the forces (the force exerted on by B on A) must be proportional to the mass of A. But if the magnitudes of these two forces have to be equal, then both forces must be proportional to both masses. That is, Newton's third law requires that the gravitational force exerted by A on B must be proportional to the mass of B and must also be proportional to the mass of A.

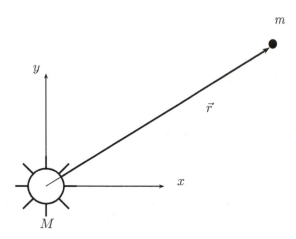

Figure 3.2: An object of mass M (shown here as the Sun) is located at the origin of a coordinate system. Another object, of mass m (here a planet), is located at position $\vec{r}$. The magnitude of the gravitational force exerted by M on m is $GMm/|\vec{r}|^2$. One can write a full vector equation for the force this way: $\vec{F} = -GMm\vec{r}/|\vec{r}|^3$.

And, of course, we have already discussed the evidence that the gravitational force between two bodies varies with the distance between those bodies as the inverse square of their separation.

The most general expression consistent with all of this is the following:

$$F = \frac{Gm_1m_2}{r^2} \tag{3.11}$$

where F is the strength (magnitude) of the gravitational force exerted by an object of mass m_1 on an object of mass m_2 located a distance r away from it. The proportionality constant G is then not dependent on the masses of the two objects involved, nor the distance between them – i.e., it is independent of all the properties the force itself depends on. In short, G is a universal constant that shall be called Newton's constant. (Incidentally, Newton himself never wrote the equation for gravitational forces in this form, and never knew the value of the constant, G, that we now name after him. We will see in the Exercises how an approximate value of G can be arrived at, and then we will study in a couple weeks – and actually reproduce in class – the famous experiment that led to the first accurate measurement of G.)

It is helpful also to write an expression for the force in full vector form. Since we will be so often concerned with the motion of the planets around the Sun, let us pick a coordinate system with the origin at the Sun, and write an expression for the force exerted by the Sun on a planet of mass m located at position $\vec{r}$.

The magnitude of the force is just $GMm/|\vec{r}|^2$. Its direction is the direction from m back toward M, i.e., just opposite the direction of the coordinate vector $\vec{r}$. A unit vector in this direction can be constructed by dividing $\vec{r}$ by its magnitude, $|\vec{r}|$, and then putting in a minus sign. That is, the force exerted by M on m is in the direction $-\vec{r}/|\vec{r}|$. And so the (full vector) force exerted by M on m is given by

$$\vec{F} = -\frac{GMm}{|\vec{r}|^3}\vec{r}. \tag{3.12}$$

We may finally use $\vec{r} = x\hat{i} + y\hat{j}$ and $|\vec{r}| = \sqrt{x^2 + y^2}$ to rewrite the force in terms of the Cartesian coordinates:

$$\vec{F} = -GMm\frac{x\hat{i} + y\hat{j}}{(x^2 + y^2)^{3/2}}.$$ (3.13)

This will be a particularly useful expression for the computer simulations we will develop during the week.

§ 3.6 Setting up Integrals and Newton's Spherical Shell Theorem

So far we've been talking about the gravitational forces between objects like baseballs, planets, and the Sun – and treating these objects as if they were pointlike particles. But, of course, such objects are not in fact pointlike particles. In the coming weeks, we will systematically develop some of the implications of Newton's laws of motion for large extended objects and other systems composed, in principle, of many pointlike particles.

Here we just dip our toes into this general topic by exploring the following question: if we take the inverse-square law as formulated in the previous section as a correct description of the gravitational force between two *point particles*, what will be the gravitational force exerted by (and/or on) an extended *collection* of point particles, such as a planet?

You should perhaps be a little puzzled about how this question could be coming up at this point. Didn't Newton get the inverse-square-law in the first place by examining (among other things) the behavior of the planets? So how could there be any question about whether it should apply to them?

It's a good question! In a way, there should be no question about whether an inverse-square-law force toward the Sun is responsible for the motions of the planets. But remember that a crucial part of Newton's theory was that this force was *gravitational* – i.e., the same sort of force that the Earth exerts on apples and (he argued) the Moon. And a big part of the argument for *that* aspect of the theory was the claim that the Earth's gravitational influence falls off as the inverse-square of the distance from its center – a point he argued for by, you'll recall, comparing the accelerations of the apple and the Moon, and noticing that their ratio was about 3600, i.e., the inverse square of the ratio of their respective distances from the Earth.

But in what sense exactly is an apple one Earth radius away from the Earth? The Earth is composed (presumably) of many tiny massive particles, distributed (presumably) more or less uniformly throughout its whole spherical volume. And the whole idea of the gravitational force is supposed to be that each one of these constituent particles exerts its own gravitational force on the apple – the *net* gravitational force on the apple then being the vector sum of all of these little forces, which have *all sorts of different magnitudes and directions*. So it is *not at all obvious* that the net gravitational force on the apple (assuming gravitation works the way Newton hypothesized, for the Earth's and apple's constituent particles) can be written as

$$F = \frac{GM_{earth}m_{apple}}{R_{earth}^2}$$ (3.14)

where M_{earth} is the total mass of the Earth (i.e., the sum of the masses of all of the Earth's little constituent particles) and R_{earth} is the Earth's radius (i.e., the distance between the apple and the *center* of the Earth).

This requires proof. And that proof requires calculus. Interestingly, there is some historical evidence that Newton had hit on the basic idea of his theory of gravitation some years prior to the publication of the *Principia*, but delayed the publication of his ideas precisely because he lacked the mathematical tools needed to complete this proof, i.e., to fill in this one logical gap in the argument as it was presented in the earlier sections. Of course, during that time (and among other things!) Newton *invented integral calculus* to solve precisely this problem. This should no doubt add to your assessment of the grandeur of Newton's achievements as a scientist. He set the bar for himself extremely high when it came to producing really rigorous, conclusive arguments, with all the i's dotted and t's crossed, for his proposed conclusions. And then he worked tremendously hard to live up to his own high expectations.

Back to the actual proof. What we want to establish is that a spherical blob of total mass M (such as the Earth) acts, gravitationally, the same as a *point mass* of mass M, located at the center of the sphere, would. We can make this slightly simpler by noting that a spherical blob of mass can be thought of as a collection of a bunch of concentric *spherical shells*. So if we can prove that a single spherical shell of mass M exerts the same gravitational force (at least on objects outside the shell) as would a point mass M located at the center of the shell, the corresponding point for spherical blobs will follow. So this is what we will prove.

Actually, this is something *you* will prove. Let's work up to it, though, by developing some techniques with a couple of simpler examples.

3.6.1 First Example

What is the magnitude of the gravitational force exerted by a narrow *stick* of length L and mass M, on a mass m located some distance d away from the segment's center (and, for simplicity, along a line perpendicular to the stick)? See Figure 3.3.

As shown in the Figure, we'll pick a coordinate system with the x-axis along the stick and with its origin at the center of the stick. Our goal is to then calculate the force exerted by the stick on a point of mass m located some distance d away along the y-axis. Pay careful attention to the way this is set up, because that is the real point of going through this (intrinsically uninteresting) example first.

Let's start by focusing our attention on one little (and non-special) piece of the stick – say, the piece of width dx located at position x. Note that, since we are treating dx as infinitesimally small, we can treat this piece as a point mass. Assuming the total mass M is uniformly distributed between $x = -L/2$ and $x = L/2$, the mass of this little piece will be the same fraction of M as dx is of L: $dM = dx\, M/L$. Since this little piece can be treated as a point, the gravitational force $d\vec{F}$ it exerts on the mass m will have

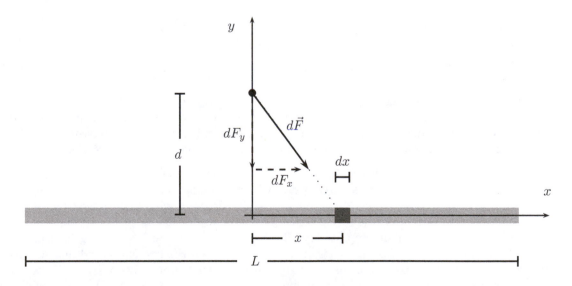

Figure 3.3: How to calculate the gravitational force exerted on a point mass by a stick of length L.

magnitude

$$dF = \frac{G\,dM\,m}{r^2} \tag{3.15}$$

where r, the distance between the two point masses, is given by $r = \sqrt{d^2 + x^2}$, and direction as shown in the figure, along the line from m to dM.

A little thought should convince you that, when we add up the forces on m due to all of the different pieces of the stick, the x-components will add up to zero and only a y-component will remain. (For each piece on the right side that gives a positive dF_x, there is a corresponding piece on the left that contributes an exactly-cancelling negative dF_x.) So to find the net force on m, we need only consider the y-component of the force $d\vec{F}$. This is given by

$$dF_y = -dF\,\frac{d}{r} \tag{3.16}$$

which, plugging everything in, reduces to

$$dF_y = -\frac{G\,dM\,m}{r^2}\frac{d}{r} = -\frac{GMmd}{L}\frac{dx}{(x^2 + d^2)^{3/2}}. \tag{3.17}$$

(The minus sign on the right hand side is there because the force $d\vec{F}$ has a negative y-component, as is evident in the Figure.)

What remains is now only to add up – by *integrating* – all of the little contributions dF_y from all of the little pieces of the stick, to find the total gravitational force exerted on the mass m by the segment.

$$F_y = \int dF_y = -\int_{-L/2}^{L/2} \frac{GMmd}{L}\frac{dx}{(x^2 + d^2)^{3/2}}. \tag{3.18}$$

The constants can be taken outside the integral, and we are left with

$$F_y = -\frac{GMmd}{L}\int_{-L/2}^{L/2}\frac{dx}{(x^2 + d^2)^{3/2}}. \tag{3.19}$$

This integral can be done by looking it up in a table or making a trig substitution. The result is:

$$F_y = -\frac{GMm}{d\sqrt{d^2 + L^2/4}} \tag{3.20}$$

It is always worth pausing at the end of a calculation like this and asking: does the result make sense? First of all, does it have the right units? We know that Newton's constant G times a mass times another mass, divided by a distance squared, will give a force – because that is the form of the basic gravitational force law postulated for point particles. And that is indeed what we have in the above expression.

So far so good. What about various physical limits in which we can intuitively reason out what the answer should be? For example, suppose the stick is really really long: $L \to \infty$. Thinking about that physically, the same total mass M is being distributed ever more "thinly" across a longer and longer line segment. Since the force between m and one of the tiny pieces of the segment falls off as $1/r^2$, only the part of the segment near $x = 0$ should contribute appreciably. But this region contains a vanishing fraction of the total mass M as we let $L \to \infty$. Plus, the parts of the segment off on the two sides tend increasingly to pull m in opposite directions, resulting in increasingly cancelling contributions to the total force. So it seems that the total force should go to zero in this limit. And that is precisely what happens if we take the $L \to \infty$ limit of Equation 3.20.

What if we consider being very very far away from the line segment, i.e., $d \to \infty$? Intuitively, if we are very very far from the line segment (much farther than it is long), the fact that it is a line segment instead of a point should stop mattering, and we should get back the inverse-square-law expression for the force between two points. And again, that is precisely what we get: for $d \gg L$, we can neglect the $L^2/4$ inside the square root sign compared to the d^2, and we get back that the force is GMm/d^2.

So it seems reasonable to believe that we calculated the force correctly.

3.6.2   Second Example

Let's practice this kind of thing with one more example. Suppose we have a thin flat *disk* of radius R, with a total mass M that is uniformly distributed over the area of the disk, and we want to know the gravitational force that would be exerted on a particle of mass m located a distance z away from the center of the disk (along its axis), as shown in Figure 3.4. We can think of the disk as a collection of concentric rings, so let us begin by working out the force $d\vec{F}$ exerted by just the one ring, of radius r and thickness dr, that is shaded more darkly in the Figure.

The force exerted by each (pointlike) piece of the ring will have a component parallel to the axis of the disk, and another component perpendicular to that axis. These perpendicular components will cancel when we add the contributions from the entire ring, so the net force $d\vec{F}$ exerted by the ring will be along the axis, back toward the center of the disk, as shown. Its magnitude will be

$$dF = \frac{G\,m\,dM}{s^2}\frac{z}{s} \tag{3.21}$$

where dM is the mass of the ring. (The "extra" factor of z/s in the formula is just the trig factor – the cosine of the angle at the top of the triangle in the Figure – needed to pick off the non-cancelling, axial component of the force from one piece of the ring.)

The mass dM of the ring can be calculated as follows. Since the mass of the disk is uniformly distributed over its area, the mass-to-area ratio of the small ring will be the same as for the disk as a whole:

$$\frac{dM}{dA} = \frac{M}{A} \tag{3.22}$$

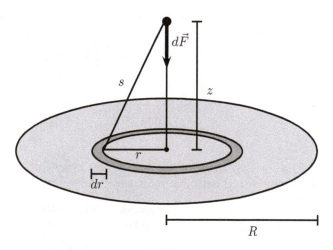

Figure 3.4: How to calculate the gravitational force exerted on a point mass by a thin disk of radius R.

or, equivalently,

$$dM = \frac{M}{\pi R^2} dA \tag{3.23}$$

where dA is the area of the ring. If we imagine cutting the ring with scissors and bending it into a rectangle (which doesn't distort the shape or change its area since its "width" dr is infinitely small!), we can see that $dA = 2\pi r \, dr$, so that

$$dM = \frac{2M}{R^2} r \, dr. \tag{3.24}$$

Hence

$$dF = \frac{2GMmz}{R^2} \frac{r \, dr}{(r^2 + z^2)^{3/2}} \tag{3.25}$$

since $s = \sqrt{r^2 + z^2}$.

To find the force exerted by the entire disk, we then simply add up the contributions from all the rings which compose it:

$$\begin{aligned} F &= \int dF \\ &= \int_{r=0}^{r=R} \frac{2GMmz}{R^2} \frac{r \, dr}{(r^2 + z^2)^{3/2}}. \end{aligned} \tag{3.26}$$

This integral can be done pretty painlessly with a u-substitution. The result is

$$F = \frac{2GMmz}{R^2} \left[\frac{1}{z} - \frac{1}{\sqrt{z^2 + R^2}} \right]. \tag{3.27}$$

Does this answer seem reasonable? Well, we know it has the right units to be a force, because the units of the z out front cancel with the units of the stuff in square brackets, to leave something with the same units as GMm/R^2.

If $z \gg R$ (i.e., the disk is very small or, equivalently, we are very far away from it), we can use the binomial approximation

$$(1 + x)^n \approx 1 + nx \tag{3.28}$$

(for small x) to simplify the second term in square brackets as follows:

$$\frac{1}{\sqrt{z^2 + R^2}} = \frac{1}{z} \left(1 + R^2/z^2 \right)^{-1/2} \approx \frac{1}{z} \left(1 - \frac{1}{2} \frac{R^2}{z^2} \right). \tag{3.29}$$

Plugging this into Equation (3.27) and simplifying gives

$$F \to \frac{GMm}{z^2} \tag{3.30}$$

which makes perfect sense: if the disk is small (or we are very far away from it) the force that it exerts is the same as the force that would be exerted by a point mass.

On the other hand, if $R \gg z$ (i.e., if the disk is very big or, equivalently, we are very close to it), the second term in square brackets is negligible compared to the first, and we are left with

$$F \to \frac{2GMm}{R^2}. \tag{3.31}$$

Interestingly, in this limit, the force doesn't depend on z – our distance from the disk – at all. That is a little surprising at first, but actually makes sense, for reasons that you can think through in one of the Exercises.

Note that, for this kind of problem, we end up doing an integral to get the answer. But the essence of the problem is not doing the integral. That's the easy part, which we could do by handing it off to a math student or looking it up in an integral table or searching online. The essence of the problem is rather: figuring out what integral we need to do. This involves, as we have laid out, drawing a clear picture of the situation, and focusing our attention on the contribution, to the thing we ultimately want to know, from just one little piece of the whole. And good problem-solving practice requires us to assess our answer at the end by confirming that it has appropriate units and simplifies to reasonable things in appropriate limits.

You'll have the opportunity to practice with many more of these "setting up integrals" type problems – including working through the full proof of Newton's spherical shell theorem – in the coming days and weeks.

Questions:

Q1. What is "centripetal acceleration"? Explain qualitatively how we know that an object moving with uniform speed in a circular orbit is accelerating toward the center. (One way to do this effectively involves using graphical vector subtraction.)

Q2. If universal gravitation implies that *all* massive objects attract one another, why don't we observe pairs of familiar household objects (cats and toasters and such) attracting one another gravitationally? According to Newton's theory, roughly how strong *is* the gravitational force between a cat and a nearby toaster? (Bonus points for finding a way to answer this quantitatively but without having to look up the value of Newton's constant, G.)

Q3. In the first example problem, we showed that the gravitational force – exerted on a point mass m by a length-L and mass-M stick a distance d away – has magnitude $GMm/d\sqrt{d^2 + L^2/4}$. What is the magnitude of the gravitational force that the point mass exerts on the stick in this same situation? Explain why we don't need to go through another complicated integral-setup procedure to answer.

Q4. When you prove Newton's spherical shell theorem, you'll discover that the gravitational force exerted by a spherical shell of mass, on a point mass that is *inside* the shell, is *zero*. Does this mean that a spherical shell functions as an anti-gravity shield? Explain.

Exercises:

E1. Rehearse a modern derivation of the "centripetal acceleration" formula, $a = v^2/R$.

E2. Here is another interesting way to derive the important formula for the magnitude of the (centripetal) acceleration associated with uniform circular motion: $a = v^2/R$. Consider a particle bouncing around on the inside of a circle, such that its trajectory is a regular polygon (an equilateral triangle, a square, a pentagon, etc.). Suppose the particle has mass m and moves with speed v and that the circle (which circumscribes the particle's trajectory) has radius R. Start with the case of a square trajectory. Find an expression for the amount by which the particle's velocity changes during one of its collisions with the circle (note that this is not zero!) and, dividing this by the time between such collisions, find an expression for the average centripetal acceleration of the particle. Now see if you can generalize to a regular polygon with N sides. You should expect that, in the limit $N \to \infty$, this average centripetal acceleration for a polygonal orbit goes into the exact expression for the instantaneous centripetal acceleration of a circular orbit. Do you indeed reproduce the familiar formula this way?

E3. Imagine you are living in a parallel universe where the laws of physics are different. An astronomer from the previous generation has discovered that the orbital periods (T) and radii (R) of all the planets in your solar system have the following curious feature: the square of the radius, divided by the period, i.e., R^2/T, is the same value for all of the planets. What would you conclude from this?

E4. Combining Newton's second law with Equation (3.14) tells us that an apple in freefall should move with an acceleration $a = GM_{earth}/R_{earth}^2$. But the freefall acceleration of an apple (or, really, any other object near the surface of the Earth) is known: $a = g = 9.8\,\text{m/s}^2$. With g and R_{earth} known, we can thus determine the value of Newton's constant G... if we also know M_{earth}. The mass of the Earth was not known accurately in Newton's time, but it is possible to estimate it by making some reasonable assumption about the Earth's average density, e.g., that it is about the same as some heavy rocky material like granite. Do this, and thereby compute an approximate value for Newton's constant G. (Interestingly, because of the equivalence of knowing G and knowing M_{earth}, the Cavendish experiment that we will discuss – and reproduce – in a couple weeks, which we think of now as a way of accurately determining G, was described at the time as a method of "weighing the Earth"!)

E5. The Astronomical Unit (AU) – i.e., the distance from the Earth to the Sun – was first determined with reasonable accuracy by the Italian astronomer Cassini during Newton's lifetime. Cassini found (by measuring the slightly-different apparent positions of Mars, against the backdrop of fixed stars, from two different locations on Earth, and using the Copernican relationship between the sizes of the orbits of Earth and Mars around the Sun) that the AU was about 22,000 times the radius of the Earth, or about 150 million kilometers. Using this and your estimated value of G from E4, explain how the mass of the Sun can be determined. (And determine it!)

E6. Two moons orbiting Mars – Phobos and Diemos – were discovered in 1877 by the astronomer Asaph Hall. Subsequent observations revealed Phobos to have an orbital radius of $9,380$ km and an orbital period of 7 hours, 39 minutes. Diemos was found to have an orbital radius of $23,460$ km and an orbital period of 1.26 days. What is the mass of Mars? (You should use your estimated value of G from E4.)

E7. The Earth's moon has mass 7.35×10^{22} kg and radius $1,737$ km. What is the local gravitational acceleration g at the surface of the Moon? How long would it take a golf ball to drop from head height to the ground if you were on the Moon? How high could you jump? (You can use the modern accepted value of Newton's constant, $G = 6.67 \times 10^{-11}\,\text{m}^3/\text{kg}\,\text{s}^2$.)

E8. Imagine a flat tabletop with fixed peg sticking up in the center. One end of an ideal, Hooke's Law spring (with spring constant k and rest length L that can be approximated as zero) is looped around the peg (so it is free to turn), and the other end of the spring is attached to a puck of mass m which can slide frictionlessly on the tabletop . It is possible to make the puck go around in a circular orbit (with $R \gg L$) with the peg at the center. What would be the analog of Kepler's *third* law for this system, i.e., what would be the relationship between the radius R and period T of the puck's "orbit" around the peg?

E9. A geo-synchronous orbit is one in which the orbiting satellite moves with the rotating Earth such that it is always in exactly the same place on the sky as seen from Earth. What must be the radius of the orbit of such a satellite? (Hint: use the fact that the Moon orbits the earth with a radius of

60 R_{earth} and a period of 27.3 days, to avoid having to use a value for G.) Are all circular orbits with that radius geo-synchronous?

E10. Most galaxies consist of a dense "nucleus" of stars near their center, with the density of stars getting much smaller out near the edges. How would you expect the velocity v (with which stars out near the edge orbit around the central nucleus) to depend on their galactic radius R (i.e., their distance from the central nucleus), if most of the mass of the galaxy is concentrated in the nucleus, where most of the visible stars are? How does this compare to observations, which suggest that (for large R) v is roughly constant, independent of R, as shown here?

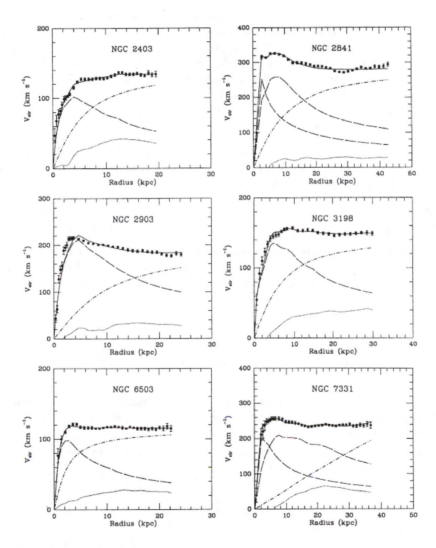

E11. Calculate the magnitudes of the forces of gravitational attraction of the earth toward (a) the Sun, (b) the Moon, (c) Jupiter, and (d) Mars. For the planets, use the force that obtains when the earth-planet distance is at its minimum, i.e., find the maximum force. (You can just look up currently-accepted values for G and these bodies' masses and orbital radii.)

E12. Explain why it makes sense that the gravitational force exerted on a point mass m, a distance z from the center of a radius-R disk of mass M, should be independent of z so long as $R >> z$. Hint:

if $R >> z$, the disk is basically equivalent to an infinite planar sheet of mass. It is then helpful to think separately about the contributions from points on the sheet outside, and inside, a cone like this:

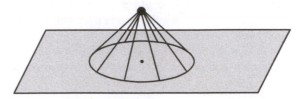

E13. Stars on the (apparent) fringes of our own galaxy, the Milky Way, appear to orbit the center with a speed of about 225 km/sec. For a star whose galactic radius is 17 kiloparsecs (about twice as far from the galactic center as the Sun), what does this imply about the total mass of the Milky Way galaxy? How does this compare to the results of statistical studies which show that the Milky Way contains about 50 Billion (5×10^{10}) stars comparable to the Sun? Based on the numbers given here, what fraction of the Milky Way's total mass is dark matter? (A "parsec" is equal to 2×10^5 AU, or equivalently 3×10^{16} meters. You may look up and use currently-accepted values for G and the mass of the Sun.)

E14. Set up an integral to calculate the gravitational force exerted by a stick of length L and mass M on a point mass m which is a distance $r > L/2$ from the center of the stick, in the direction the stick is aligned along.

E15. Set up an integral to calculate the gravitational force exerted by an infinite planar sheet (with constant mass-per-unit-area σ) on a point mass m a distance z away from the sheet.

E16. Set up an integral to calculate the gravitational force exerted by a hemispherical shell dome (of mass M and radius R) on a point mass m located at the center of the dome's base (i.e., at the point that would be the center of the sphere if it were a full sphere instead of just a hemisphere).

Projects:

P1. Learn how to numerically solve $F = ma$ using a computer, and simulate projectile motion with air drag.

P2. Use a computer to solve for the trajectory of a planet moving under the influence of an inverse-square-law gravitational force from the Sun (fixed at the origin). (Note that it is helpful to work in units of "AU" (for distances) and "years" (for time). Since the accelertation of the Earth toward the Sun can be written either as $GM_{sun}/(1\,\text{AU})^2$ or $4\pi^2(1\,\text{AU})/(1\,\text{year})^2$, we have the numerically convenient fact that

$$GM_{sun} = 4\pi^2(\text{AU})^3/(\text{year})^2.$$

Also, the speed v with which the Earth moves around the Sun is roughly $v = 2\pi\text{AU}/\text{year}$.) For initial conditions that produce a nice elliptical orbit, verify that the orbit respects Kepler's first two laws: equal areas are swept out in equal times, and the points on the orbit form an ellipse with the Sun at one focus.

P3. Explore, again numerically using a computer, the nature of orbits with different (non-inverse-square) force laws. For example, suppose the planets were connected to the Sun by giant invisible springs (with, for simplicity, $L = 0$). What would be the shape of (non-circular) orbits? Is Kepler's first law respected? Kepler's second law? What about other possible force laws, e.g., $F \sim 1/r$ or $F \sim 1/r^3$ or $F \sim 1/r^{2+\delta}$ where δ is small? (This is meant to help you appreciate something that Newton proved, namely that *only* an inverse-square-law force would produce planetary orbits that respect Kepler's laws. But of course Newton proved this in a very different way!)

P4. Explore, again numerically using a computer, the effects of inter-planetary gravitational forces. That is, modify your previous orbital simulation code to include not only the force exerted on your planet by the Sun, but also a gravitational force exerted by some heavy planet (like Jupiter) that lies outside the orbit of the planet you are studying. (For simplicy, you could just plunk Jupiter down at some fixed location and have it stay there instead of having it move... This is a reasonable approximation to what really happens, since Jupiter's orbital period is much longer than that of the inner planets.) Qualitatively, what sort of deviations (from ideal Keplerian orbits) do inter-planetary gravitational forces produce?

P5. Set up an integral to calculate the force that a thin spherical shell (of radius R and total mass M uniformly distributed over its area) would exert on a point mass m a distance r away from its center. The following figure, showing how the shell can be thought of as a collection of rings, should help you get started in a good direction:

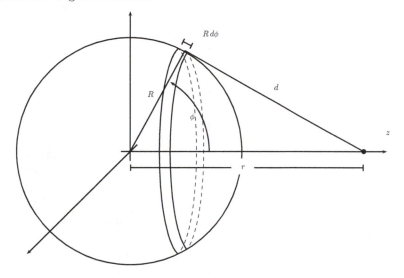

You should tackle at least a couple simpler problems of this type (e.g., Exercises 14-16) first. One mathematical hint that will help you on the home stretch: $\sqrt{(r-R)^2}$ should be understood as $|r-R|$, which is $(r-R)$ if $r > R$, but is instead $(R-r)$ if $r < R$. And here is a relevant quote from Newton to help inspire you:

> "After I had found that the gravity toward a whole planet arises from and is compounded of the gravities toward the parts and that toward each of the individual parts it is inverse[ly] proportional to the squares of the distances from the parts, I was still not certain whether that proportion of the inverse square obtained exactly in a total force compounded of a number of forces, or only nearly so. For it could happen that a proportion which holds exactly enough at very great distances might be markedly in error near the surface of the planet, because there the distances of the particles may be unequal and their situations dissimilar. But at length ... I discerned the truth of the proposition dealt with here."

CHAPTER 4

Systems of Particles

THIS week, we continue our exploration of Newtonian mechanics and in particular pick up the thread we began in the final section of Chapter 3. There, we tackled the problem of calculating the gravitational forces that would be exerted by extended objects (such as sticks, disks, and spherical shells), using the technique of treating the object as an infinite collection of infinitely small points and then adding up the forces contributed by each pointlike piece using integral calculus.

Note that the perspective here involves the idea that matter is composed of some elementary corpuscles or "atoms". We will study the development of the modern atomic theory of matter in Part III of the book. As we'll see, although the idea of atoms goes all the way back to the ancient Greeks, the modern scientific version of atomism grew out of experimental chemistry in the early 1800s. But already in Newton's time, the atomic worldview – according to which the fundamental laws of nature were understood as applying primarily to the elementary corpuscles – was "in the scientific air". And we can think of what we were doing last week in this light: given that the gravitational interaction between two elementary corpuscles is described by Newton's inverse square law, we can infer the gravitational behavior of a large extended object (composed of many such elementary corpuscles) by literally adding up their individual contributions.

This week we focus on a related problem: given that the fundamental laws of motion are postulated to apply to the elementary pointlike constituents of extended bodies, what laws of motion will govern the behavior of the bodies as wholes? And furthermore, what do the fundamental laws of motion imply about the behavior of systems composed of two or more such extended bodies? We will develop and explore some (probably familiar) concepts such as "center of mass" and "momentum" which help answer these questions, continue developing our skills at setting up integrals to solve problems, and apply these ideas to some interesting (astronomical and other) systems.

§ 4.1 Newton's Laws for Extended Bodies

Suppose we have an extended object composed of several elementary massive corpuscles. For simplicity, let's assume first that the object is composed of only two such particles – e.g., a barbell as shown in Figure 4.1. We'll then cycle back at the end and show how everything we've said about the barbell applies also to any arbitrary extended object.

The Figure shows the two mass points that compose the barbell, and also the forces acting on each of them. The forces can be classified as either "internal" or "external" depending on whether the force in question is exerted by an object that is part of, or outside of, the extended object in question. Specifically the Figure shows, for each particle, the net "external" force (i.e., the sum of all the forces exerted on that

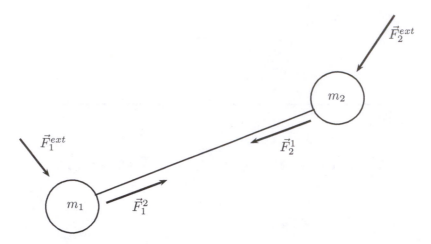

Figure 4.1: *A simple example extended body, composed of two massive corpuscles labeled m_1 and m_2. (Never mind what the "rod" connecting them is made of! It is treated here as massless.) This diagram is also double-tasking as a free-body-diagram, so the forces acting on each corpuscle are also shown. These are classified as internal or external forces depending on whether or not they are exerted by other corpuscles that are part of the extended body in question. For the internal forces, our labeling convention is that $\vec{F}_1^2$ denotes the force exerted on mass 1 by mass 2.*

object by other particles in the universe, not pictured) and then also the "internal" force exerted by the *other* pictured mass.

Let's assume we're using an inertial reference frame, so that Newton's first law holds. Then, we are positing that Newton's second law applies to each of the two masses, and also that Newton's third law correctly describes the relationship between the forces the two masses exert on each other. In equations, these assumptions read

$$\vec{F}_1^{ext} + \vec{F}_1^2 = m_1 \vec{a}_1 \tag{4.1}$$

and

$$\vec{F}_2^{ext} + \vec{F}_2^1 = m_2 \vec{a}_2 \tag{4.2}$$

where $\vec{a}_1$ and $\vec{a}_2$ are the accelerations of the two corpuscles, and

$$\vec{F}_1^2 = -\vec{F}_2^1. \tag{4.3}$$

Let us now introduce the concept of the "center of mass" of this extended object. This is simply a point in space that in some way represents the position of the object as a whole. To be precise, it represents the *average* position of all the component corpuscles, with the masses of the corpuscles used as a weighting for the average. Mathematically, the definition is

$$\vec{R}_{CM} = \frac{m_1 \vec{r}_1 + m_2 \vec{r}_2}{m_1 + m_2} \tag{4.4}$$

where $\vec{r}_1$ and $\vec{r}_2$ are the position vectors for the individual corpuscles. Since the masses m_1 and m_2 are just constants independent of time, it is easy to derive (by taking derivatives with respect to time) expressions for the center of mass velocity

$$\vec{V}_{CM} = \frac{m_1 \vec{v}_1 + m_2 \vec{v}_2}{m_1 + m_2} \tag{4.5}$$

and acceleration

$$\vec{A}_{CM} = \frac{m_1 \vec{a}_1 + m_2 \vec{a}_2}{m_1 + m_2} \tag{4.6}$$

where $\vec{v}_1$ is the velocity of particle 1, $\vec{a}_2$ is the accleration of particle 2, etc.

Now let us put all of this together. The idea is to simply *add* together Equations (4.1) and (4.2):

$$m_1 \vec{a}_1 + m_2 \vec{a}_2 = \vec{F}_1^{ext} + \vec{F}_1^2 + \vec{F}_2^{ext} + \vec{F}_2^1. \tag{4.7}$$

But now we can use Equation 4.3 to cancel the two (equal and opposite) internal forces which appear added on the right hand side. The result is:

$$m_1 \vec{a}_1 + m_2 \vec{a}_2 = \vec{F}_1^{ext} + \vec{F}_2^{ext}. \tag{4.8}$$

Finally, we can recognize the left hand side as the numerator from the right hand side of Equation 4.6. This allows us to write

$$\vec{F}^{ext} = M \vec{A}_{CM} \tag{4.9}$$

where $\vec{F}^{ext}$ without any subscripts refers to the *total* external force $\vec{F}_1^{ext} + \vec{F}_2^{ext}$, and we define the total mass $M = m_1 + m_2$.

The final result here looks just like Newton's second law ($F = ma$) but is a description of the two-particle-system treated, in some sense, as a whole, as a single object. In words, what we proved is that if the constituent particles of this barbell object obey Newton's laws of motion individually, then the object as a whole will, too – with (not surprisingly) the total mass functioning as "the mass of the whole," the total external force acting as "the net force acting on the whole," and the acceleration of the center of mass point functioning as "the acceleration of the whole."

It shouldn't be surprising that this works out. After all, it was empirical observation of big extended objects that led Newton to posit his laws of motion in the first place. So there would be a pretty serious problem if positing those laws for the small constituent particles resulted in anything else for the motion of the whole. Still, it's nice to see that – and how – it works out.

Before going on, we should check explicitly that the same method works for a more general extended object, composed of some arbitrary number of elementary massive particles. The logic will be exactly the same, so let's just breeze through this quickly for the record. (If you understood the discussion of the barbell, there's really nothing new here except bigger numbers, and we won't even see those since we'll write things more abstractly, as sums.) Thus, suppose an extended object is composed of N massive particles with masses m_i and positions $\vec{r}_i$. We can then define the total mass and center of mass position by

$$M = \sum_{i=1}^{N} m_i \tag{4.10}$$

and

$$\vec{R}_{CM} = \frac{1}{M} \sum_{i=1}^{N} m_i \vec{r}_i \tag{4.11}$$

The CM point's velocity and acceleration are then defined by differentiation:

$$\vec{V}_{CM} = \frac{1}{M} \sum_{i=1}^{N} m_i \vec{v}_i \tag{4.12}$$

and

$$\vec{A}_{CM} = \frac{1}{M} \sum_{i=1}^{N} m_i \vec{a}_i. \tag{4.13}$$

We are assuming that the motion of each individual particle is governed by Newton's laws, i.e.,

$$m_i \vec{a}_i = \vec{F}_i^{net} = \vec{F}_i^{ext} + \sum_{j \neq i} \vec{F}_i^j \tag{4.14}$$

and that Newton's third law holds:

$$\vec{F}_i^j = -\vec{F}_j^i. \tag{4.15}$$

Now, just following what we did for the case $N = 2$ above, we may simply sum Equation 4.14 for all the particles:

$$\sum_{i=1}^{N} m_i \vec{a}_i = \sum_{i=1}^{N} \left(\vec{F}_i^{ext} + \sum_{j \neq i} \vec{F}_i^j \right). \tag{4.16}$$

The internal forces will cancel pairwise, leaving, on the right hand side, only the total external force:

$$\vec{F}^{ext} = \sum_{i=1}^{N} \vec{F}_i^{ext}. \tag{4.17}$$

Equation (4.16) can then be re-written, using Equation (4.13), as

$$\vec{F}^{ext} = M \vec{A}_{CM} \tag{4.18}$$

as expected. So what we found above for the barbell is really general: any multiple-particle system (whose individual particles all obey Newton's laws of motion) will obey Equation (4.18).

§ 4.2 Finding the Center of Mass

Before exploring some interesting examples of the principle developed in the first section, let's pause to (again) practice "setting up integrals", this time to find the location of the center of mass of some given object. The main formula we'll be using is just the definition of the center of mass point,

$$\vec{R}_{CM} = \frac{1}{M} \sum_{i=1}^{N} m_i \vec{r}_i \tag{4.19}$$

or, rewriting the same thing as a sum over an infinite number of infinitely small pieces (rather than a sum over a finite number of finite-sized pieces),

$$\vec{R}_{CM} = \frac{1}{M} \int \vec{r} \, dm. \tag{4.20}$$

This implies that each Cartesian coordinate of the center of mass point can be found by taking the mass-weighted average of the corresponding Cartesian coordinates of the small pieces that compose the body. For example,

$$X_{CM} = \frac{1}{M} \int x \, dm, \tag{4.21}$$

and similarly for Y_{CM} and Z_{CM}.

We should think of Equation (4.21) as saying the following: to find the x-coordinate of the center of mass of an object, we should imagine slicing the object up into an infinite number of infinitely small pieces; for each piece, we multiply the mass (dm) of the piece by its x-coordinate (x); we add this product, $x\,dm$, up for all of the pieces that compose the object, and divide by the total mass M of the object which, incidentally, can be written as the sum of the masses of all the pieces:

$$M = \int dm. \tag{4.22}$$

Let's illustrate the process with a couple of examples.

4.2.1   First example

Let's find the location of the center of mass point of a stick of length L whose mass-per-unit length λ is given by

$$\lambda(x) = cx \tag{4.23}$$

where c is a constant. Figure 4.2 shows the stick, situated with respect to a coordinate system, and chopped up into little pieces of width dx.

We begin, as always, by focusing our attention on one generic piece, like the one highlighted with cross-hatching in the figure. This piece has a position x along the x-axis, and a length dx. To find its mass dm, we can multiply the mass-per-unit-length at its location by its length, i.e.,

$$dm = \lambda(x)\,dx = c\,x\,dx. \tag{4.24}$$

We are then ready to plug into Equation (4.21):

$$X_{CM} = \frac{1}{M}\int x\,dm = \frac{1}{M}\int_{x=0}^{x=L} cx^2\,dx = \frac{cL^3}{3M}. \tag{4.25}$$

In this form, however, it is very hard to decide whether or not the answer makes sense.

To put it in a more transparently comprehensible form, we can figure out how the total mass M of the stick relates to the constant c from the given expression for the stick's linear mass density. Using Equation (4.22),

$$M = \int dm = \int_{x=0}^{x=L} c\,x\,dx = \frac{cL^2}{2}. \tag{4.26}$$

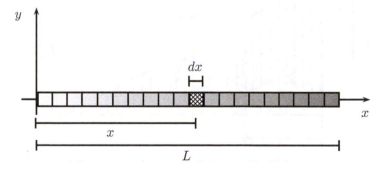

Figure 4.2: A stick of length L has a non-constant mass-per-unit-length $\lambda(x) = cx$, indicated by the shading. To find the location of its center of mass, we divide the stick up into little pieces of (infinitessimal) length dx, add up (by integrating) the mass dm times the position x of each piece, and finally divide by the total mass M.

Plugging this expression for M into Equation (4.25) gives

$$X_{CM} = \frac{2}{3}L \tag{4.27}$$

which is much easier to make sense of! In particular, it is clear that our answer for X_{CM} has appropriate (position) units, because it is just a pure fraction times the length L of the stick. And the particular fraction seems quite reasonable: the center of mass point *should* be closer to the end of the stick where the mass density is higher.

4.2.2 Second example

As a second example, let's consider the triangle shown in Figure 4.3. To find the x-coordinate of the center of mass point, we need to follow the recipe described by Equation (4.21), which remember means that we add up, for each little piece of the triangle, the x-coordinate of the piece times the mass of the piece. But following this recipe requires that each piece *have* a well-defined x-coordinate. Thus, to compute X_{CM} we need to slice the triangle as shown in the Figure – that is, into a bunch of narrow vertical strips, each of which (because its width dx is supposed to be infinitely small) *has* a sharply-defined x coordinate despite being extended in the y-direction.

We focus our attention, to begin with, on one generic piece of the triangle (highlighted by darker shading in the Figure). It is essentially a rectangle of width dx and height $y = H - (H/W)x$. (This is just the equation for the line that defines the top edge of the triangle and hence the height of the little piece at horizontal position x.) The area of our little piece is therefore

$$dA = dx\, y = \left(H - \frac{H}{W}x \right) dx. \tag{4.28}$$

This is helpful to know because now we can write the mass of the little piece using "mass equals mass-per-unit-area times area":

$$dm = \sigma\, dA = \frac{M}{A}\, dA = \frac{2M}{HW} \left(H - \frac{H}{W}x \right) dx \tag{4.29}$$

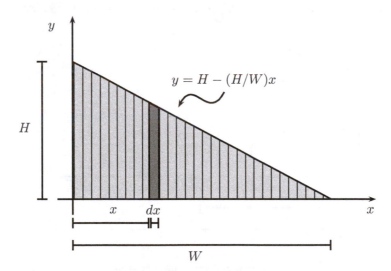

Figure 4.3: *A right triangle of width W and height H has a uniform mass-per-unit-area $\sigma = M/A = 2M/HW$. To find the x-coordinate of its center of mass point, we need to slice the triangle up into narrow vertical strips, each of which has a sharply-defined x-coordinate.*

where we have used the fact that the total area of the triangle is $A = HW/2$.

We are now ready to plug the expression for dm into Equation (4.21):

$$X_{CM} = \frac{2}{HW} \int_{x=0}^{x=W} x \left(H - \frac{H}{W} x \right) dx. \tag{4.30}$$

The integral is readily carried out and gives the result

$$X_{CM} = \frac{1}{3} W \tag{4.31}$$

which seems very reasonable: the x-coordinate of the center of mass point *should* be proportional to the width W (i.e., the units are correct), and it makes sense that the center of mass point is less than half of the way to the right-hand edge since the triangle is taller on the left. So the answer makes sense.

We can also calculate the y-coordinate of the center of mass of this triangle, but a crucial point is that doing so requires slicing it up in a new way! To see why, look at the formula that we will use to calculate the y-coordinate of the CM point:

$$Y_{CM} = \frac{1}{M} \int y \, dm. \tag{4.32}$$

This says: add up, for each little piece of the triangle, the y-coordinate of the piece times the mass of the piece; then divide the result by the total mass M. But look at the generic piece of the triangle highlighted in Figure 4.3. This piece *does not have* a sharply-defined y-coordinate. It is, so to speak, all over the place in y! So we just cannot implement the recipe that is being described in Equation (4.32) if we slice the triangle up as depicted in Figure 4.3.

It should be clear, though, that we can implement the recipe if we instead slice the triangle up as depicted in Figure 4.4. That is, to compute the y-coordinate of the center of mass point, we need to slice the triangle into thin horizontal strips, each of which has a sharply-defined y-coordinate.

The procedure is then basically the same as before. The area of the highlighted generic piece can be written

$$dA = x \, dy = \left(W - \frac{W}{H} y \right) dy \tag{4.33}$$

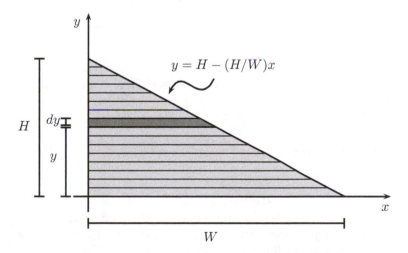

Figure 4.4: To find the y-coordinate of the center of mass point of the same triangle, we need to instead slice it up into thin horizontal strips, each of which has a sharply-defined y-coordinate.

where we have expressed the width x of the piece in terms of its y-coordinate y using, again, the equation for the line that defines the top/right edge of the triangle. We thus have

$$dm = \sigma \, dA = \frac{2M}{HW} \left(W - \frac{W}{H}y \right) dy \tag{4.34}$$

and therefore

$$Y_{CM} = \frac{2}{HW} \int_{y=0}^{y=H} y \left(W - \frac{W}{H}y \right) dy = \frac{1}{3}H \tag{4.35}$$

which makes perfect sense, for basically the same reasons as before: the y-coordinate of the center of mass point should be a little lower than $y = H/2$ since the triangle is wider at the bottom than the top, i.e., more of its mass is closer to $y = 0$ than to $y = H$.

You'll have the opportunity to practice with several more examples like these in class during the week.

§ 4.3 Isolated Binary Systems

One important implication of Equation (4.18) is that the center of mass point of a system on which no external forces act, will not accelerate. Hence, if it is initially at rest, it will remain at rest. But the center of mass point being at rest does not imply that the separate parts of the system are all at rest. The parts can have an interesting, non-trivial motion even as the center of mass point remains motionless.

An important astrophysical example is provided by an isolated system of two gravitationally-bound objects, for example, a binary star system. We sometimes say that the two stars in such a system orbit one another, but it is more accurate to say that each star orbits around the center of mass point of the two-star system, as shown in Figure 4.5. Importantly, both stars move with the same angular velocity ω relative to the center of mass point.

Observing binary star systems like this allows astronomers to determine the masses of distant stars. Let's understand how. To begin with, if we take the center of mass point – the black dot in Figure 4.5 – as the origin of our coordinate system, it follows from the definition of the center of mass, $X_{CM} = (m_1 x_1 + m_2 x_2)/M$, that $0 = m_1 R_1 - m_2 R_2$, i.e.,

$$\frac{m_1}{m_2} = \frac{R_2}{R_1}. \tag{4.36}$$

Thus, if the orbital radii of the two stars can be determined from observation, the ratio of the two stars' masses can be found.

But the *sum* of the two stars' masses can also be determined. To see how, apply Newton's second law to both orbits. Each star experiences a (centripetal) gravitational force from the other star. Thus we have, for the star of mass m_1,

$$\frac{Gm_1 m_2}{(R_1 + R_2)^2} = m_1 \omega^2 R_1 \tag{4.37}$$

(where we have re-written the centripetal acceleration $a = v^2/R$ using $v = \omega R$ as $a = \omega^2 R$). Similarly, for the other star, we have

$$\frac{Gm_1 m_2}{(R_1 + R_2)^2} = m_2 \omega^2 R_2. \tag{4.38}$$

The left hand sides of these two equations are the same, which implies that their right hand sides are equal to each other. That just reproduces Equation (4.36). But if we instead solve the first equation for m_2 and the second for m_1 and add, we find that

$$m_1 + m_2 = \frac{\omega^2}{G}(R_1 + R_2)^3 \tag{4.39}$$

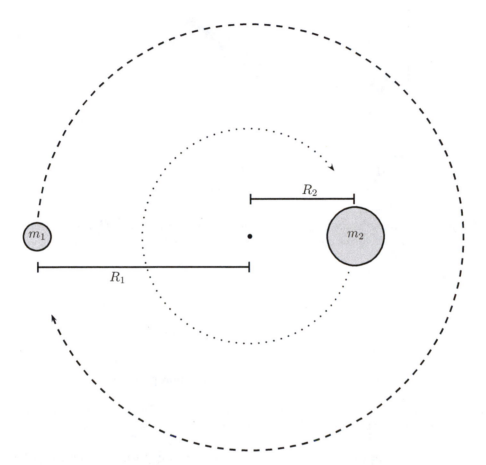

Figure 4.5: Two objects (e.g., the two stars in a binary star system) that attract one another can each orbit around their mutual center of mass point (such that the center of mass point – the black dot here – remains at rest).

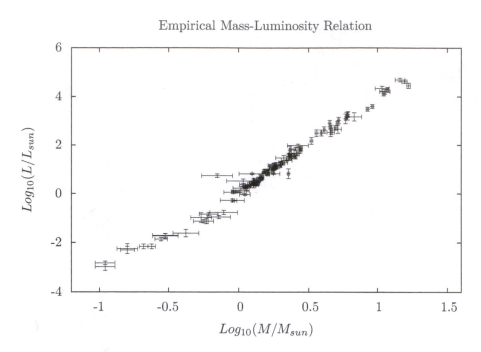

Figure 4.6: *Graph of the mass-luminosity relation for main sequence stars. (Data are from D.M. Popper, "Stellar Masses", Ann. Rev. Astron. Astrophys., 1980, 18:115-64.)*

or, if we re-write in terms of the (shared) orbital period $T = 2\pi/\omega$,

$$m_1 + m_2 = \frac{4\pi^2}{GT^2}(R_1 + R_2)^3. \tag{4.40}$$

which can be understood as a version of Kepler's Third Law appropriate for binary systems.

Anyway, with both the ratio m_1/m_2 and sum $m_1 + m_2$ of the two masses known, it is clear that each mass can be independently determined from observation.

This technique of determining the masses of distant stars (which happen to be part of a binary system) has allowed astronomers to discover important correlations between mass and other stellar properties. In particular, it seems that the intrinsic brightness (aka "luminosity") of a star, and also its surface temperature (which we can see as its color), are uniquely determined by the total mass of the star, at least for so-called "main sequence" stars. See Figure 4.6, for example, for a plot of the Mass-Luminosity relation, as determined directly from observation of nearby binary star systems.

Let us finally mention that, although we have been focusing on the case in which both stars in a binary star system move in perfectly-circular orbits (about their mutual center of mass point), non-circular orbits are also possible. In fact, not too surprisingly given what we've seen in recent weeks, the trajectories of the individual stars in a binary system will, in general, be *ellipses*, with the center of mass point occupying one of the focal points of both elliptical orbits, as sketched in Figure 4.7. In class you may have the chance to simulate a two-body system and explore different possible orbits like the one shown here.

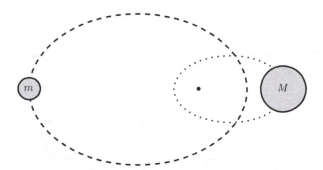

Figure 4.7: In general, the orbits of the two stars in a binary star system are not perfectly circular, but rather elliptical.

§ 4.4 Momentum

Let us consider a different sort of isolated two-body system: suppose there is a large rock of mass m_2 just floating at rest in outer space, far from any other nearby stars or planets that would exert significant gravitational forces on it; and suppose a second, smaller rock of mass m_1 is headed toward it with speed v_0 as shown in the top frame of Figure 4.8

The center of mass position of the two-rock system, represented by the black dot in the Figure, will be somewhere along the line connecting the two rocks, closer to the heavier one, in accordance with Equation (4.11). And as the moving rock approaches the initially-stationary one, the center of mass point will be moving with velocity given by Equation (4.12):

$$\vec{V}_{CM}^{before} = \frac{m_1 \cdot v_0 \hat{i} + m_2 \cdot \emptyset \hat{i}}{m_1 + m_2} = \frac{m_1}{m_1 + m_2} v_0 \hat{i}. \tag{4.41}$$

If the two rocks collide and stick together, they will evidently move off together, as a single body of mass $m_1 + m_2$, with some final speed v_f. The center of mass point of this now single body will, of course, just be located at its position, so the velocity of the center of mass point, after the collision, will just coincide with the velocity of the body itself:

$$\vec{V}_{CM}^{after} = v_f \hat{i}. \tag{4.42}$$

But if the two-rock system is isolated from its environment – i.e., if no external forces act on the system during the time period in question – then it follow from Equation (4.18) that the acceleration of the center of mass point, $\vec{A}_{CM}$, is zero, i.e., that $\vec{V}_{CM}$ is *constant*. And so, in particular, the values before and after the collision, as given in Equations (4.41) and (4.42), should match. It follows that the speed with which the now-combined rocks must move off together after the collision should be given by

$$v_f = \frac{m_1}{m_1 + m_2} v_0. \tag{4.43}$$

You have probably seen this same result derived, in a different way, using the idea of "momentum conservation". We came at it here from a slightly different direction in order to emphasize the fact that the following two ideas are completely equivalent:

1. The center of mass of an isolated system moves with constant velocity.

2. The total momentum of an isolated system is constant.

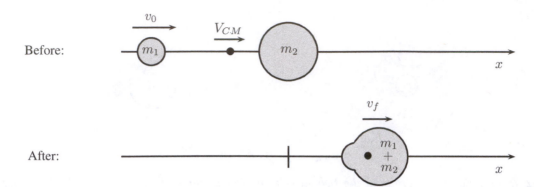

Figure 4.8: A one-dimensional "sticky" collision between two rocks in space. Assuming the two-rock system is isolated (no external forces act on it), the velocity V_{CM} of the center of mass point will remain constant throughout the collision process. This allows us to determine the final speed v_f (with which the two stuck-together rocks move after the collision) in terms of the initial speed v_0 and the two masses.

We can see this equivalence more explicitly by recalling, from Equation (4.12), that the velocity of the center of mass point is given by

$$\vec{V}_{CM} = \frac{1}{M}\sum_i m_i \vec{v}_i. \tag{4.44}$$

Defining the momentum of a particle to be the product of its mass and velocity

$$\vec{p}_i = m_i \vec{v}_i \tag{4.45}$$

and moving the total mass M to the other side, the definition of $\vec{V}_{CM}$ is thus equivalent to

$$M\vec{V}_{CM} = \sum_i \vec{p}_i. \tag{4.46}$$

The constancy of $\vec{V}_{CM}$ – which follows from Equation (4.18) for isolated systems – is thus equivalent to the statement that the *total momentum*, $\sum_i \vec{p}_i$, remains constant in time, or, as it is more commonly said, the isolated system's momentum is conserved.

Let's quickly rehearse how we could use this idea of momentum conservation as an alternative way of thinking about how to derive Equation (4.43). Considering again the situation depicted in Figure 4.8, we would say that, before the collision, the total momentum of the two-rock system is given by

$$\vec{P}_{total}^{before} = m_1 v_0 \hat{i} + m_2 \cancel{0}\hat{i} = m_1 v_0 \hat{i} \tag{4.47}$$

since the rock with mass m_2 is initially at rest and therefore has momentum equal to zero. After the collision, the total momentum of the system is just equal to the momentum of its now-only body:

$$\vec{P}_{total}^{after} = (m_1 + m_2)v_f \hat{i}. \tag{4.48}$$

But then, the constancy of the total momentum throughout the collision process – i.e., the fact that, because the system is isolated, its momentum is conserved – implies that $\vec{P}_{total}^{before} = \vec{P}_{total}^{after}$, so that

$$v_f = \frac{m_1}{m_1 + m_2}v_0 \tag{4.49}$$

as before.

Note that although we have focused here on the simplest possible example of a process in which the parts of an isolated system have some nontrivial interaction – namely, a "sticky collision" in which one of the objects is initially at rest – the principle applies quite generally. So, for example, the total momentum of an isolated two-body system remains constant during a collision (or, equivalently, the velocity of the center of mass point of the system remains constant) even if both bodies are initially moving, even if the bodies do not stick together afterwards, even if the collision is irreducibly two-dimensional (rather than one-dimensional), etc. And the same principles apply as well for interactions that one would not describe as "collisions" (e.g., the mutual orbiting that we discussed previously), and even if the number of bodies in the system is larger than two.

All that is required for momentum conservation is that the system be *isolated*, meaning that there is *no external force* on the system, so that the only relevant forces are *internal* forces – i.e., forces that the several parts of the system exert on one another. Momentum conservation for isolated systems is thus ultimately a consequence of Newton's Third Law, which states that these internal forces come in equal-and-opposite *pairs*, whose effects on the momentum of the system as a whole *cancel*.

§ 4.5 Non-Isolated Systems

We have been focusing on the important special case of isolated systems. As we have seen, for such systems, the center of mass point moves with a constant, unchanging velocity and, equivalently, the total momentum of the system is conserved.

But the close relationship between the motion of the center of mass point, and the purely external forces acting on the system, as expressed in Equation (4.18), remains true whether the system is isolated or not. This provides a powerful technique for analyzing and understanding the behavior of systems which *do* experience external forces. Let us first illustrate with a simple example, and then discuss a more complicated (but more important and interesting) case.

4.5.1   The simple example

Consider again the rigid barbell with which we began in Section 1. If this object were floating in splendid isolation in outer space, such that the only force acting on either ball was the force exerted on it (via the mysterious massless rod connecting them) by the other ball, it is pretty clear what sorts of motion are possible. If the center of mass point of the barbell is at rest (or, equivalently, if we adopt a frame of reference in which the center of mass point is at rest), then basically the barbell can simply *rotate*, with each ball making a circular orbit like those pictured in Figure 4.5. On the other hand, if the center of mass point is moving with some nonzero constant velocity (or, equivalently, if we adopt a frame of reference in which the center of mass point is moving), then the two individual balls will move in uniform circular motions *relative* to the now-moving center of mass point.

The resulting trajectories of the individual balls will be a little more interesting in this case, but overall the situation is still basically straightforward, because the motion of the balls relative to the center of mass point, and the motion of the center of mass point itself, are decoupled (i.e., not in any way interacting with each other) and each of these motions is relatively simple.

Now imagine that instead of floating in outer space, our barbell has been thrown into the air near the surface of the earth. (But let us assume that air drag is negligible.) That is, suppose that instead of being an isolated system, the barbell is now subject to nonzero external forces – in particular, the downward gravitational forces exerted on each ball.

The situation is depicted in Figure 4.9. The interesting thing here is that although the trajectories of the individual balls can be rather complicated (as suggested by the dotted curve in the Figure), this

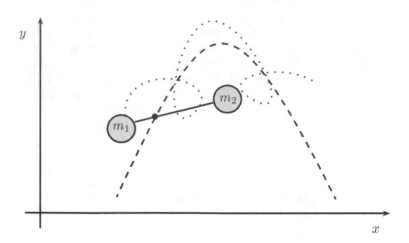

Figure 4.9: A barbell is thrown near the surface of the earth. If the barbell is rotating as it moves, the trajectories of the individual balls can be rather complicated, as suggested by the dotted curve. But this complicated motion can be understood as nothing but uniform circular motion, of each ball, relative to the center of mass point (the black dot), while the center of mass point itself moves along the familiar parabolic trajectory that an equivalent pointlike particle (with total mass $M = m_1 + m_2$) – a kind of proxy for the barbell system as a whole – would move along.

motion can be analyzed into two simple components, just as in the case where the system is isolated. In particular, each ball here simply moves with uniform circular motion about the (moving) center of mass point (represented by the black dot in the Figure). And, simultaneously, the center of mass point itself moves along the familiar parabolic trajectory indicated by the dashed line.

In short, the behavior of this system can be completely understood using the following scheme.

First, we analyze the behavior of the system as a whole, by treating it as an equivalent particle located at the center of mass point of the real system. The total *external* force on the real system is given by the sum of the gravitational forces on the two balls, that is,

$$\vec{F}_{ext} = -m_1 g\hat{j} - m_2 g\hat{j} = -(m_1 + m_2)g\hat{j}. \tag{4.50}$$

This is the same net force that would be experienced by a particle of mass $M = m_1 + m_2$ located at the position of the center of mass of the actual barbell. How would this equivalent particle move? Well, applying Newton's Second Law tells us that it would have acceleration given by

$$\vec{A} = -g\hat{j} \tag{4.51}$$

which means it would move with the familiar sort of parabolic trajectory characteristic of projectiles (when air drag is negligible):

$$\vec{R}(t) = [X_0 + V_0^x t]\,\hat{i} + \left[Y_0 + V_0^y t - \frac{1}{2}gt^2\right]\hat{j} \tag{4.52}$$

where X_0, V_0^x, Y_0, and V_0^y are constants that depend on the initial conditions.

So, by treating the real system as an equivalent particle, subject to the same external force as the real system is subject to, we can first calculate the trajectory of the center of mass point of the real system.

Then, second and separately, we can analyze the motion of the individual particles relative to one another, ignoring the motion of the center of mass. For our rigid barbell system, this just means that the individual balls may be "orbiting" about their mutual center of mass point. That is, if the barbell is set spinning when it is launched, each ball will undergo uniform circular motion about the center of mass point. The apparently-complicated motion of each ball can thus be understood as a simple compounding of two quite simple and comprehensible motions.

4.5.2   The Earth-Moon System

The analysis scheme we have just been describing is powerful and important. But one of the main reasons for going into it, is to appreciate that it does not always work exactly! Let us develop this point by considering the Earth-Moon system and its orbit about the Sun.

If the Earth-Moon system were completely isolated, it is clear that its behavior would be just like the binary star system we considered in Section 4.3 – in particular, the two objects, the Earth and the Moon, would orbit about their mutual center of mass point, in (roughly circular but in fact slightly elliptical) orbits. But of course – and thankfully! – the Earth-Moon system is not isolated. Both objects experience not only gravitational attractions toward one another (these being internal forces as far as the Earth-Moon system is concerned) but also gravitational attractions toward the Sun. (We set aside here the additional gravitational forces that are in fact exerted on the Earth and Moon by other, less massive, planets and moons in the solar system.)

The discussion in the previous sub-section might suggest that you could analyze the three-body Sun-Earth-Moon system in the way that is illustrated in Figure 4.10. First, treat the Earth-Moon system as an equivalent particle (whose mass is equal to the total mass of the Earth and Moon combined and whose position is just the position of the center of mass of the Earth-Moon system). As we have seen

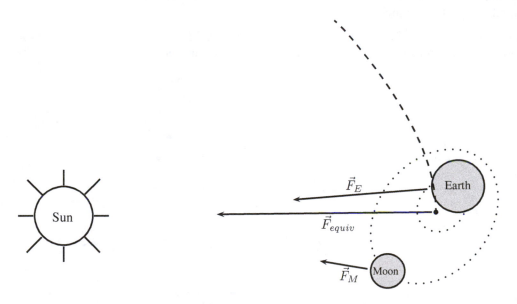

Figure 4.10: The motion of the Earth and Moon can be approximately – but only approximately! – understood by assuming that each body moves in a Keplerian elliptical orbit about their mutual center of mass point (the black dot), which in turn follows the Keplerian orbit about the Sun that an equivalent particle (with the same mass as the Earth-Moon system, and located at its center of mass point) would follow. (Note that the objects and distances in the Figure are not at all to scale!)

previously, this equivalent particle should move in a nice Keplerian elliptical orbit about the Sun. (The dashed curve in the Figure represents this motion.) Second, just compound this Keplerian elliptical orbit of the center of mass point, with the (also elliptical) orbits that we know the Earth and Moon would each have, relative to their mutual center of mass point, if the Earth-Moon system were isolated. (The dotted curves in the Figure represent this motion.) Thus, the complicated-looking trajectories of the Earth and Moon separately, could be understood as nothing but the combination of two rather simple (Keplerian elliptical) motions.

This way of analyzing things certainly provides a helpful perspective and a decent approximation to the actual behavior. But, importantly, it is not quite right. The reason it is not quite right is that – unlike the simpler case of the barbell near the surface of the Earth – the effect of the external forces here is not quite perfectly equivalent to a simple translational motion of the center of mass of the system.

One aspect of this is fairly straightforward to understand. The external forces on the Earth-Moon system are just the gravitational forces exerted by the Sun on the Earth and Moon, respectively. But the gravitational force exerted on an object by the Sun will be different, depending on the precise location of that object. For example, the forces exerted by the Sun on the Earth and on the Moon will be in slightly different directions, as shown in Figure 4.10. But this, combined with the inverse-square-law distance dependence of the forces, turns out to imply that the actual exernal force on the Earth-Moon system

$$\vec{F}_{ext} = \vec{F}_E + \vec{F}_M \tag{4.53}$$

is *not quite the same as* the gravitational force that would be exerted, by the Sun, on the "equivalent particle" whose mass is equal to the combined mass of the Earth and Moon and whose position is just the location of the center of mass of the Earth-Moon system. That is,

$$\vec{F}_{equiv} \neq \vec{F}_E + \vec{F}_M \tag{4.54}$$

and so the motion of the center of mass point of the actual Earth-Moon system is *not* exactly the perfect Keplerian ellipse that the equivalent particle would follow.

The other aspect of the failure of our simple analysis scheme is a little harder to understand, but is quite interesting and important. Consider first Figure 4.11, which shows a rough map of the acceleration

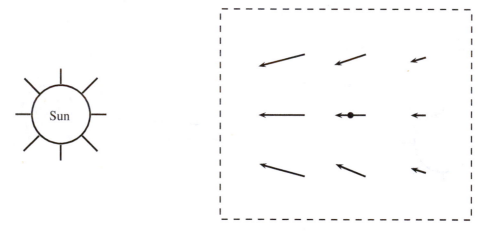

Figure 4.11: A rough map of the acceleration that an object would experience, due to its gravitational attraction toward the Sun, at several different locations near the center of mass point of the Earth-Moon system. Because the gravitational force falls off with increasing distance from the Sun, and is always directed toward the Sun, the accelerations have slightly different directions and magnitudes at the different points. (These differences have been somewhat exaggerated in the Figure.)

that an object would experience, due to its gravitational attraction toward the Sun, at several different locations near the center of mass point of the Earth-Moon system. The important point is that the magnitude and direction of the acceleration are different at different locations. (This is precisely the thing that was *not* the case near the surface of the Earth, where, at least to an excellent approximation, the acceleration that an object in freefall would have is just $\vec{a} = -g\hat{j}$ regardless of its position.)

Now, if we are trying to let a hypothetical "equivalent particle" at the center of mass location proxy for the real Earth-Moon system, it is interesting to consider the *difference* between the acceleration that the actual Earth and Moon would have (as a result of their attractions toward the Sun), and the acceleration that this hypothetical particle located at the center of mass point would experience. It is this difference which tells us something about the *error* involved in using the "equivalent particle" as a stand-in for the real system. A rough map of this difference is shown in Figure 4.12. Let's think about the implications of this "relative acceleration" map for several different possible Earth-Moon orientations.

Suppose first that it is Full Moon, so that the Earth is located a little to the left of the black dot and the Moon is a little to the right. The "equivalent particle" scheme would imply that the effect of the gravitational attraction toward the Sun is *merely* to pull the Earth and Moon equally toward the Sun, so that the center of mass point accelerates toward the Sun but without any effect on the *relative* positions of the Earth and Moon. But looking at the "relative acceleration" map, it is clear that, around Full Moon, the Earth accelerates toward the Sun, and the Moon accelerates *away* from the Sun, *relative to what the "equivalent particle" would do*. That is, in addition to accelerating toward the Sun, the Earth-Moon system is "stretched out" somewhat away from the center of mass point. The same thing would happen at the New Moon configuration as well.

But at the two Half Moon configurations, when the Moon and Earth are a little above and a little below the black dot, the effect is just the opposite: (in addition to accelerating toward the Sun,) the Earth-Moon system is now "squished in" somewhat toward the center of mass point.

And in between, when the line connecting the Earth and Moon makes a diagonal in the Figure, the Earth-Moon system is (in addition to accelerating toward the Sun) *rotated* by the gravitational forces

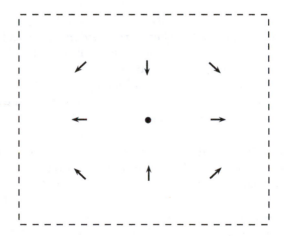

Figure 4.12: A similar rough map, over the same region as in Figure 4.11, but this time showing the acceleration that an object at various points would feel relative to the acceleration that would be felt by an object at center of mass of the Earth-Moon system, represented here by the black dot. It is this relative acceleration which tells us about the error associated with pretending that the gravitional effects, on the Earth-Moon system, of the attraction toward the Sun, are perfectly captured by the translational motion of the center of mass point.

from the Sun – clockwise if the Earth and Moon are along the diagonal that goes up and to the right, and counterclockwise if the Earth and Moon are along the diagonal that goes down and to the right.

To summarize, as the Earth and Moon orbit about their mutual center of mass point, the center of mass point roughly orbits the Sun as an "equivalent particle" located at that center of mass point would do. But only roughly, because the total force exerted by the Sun on the Earth-Moon system is not quite the same as the total force that would be exerted on that "equivalent particle". And then, in addition, there is a residual effect, of the two bodies' attractions toward the Sun, on the *relative* motion of the Earth and Moon. It is as if the Earth-Moon system were being "kneaded" slightly (stretched one way, then twisted, then squished in, then twisted back the other way, etc.) as it orbits the Sun. And needless to say, this "kneading" produces somewhat different motions, of the two objects, relative to their mutual center of mass, compared to the simple Keplerian orbits that would obtain for a truly isolated two-body system like the ones we studied in Section 4.3.

All of this explains why it was rather difficult to extract, from Newton's theory of gravity, accurate predictions for the motion of the Moon relative to the Earth. The theory does, at the end of the day, account for those motions correctly, but it is more complicated than analyzing the motion of a single planet around the Sun.

We are also seeing here that, although certain approximation schemes may provide insight and predictions that are good enough for many purposes, a true 3-body system (like that comprising the Sun, Moon, and Earth) cannot be reduced to two easily-solvable two-body systems. Indeed, modern chaos theory in some sense grew out of the recognition, by Henri Poincaré and others starting in the 1800s, that 3-body systems allowed for qualitatively novel behaviors such as non-periodic, never-repeating orbits.

The Sun-Earth-Moon system is thus a nice concrete illustration of one of the things that makes physics really interesting and beautiful: even very simple laws, which give rise to simple behavior in simple situations, can produce complicated behavior, and often qualitative surprises, in only-slightly-less-complicated situations.

Note, finally, that what we have been describing here – the residual effect on the Earth-Moon system that is not captured by treating the Earth-Moon system as an "equivalent particle" and by assuming that it is only the internal forces which determine their relative motion – is usually discussed by saying that in addition to exerting an attractive force on the Earth-Moon system, the Sun also exerts *tidal* forces on that system. These tidal forces tend to stretch the system out along the line connecting the system with the Sun and squish it along the direction perpendicular to that line.

The reason for the name – "tidal" – is that this is exactly the same effect that explains the phenomenon of ocean tides on the Earth. If you treat the Earth as an equivalent particle, located at the Earth's center of mass, you would say that the effect of the Moon's gravitational attraction is to make the Earth accelerate toward the Moon. And that is true: it is this attraction which provides the centripetal acceleration associated with the Earth's orbit around the Earth-Moon system's center of mass point. But in addition to this overall acceleration toward the Moon, there are also residual effects which are different at different locations on Earth. In particular, the tidal forces from the Moon stretch the Earth along the line connecting the Earth with the Moon, (producing two high-tide bulges in the oceans around the points that are along this line), and squish it perpendicular to this line (low tide).

So that, in a quick nutshell, is how Newton's theory of gravity accounts for the ocean tides – something that was mentioned, but not really explained, in the previous Chapter.

Questions:

Q1. Is the center of mass point of an object necessarily a point *on* the object? For example, the center of mass of a solid (uniform density) disk is just the center point of the disk: you could draw a dot there. Is it always possible to draw a dot on the center of mass point of an object? Explain.

Q2. Imagine that you have been asked to find the y-coordinate of the center of mass point of the half-disk shown here:

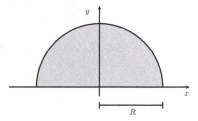

You know that to do the calculation you have to imagine chopping the half-disk up into small pieces. Three possible methods of chopping occur to you: (1) you could chop it into narrow vertical strips, (2) you could chop it into long skinny horizontal slices, or (3) you could chop it into thin half-circle-shaped arcs like the different colors in a rainbow. Which method should you use, and why? You should explain, by reference to the formula you will use to make the calculation, why the other two methods are not viable.

Q3. Astronomers observe a distant star whose luminosity is about ten thousand times (i.e., 10^4) that of our Sun. This star is *not* part of a binary star system. What do you think is the mass of the star? Explain the reasoning carefully.

Q4. Summarize, in your own words, why (for example) the relative motion of the Earth and the Moon is not quite what you would expect if you ignored the external forces acting on the Earth-Moon system.

Q5. The Earth's oceans have two "high tide bulges" (one just under the Moon, and one ont he opposite side of the Earth). Sketch an (exaggerated) diagram showing the shape of the Earth's oceans and the Moon. Indicate on your sketch where, on the Earth, it is low tide. Use your diagram to explain why (for most places on Earth) the tide is observed to rise and fall twice per day. Where on Earth (and/or under what astronomical conditions) could qualitatively different behavior be observed?

Exercises:

E1. The rectangle of height H and width W has a non-constant mass-per-unit area given by $\sigma = cx$ where x is the distance from the left edge. Find the x- and y-coordinates of its center of mass point.

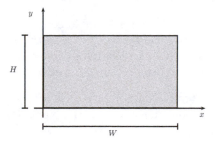

E2. The center of mass point of this (uniform density, radius-R) semi-circle lies somewhere along the y-axis, i.e., $X_{CM} = 0$. What is Y_{CM}?

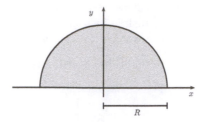

E3. A solid, uniform density cone has a height H and has its radius-R circular base in the x-y-plane. The center of mass point is, by symmetry, located somewhere along the z-axis. What is Z_{CM}? (It may be helpful to recall – or, better, derive – the fact that the volume of a cone is one-third the area of its base times its height.)

E4. A solid, uniform-density hemisphere of radius R rests with its circular face on the floor. At what height above the floor is its center of mass?

E5. Imagine a solar system just like ours, but with all of the planets other than Jupiter removed so that we are left with an isolated two-body system comprising the Sun and Jupiter. As we have seen before, Jupiter orbits the Sun with a period $T = 11.86$ years and a radius $R = 5.2$ AU. The mass of Jupiter is 1.9×10^{27} kg, and the mass of the Sun is 2.0×10^{30} kg. What is the speed v with which the Sun circles the mutual center of mass point of the Sun-Jupiter system?

E6. Is the center of mass point of the Earth-Moon system inside, or outside, the body of the Earth? (You can look up relevant data such as the mass of the Earth, the mass of the Moon, the Earth-Moon separation, and the radius of the Earth.)

E7. Develop, in a slightly different way, the connection between Newton's Third Law and momentum conservation that was mentioned at the end of Section 4. In particular: (a) Argue that Newton's Second Law, $\vec{F}_{net} = m\vec{a}$ can be re-written in terms of the momentum $\vec{p} = m\vec{v}$ as $\vec{F}_{net} = d\vec{p}/dt$. (b) Write down Newton's Second Law, in this re-written form, for a set of N particles, notationally distinguishing, for each particle, the internal and external contributions to the net force. (c) Add your N equations together and show that, because of Newton's third law, the total momentum of the system will be unchanging if the external forces vanishes.

E8. A car of mass m_1 is traveling due north through an icy intersection at speed v_1. At the same time, another car of mass m_2 is traveling due east through an icy intersection at speed v_2. Unfortunately, they are in the same intersection, so the cars collide and stick together. Momentum conservation implies that, immediately after the collision, the now-stuck-together cars are moving with a speed $v_f = \sqrt{m_1^2 v_1^2 + m_2^2 v_2^2}/(m_1 + m_2)$. Don't take my word for it, though – confirm for yourself that this is true. Now explain/derive this expression a different way, by calculating the speed with which the two cars' mutual center of mass point is moving prior to the collision.

E9. Prove, by choosing and then analyzing a suitable concrete example, that the gravitational force exerted by the Sun on a two-particle system (such as the Earth-Moon system) is not necessarily exactly equal to the force that would be exerted on an equivalent particle (whose mass is the total system mass and whose position is that of the system's center of mass).

E10. Confirm, using graphical vector subtraction, that the *difference* between the acceleration vector at a given point in Figure 4.11, and the vector at the black dot, looks like the vector drawn at that same point in Figure 4.12.

Projects:

P1. Make a barbell-type object (e.g., by tying two different-mass balls together with a short piece of light string). Capture a video of this object flying, while spinning in some complicated way, across the room. Use computer video analysis software to mark the locations of each ball individually in each frame of the video, and plot their individual trajectories through space. Then have the computer calculate the position of the center of mass point at each moment, and plot the trajectory of this center of mass point through space. It is also illuminating to calculate, and then plot, the trajectory of each ball *relative* to the center of mass point.

P2. Write a computer simulation to calculate the trajectories of two objects, with different masses, that are gravitationally bound. By arranging the initial conditions appropriately, you should be able to reproduce circular trajectories like those shown in Figure 4.5 and elliptical trajectories like those shown in Figure 4.7. You should also find that if you are not careful to ensure that your initial conditions imply that $\vec{V}_{CM} = 0$, your binary system will drift through space, with constant but nonzero $\vec{V}_{CM}$, while the objects orbit the (moving!) center of mass point.

P3. Adapt your simulation from P2 to include a fixed central gravitating body around which your gravitationally-bound two-particle system can orbit. That is, in effect, simulate the Sun-Earth-Moon system (but treating the Sun, for simplicity, as fixed in place). By calculating and plotting the trajectory of the Moon relative to Earth, you should be able to see that the orbit does *not* simply obey Kepler's laws, due to the tidal effects discussed in Section 5. This same code (perhaps with tweaked initial conditions and mass values) can then also be used to simulate a number of other interesting things such as the effect of a Jupiter-like planet on the orbit of a lighter, interior planet; the motion of a light comet whose orbit about the Sun is (perhaps catastrophically!) influenced by an orbiting planet; etc.

P4. Following up on E5, the line-of-sight velocity component of a star can be observed to oscillate sinusoidally, due to the gravitational influence of a large (but not directly observable) orbiting planet. The first planets outside of our own solar system ("extra-solar planets") were discovered, just in recent decades, in this method. You will be given some data for the line-of-sight velocity component of a star, and can attempt to fit a sinusoidal curve to it; the amplitude and period of this sinusoidal oscillation should then allow you to determine the mass and orbital radius of an extra-solar planet.

CHAPTER 5

Rotational Motion

THIS week we tease out the implications of Newton's laws for rotational motion. We begin by developing a new perspective on planetary orbits which allows us to see, more deeply, why Kepler's Second Law – the rule that planets "sweep out equal areas in equal times" – holds. This leads to a more formal development of the rotational-motion analogs of "force", "mass", and "momentum" – namely the concepts of "torque", "moment of inertia", and "angular momentum". After yet another interlude on setting up integrals (this time to calculate the moments of inertia for various different objects) we apply some of the ideas we've developed to a couple of important loose ends in the story of the historical development of Newton's theory of universal gravitation.

§ 5.1 Newton and Kepler's Area Law

One of the crucial arguments Newton developed in support of his theory was a proof that the forces exerted on the planets are directed *toward the Sun*, and hence (like the gravitational forces exerted by the Earth) are associated with a real physical object and not some mere mathematical point (like the "Mean Sun" in Copernicus' model). Let us develop this argument using contemporary concepts and notation.

Here is a simple theorem in Newtonian mechanics. Consider a particle of mass m moving with velocity $\vec{v}$. Its momentum is then $\vec{p} = m\vec{v}$. Let us define a new quantity, its *angular momentum*, as follows:

$$\vec{L} = \vec{r} \times \vec{p} \tag{5.1}$$

where $\vec{r}$ is the position vector for the particle (i.e., the vector that goes from the origin of our coordinate system out to the location of the particle) and the "$\times$" in the equation denotes the vector cross product. For reasons that will emerge as we get deeper into the material this week, this quantity, the "angular momentum", is a kind of rotational-motion analog of (regular old, translational) momentum, in the same way that (for example) the "angular velocity" $\omega = \dfrac{d\phi}{dt}$ is a rotational-motion analog of (regular old, translational) velocity $v = \dfrac{dx}{dt}$.

Let us also define the *torque* – a rotational analog of *force* – on the particle as follows:

$$\vec{\tau} = \vec{r} \times \vec{F} \tag{5.2}$$

where $\vec{F}$ is the force acting on the particle.

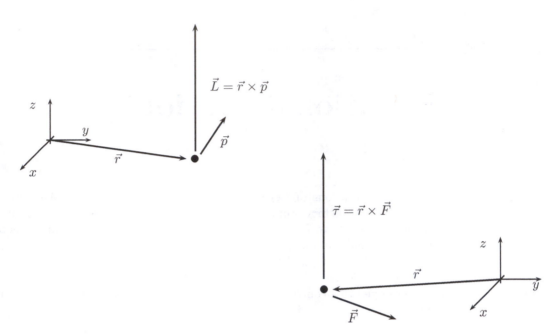

Figure 5.1: *Illustrations of the vector relationships defined in Equations 5.1 and 5.2. The angular momentum $\vec{L}$ of a particle is the vector cross product of its position $\vec{r}$ and momentum $\vec{p}$. The direction of $\vec{L}$ is found by the right hand rule: orient your right hand so your (straightened) fingers point in the direction of $\vec{r}$, then rotate your hand so your fingers can (non-painfully) bend to point in the direction of $\vec{p}$. Your thumb will then point in the direction of $\vec{L}$. Note that the cross product of two vectors is orthogonal to each of the two vectors – so the right hand rule is really just a "rule of thumb" to decide between the two possible directions that are orthogonal to the plane defined by the two original vectors. In the example shown, the idea is that $\vec{r}$ and $\vec{p}$ are in the x-y plane, so their cross product is in the z-direction. The second picture just shows an example of the relationship expressed by Equation 5.2: the torque $\vec{\tau}$ produced by a force $\vec{F}$ exerted on a particle at position $\vec{r}$ is given by the cross product. The direction of the torque is orthogonal to the plane spanned by $\vec{r}$ and $\vec{F}$, as picked out by the right hand rule. It is also useful to use a right hand rule to think qualitatively about the meaning of the vectors $\vec{L}$ and $\vec{\tau}$ which can admittedly be a bit puzzling. If you point the thumb of your right hand in the direction of the vector $\vec{L}$, your fingers will naturally "curl" in a certain "sense" (clockwise or counter-clockwise) in the plane perpendicular to the vector. This "sense" (in which your fingers curl) is a good way to think about the meaning of angular momentum. For example, in the first figure, the angular momentum vector $\vec{L}$ is in the positive z-direction; if you point your right thumb parallel to this vector, your fingers indicate a counter-clockwise orbital motion in the x-y plane, which is precisely what the particle is doing. The meaning of the direction of the torque vector $\vec{\tau}$ can be understood in the same way. In the example shown, if you orient your right thumb parallel to $\vec{\tau}$, your fingers naturally curl to indicate (again) a counter-clockwise "sense" in the x-y plane. This is precisely what a torque in the z-direction means: the force tends to "twist" the particle counter-clockwise in the x-y plane.*

All this talk of "rotational analogs" suggests there might be some connection between torque and angular momentum. In particular, from this formulation of Newton's second law

$$\vec{F} = \frac{d\vec{p}}{dt} \tag{5.3}$$

(where here $\vec{F}$ represents the net force, the sum of all the individual forces that act) one might guess the "analogous" rotational formula:

$$\vec{\tau} = \frac{d\vec{L}}{dt}. \tag{5.4}$$

(where $\vec{\tau}$ is the net torque on the particle, i.e., the sum of the torques produced by all of the individual forces that act).

This turns out to be precisely right. It is not a new postulate, but instead a theorem based on the earlier definitions of torque and angular momentum:

$$\frac{d\vec{L}}{dt} = \frac{d}{dt}\left(\vec{r} \times \vec{p}\right) = \frac{d\vec{r}}{dt} \times \vec{p} + \vec{r} \times \frac{d\vec{p}}{dt} = \vec{r} \times \vec{F} = \vec{\tau} \tag{5.5}$$

where we have used the facts that $d\vec{r}/dt = \vec{v}$ and $\vec{v} \times \vec{p} = 0$ since the two vectors are necessarily parallel.

Let us focus to begin with on an important special case: if the angular momentum is constant, so that $d\vec{L}/dt = 0$, then the net torque $\vec{\tau}$ must vanish.

You maybe already have noticed (back in the Chapter 3 Projects) that Kepler's second law is mathematically equivalent to the statement that

$$\vec{r} \times \vec{v} = \text{constant} \tag{5.6}$$

where $\vec{r}$ is the position vector of the planet relative to the Sun, i.e., we take the Sun as the origin of our coordinate system. Figure 5.2 and its caption explain this connection in a little more detail. But the basic idea is pretty simple: since the left hand side ($\vec{r} \times \vec{v}$) differs from a planet's angular momentum only by a factor of the mass m of the planet, it should be clear that Kepler's second law is also equivalent to the statement that, for each planet,

$$\vec{L} = \text{constant}. \tag{5.7}$$

And this, according to the theorem, implies that the torque exerted on the planet (with the origin taken as the Sun) must vanish. What could ensure this? The crucial point is that (with the Sun as the origin, so $\vec{r}$ points along the line connecting the Sun to the planet in question) the force $\vec{F}$ must point directly back toward the Sun. Then $\vec{r}$ and $\vec{F}$ are (anti-) parallel, so their cross-product, i.e., the torque associated with this force, necessarily vanishes.

In fact, Newton stressed that not only do the planets sweep out equal areas in equal times with respect to the Sun, but the Earth's moon sweeps out equal areas in equal times with respect to the (center of the) Earth (but only approximately, due to the additional "tidal" contributions from the Sun), and the moons of Jupiter and Saturn sweep out equal areas in equal times with respect to those planets. Thus, wherever we have one body orbiting another in the heavens, the orbiting body sweeps out area (relative to the central body) at a constant rate. And so, according to the above argument, in all of these cases, the force causing the orbit is directed precisely toward the central body that is being orbited around. It then seems irresistible to conclude that the orbits are produced by forces *exerted by* that central body.

In short, Kepler's Second Law requires that the force which holds a given planet in its orbit point directly toward the Sun. This was obviously a powerful piece of evidence for the idea that the force was exerted *by* the Sun and, more generally, in support of Newton's theory of universal gravitation.

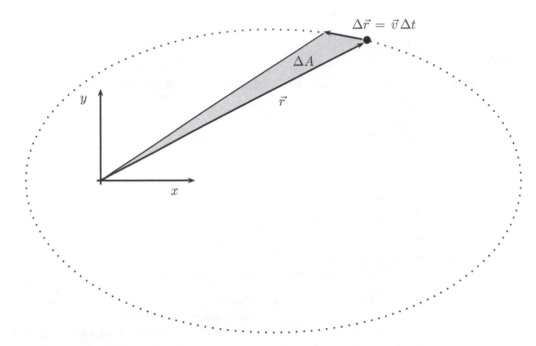

Figure 5.2: The vector $\vec{r}$ represents the position of a planet, with the Sun taken as the origin. $\Delta\vec{r}$ represents its displacement during a short time period Δt. For sufficiently small Δt, the average velocity during the time interval is close to the instantaneous velocity at the moment shown, so $\Delta\vec{r} = \vec{v}\,\Delta t$. The cross-product $\vec{r} \times \Delta\vec{r}$ is a vector pointing, by the right hand rule, out of the page. Its magnitude is the area of the parallelogram spanned by $\vec{r}$ and $\Delta\vec{r}$. This area is just twice the shaded area ΔA. Thus, $|\vec{r} \times \vec{v}| = 2\Delta A/\Delta t$. And so the magnitude of the planet's angular momentum is $|\vec{L}| = |\vec{r} \times \vec{p}| = m|\vec{r} \times \vec{v}| = 2m\Delta A/\Delta t$. Hence, the area swept out by the planet per unit time – $\Delta A/\Delta t$ – is proportional to the magnitude of the planet's angular momentum about the Sun. The constancy of $\Delta A/\Delta t$ (coupled with the fact that the orbit takes place in a fixed plane, so the direction of $\vec{L}$ is constant) thus implies the constancy of $\vec{L}$ which, as discussed in the text, implies $\vec{\tau} = 0$. The vanishing of the torque is naturally explained by $\vec{r}$ being parallel (or anti-parallel!) to $\vec{F}$, i.e., a force that is directed precisely toward the Sun.

§ 5.2 Rotational Dynamics of a Rigid Body

In the previous section we developed the rotational-motion analog of Newton's Second Law, $\vec{F}_{net} = d\vec{p}/dt$, namely

$$\vec{\tau}_{net} = \frac{d\vec{L}}{dt} \tag{5.8}$$

for the simplest possible system: a single particle.

Last week we showed that multi-particle systems obey a kind of holistic version of Newton's second law (with the center of mass point functioning as "the position" for the system as a whole), if each particle composing the system obeys Newton's Laws. Let us explore a similar path here, and see if we can develop a "holistic" version of the rotational-motion analog of Newton's Second Law, Equation (5.8), which applies to a "rigid body", i.e., an object which can be understood as many particles "glued" together.

5.2.1   Pure Rotation about the CM point

Let's begin with a simple case: our old friend the barbell, which we now take to be pinned in place (but free to rotate about) its center of mass point, as shown in Figure 5.3.

Let us allow that each of the two balls is subject to some external force as well as a force exerted on it by the other ball (via the massless rod). The net torque exerted on the two-particle barbell system will then be

$$\vec{\tau}_{net} = \vec{\tau}_1 + \vec{\tau}_2 \tag{5.9}$$

where $\vec{\tau}_i$ is the net torque exerted on particle i. With our origin at the center of mass point, and with the internal forces that the balls exert on each other (via the massless rod) along the line connecting them, the internal forces will produce no torque, and the torque on each particle (and hence on the barbell system as a whole) will be due exclusively to external forces.

Assuming each particle individually obeys Newton's laws, we may apply Equation (5.8) to each particle and arrive at

$$\vec{\tau}_{net} = \frac{d\vec{L}_1}{dt} + \frac{d\vec{L}_2}{dt} = \frac{d}{dt}\vec{L} \tag{5.10}$$

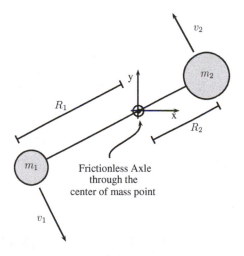

Figure 5.3: A barbell, composed of particles of mass m_1 and m_2 (and a massless rod connecting them), which is pinned in place at and free to rotate about its center of mass point.

where $\vec{L} = \vec{L}_1 + \vec{L}_2 = \vec{r}_1 \times \vec{p}_1 + \vec{r}_2 \times \vec{p}_2$ is the total angular momentum of the barbell. This is already a kind of rotational-motion analog to Newton's second law for the barbell as a whole, but we can process it into a somewhat simpler form for this special case.

Let us assume, for simplicity, that all the external forces lie in the x-y-plane, so that the net torque is purely in the z-direction. With the barbell constrained to rotate about an axle perpendicular to the x-y-plane, the angular momenta $\vec{L}_i$ of each ball will also be purely in the z-direction. So we can drop the vector signs (i.e., strictly speaking, just talk about the z-components of all the vectors, because those are the only components that are nonzero) and just write

$$\tau_{net} = \frac{dL}{dt}. \tag{5.11}$$

Furthermore, for each ball, the angular momentum L_i can be written as

$$L_i = R_i p_i = R_i m_i v_i = m_i R_i^2 \omega \tag{5.12}$$

where, for each particle, the (translational) velocity v_i has been re-written as ωR_i using the angular velocity ω with which both particles rotate about the center of mass point.

Putting the pieces together, we see that we can write

$$\tau_{net} = \frac{d}{dt} \left[\left(m_1 R_1^2 + m_2 R_2^2 \right) \omega \right] \tag{5.13}$$

or, more simply,

$$\tau_{net} = I\alpha \tag{5.14}$$

where $\alpha = d\omega/dt$ is the *angular acceleration* of the barbell (about its center of mass point), and

$$I = m_1 R_1^2 + m_2 R_2^2 \tag{5.15}$$

is called its "moment of inertia" (again, about the center of mass point).

Equation (5.14) is in some sense the simplest version of the rotational-motion analog of Newton's second Law, and there is a perfect correspondence between the three quantities involved and their translational-motion partners: the net torque τ_{net} (due, here, exclusively to *external* forces) is the rotational-motion analog of the net force F_{net}; the moment of inertia I is the rotational-motion analog of the mass m; and the angular acceleration α is the rotational-motion analog of the (regular, translational) acceleration a.

Not surprisingly, this same result applies to any object (not just a two-particle barbell) which is rotating about an axle that goes through its center of mass point. See, for example, Figure 5.4 which depicts a random blob, pinned in place and free to rotate about its center of mass point. The total angular momentum of the object (technically just its z-component, but that is the only nonzero component here) can be written

$$L = \sum_i r_i p_i = \sum_i r_i m_i v_i = \sum_i m_i r_i^2 \omega = I\omega \tag{5.16}$$

where

$$I = \sum_i m_i r_i^2 \tag{5.17}$$

is the object's moment of inertia.

Assuming that each constituent particle obeys Newton's laws then implies that

$$\tau_{net} = \frac{dL}{dt}. \tag{5.18}$$

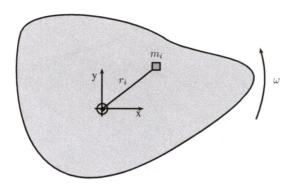

Figure 5.4: An arbitrary random-blob-shaped object, which is pinned in place through, and free to rotate about, its center of mass point. A generic piece of the blob can be thought of as a particle of mass m_i whose distance from the center of mass point is r_i.

so that, again,

$$\tau_{net} = I\alpha. \tag{5.19}$$

Note again that the torques associated with internal forces cancel one another out, so that the net torque τ_{net} results exclusively from external forces. We will discuss the reasons for this in more detail in class since they are somewhat subtle and interesting. Note also that Equation (5.18) reduces to (5.19) only for a *rigid body*, i.e., one for which the relative positions of all the particles – and in particular the distances r_i of each particle from the center of mass point – remain fixed.

5.2.2   The fully general case

In the previous subsection, we assumed that our object was pinned in place and free to rotate about an axle through its center of mass. This is obviously a fairly special situation! So let us step back and consider the fully general situation, in which an object is rotating (but not necessarily around any fixed axle) and simultaneously moving, translationally.

We'll skip the barbell and jump straight in with the generic random blob, as shown in Figure 5.5. As indicated in the Figure, the position $\vec{r}_i$ of each of the blob's constituent particles can be written in terms of the position of the center of mass point $\vec{R}_{CM}$ and the position $\vec{r}_{i,CM}$ of the particle *relative to* the center of mass point:

$$\vec{r}_i = \vec{R}_{CM} + \vec{r}_{i,CM}. \tag{5.20}$$

Similarly (or actually this follows by just differentiating this last equation with respect to time), the velocity $\vec{v}_i$ of each constituent particle can also be thought of as the velocity of the CM point *plus* the velocity of the given particle *relative to* the CM point:

$$\vec{v}_i = \vec{V}_{CM} + \vec{v}_{i,CM}. \tag{5.21}$$

For a rigid body, the velocity of a given particle relative to the CM point will just be the tangential velocity associated with circular motion about the CM point (because the distance of the particle from the CM point cannot change).

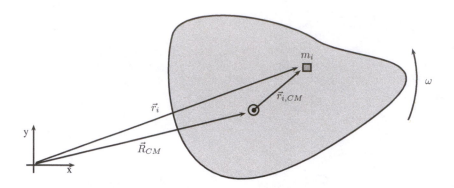

Figure 5.5: An arbitrary random-blob-shaped object, which is now free to not only rotate, but also to move translationally. We can think of the position vector $\vec{r}_i$ of one of its generic constituent particles, relative to the origin of our coordinate system (which is now no longer required to be at the center of mass point), as the sum of the position vector $\vec{R}_{CM}$ of the center of mass point, plus the position vector $\vec{r}_{i,CM}$ of the generic particle relative to the center of mass point. Similarly (but not depicted), we can think of the velocity of the generic particle as the velocity of the center of mass point plus the velocity of the particle relative to the center of mass point. This allows the beautiful way of expressing the total angular momentum that is developed in the text.

Let us define an angular velocity vector $\vec{\omega}$, whose magnitude is just the angular velocity of the object as a whole (i.e., the angular velocity that each of its constituent particles moves with, about the CM point), and whose direction is perpendicular to the plane in which that rotational motion occurs, with a direction given by the right hand rule. So, for example, if the object is rotating counter-clockwise in the x-y-plane, as shown in the Figure, we would say that $\vec{\omega} = \omega\hat{k}$, i.e., the angular velocity vector points out of the page toward us.

We can then say that the velocity of a constituent particle relative to the CM point, $\vec{v}_{i,CM}$, is perpendicular to both ω and $\vec{r}_{i,CM}$. A little contemplation (and fiddling with the right hand rule for cross-products) reveals that it can be written

$$\vec{v}_{i,CM} = \vec{\omega} \times \vec{r}_{i,CM} \tag{5.22}$$

so that

$$\vec{v}_i = \vec{V}_{CM} + \vec{\omega} \times \vec{r}_{i,CM}. \tag{5.23}$$

All right. Now let us write down an expression for the total angular momentum of our moving blob:

$$\vec{L} = \sum_i \vec{L}_i = \sum_i m_i \vec{r}_i \times \vec{v}_i. \tag{5.24}$$

Using Equations (5.20) and (5.23) gives

$$\begin{aligned}
\vec{L} &= \sum_i m_i \left(\vec{R}_{CM} + \vec{r}_{i,CM} \right) \times \left(\vec{V}_{CM} + \vec{v}_{i,CM} \right) \\
&= \sum_i m_i \left[\vec{R}_{CM} \times \vec{V}_{CM} + \vec{r}_{i,CM} \times \vec{V}_{CM} + \right. \\
&\qquad\qquad \left. \vec{R}_{CM} \times (\vec{\omega} \times \vec{r}_{i,CM}) + \vec{r}_{i,CM} \times (\vec{\omega} \times \vec{r}_{i,CM}) \right].
\end{aligned} \tag{5.25}$$

Let's consider each of the four terms separately.

Since $\vec{R}_{CM}$ and $\vec{V}_{CM}$ are constants, the first term can be written as

$$\left(\sum_i m_i \right) \vec{R}_{CM} \times \vec{V}_{CM} = M\vec{R}_{CM} \times \vec{V}_{CM} \tag{5.26}$$

where $M = \sum_i m_i$ is just the total mass of the object. With $\vec{P}_{CM} = M\vec{V}_{CM}$, this is just the angular momentum we would expect for a single particle whose mass is the same as the total mass of our blob and whose position and velocity matched the position and velocity of the center of mass point of our blob.

The second term can be thought of like this:

$$\left(\sum_i m_i \vec{r}_{i,CM} \right) \times \vec{V}_{CM}. \tag{5.27}$$

But the sum in parentheses is necessarily zero, by definition of the center of mass point! (The mass-weighted average position of all the particles just is the center of mass position. So the mass-weighted average position of all the particles, relative to the center of mass position, must vanish.) Therefore this entire term is zero.

The third term, which can be written

$$\vec{R}_{CM} \times \left(\vec{\omega} \times \left(\sum_i m_i \vec{r}_{i,CM} \right) \right), \tag{5.28}$$

is zero for the same reason.

So that leaves only the fourth term:

$$\sum_i m_i \vec{r}_{i,CM} \times (\vec{\omega} \times \vec{r}_{i,CM}). \tag{5.29}$$

But it is fairly straightforward to show that this can be written as just

$$I\vec{\omega} \tag{5.30}$$

where

$$I = \sum_i m_i (r_i^{\perp})^2 \tag{5.31}$$

is just the moment of inertia of the object about an axis (parallel to $\vec{\omega}$) through the center of mass. (The notation $r_i^{\perp}$ just means the distance of the ith constituent particle, not from the center of mass point itself, but from the rotation axis through the center of mass point. This distinction will be slightly hard to understand since we have been thinking about flat 2D objects which are easy to draw. But we'll have the opportunity to see exactly what $r_i^\perp$ means in Section 5.3 when we calculate the moments of inertia for some 3D objects.)

So, to summarize, the total angular momentum of our flying blob can be written in the following very elegant way:

$$\vec{L} = \vec{R}_{CM} \times \vec{P}_{CM} + I\vec{\omega}. \tag{5.32}$$

The first term is often called the "orbital angular momentum". It is the angular momentum that the object has by virtue of its overall translational motion about the origin. If we apply this to the Earth (and use the Sun as the origin), we would say that the Earth has orbital angular momentum (about the Sun) because it is *orbiting* the Sun. The second term, associated not with the overall translational motion of the object but rather with its purely rotational motion, is called the "spin angular momentum". The Earth has spin angular momentum (in addition to its orbital angular momentum) by virtue of its daily rotation.

Now, for the same reasons we have already seen, the net torque on an object will equal the rate of change of its angular momentum:

$$\vec{\tau}_{net} = \frac{d\vec{L}}{dt}. \tag{5.33}$$

That's the most fundamental way to express the rotational-motion analog of Newton's Second Law. The interesting thing here is that, in the most general possible case (an extended rigid body which simultaneously moves translationally and rotates), the right hand side can be understood as just the sum of the two simple expressions we encountered in the simple examples we began with: a single particle (for which $\vec{L} = \vec{r} \times \vec{p}$) and an object that is constrained to rotate about a fixed axle (for which $\vec{L} = I\vec{\omega}$).

5.2.3 Example

Let's apply some of these ideas to a simple and familiar example situation to help concretize everything.

Let's take the case of a solid (uniform density) ball, of mass M and radius R, rolling without slipping down a ramp that makes an angle θ with the horizontal, as shown in Figure 5.6. The figure shows the

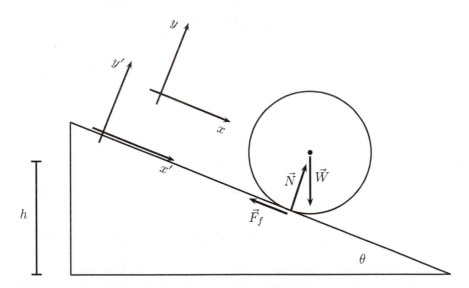

Figure 5.6: A ball rolling down a ramp, with forces shown as in a free body diagram.

three forces that act on the ball: the weight (i.e., gravitational) force $\vec{W}$, the normal force $\vec{N}$, and a friction force $\vec{F}_f$.

We will analyze the situation using two different coordinate systems, to illustrate how certain things (like the net torque) are different depending on what origin point one picks but how, nevertheless, the predictions come out the same in the end.

Let's start with the x-y-system shown in the Figure.

- Newton's (regular, translational) 2nd law, $\vec{F}_{net} = m\vec{a}$, should apply. Since the ball will have no motion in the y-direction, the y-component of this equation just tells us the size of the normal force: $|\vec{N}| = |W_y| = Mg\cos(\theta)$. The x-component tells us that

$$Mg\sin(\theta) - F_f = Ma_x. \tag{5.34}$$

- Let's then consider the rotational-motion analog of Newton's 2nd law. For this system and using this coordinate system, $\vec{R}_{CM}$ and $\vec{V}_{CM}$ will always be perfectly parallel, so the ball will have no orbital angular momentum. Thus $L = I\omega$ and $\tau_{net} = I\alpha$. With this reference frame, all three forces produce a nonzero torque, but the torques due to the normal force ($\vec{N}$) and the weight force $\vec{W}$ add up to zero. So the net torque is just the torque due to the friction force $\vec{F}_f$, which is: $\tau_f = RF_f$ and therefore the rotational-motion analog of Newton's 2nd law thus tells us that

$$RF_f = I\alpha. \tag{5.35}$$

- Finally, we can impose the "rolling-without-slipping" condition,

$$a_x = \alpha R. \tag{5.36}$$

Algebraically eliminating α and F_f from the previous three equations, we can solve for the (translational) acceleration of the ball:

$$a_x = \frac{g\sin(\theta)}{1 + \frac{I}{MR^2}} \tag{5.37}$$

which makes sense. An object which just slides frictionlessly down the ramp will move with acceleration $a_x = g\sin(\theta)$. The ball – which *rolls* because of the friction force which prevents it from sliding – moves with a smaller acceleration. Incidentally, you will have the opportunity to derive, in class, the formula for the moment of inertia of a solid ball: $I = \frac{2}{5}MR^2$. So the expression for the translational acceleration of the ball can be simplified to $a_x = \frac{5}{7}g\sin\theta$.

Now, let's go through that same analysis, but using the x'-y'-coordinate system from the Figure, to see what is different and what is the same.

- The regular translational version of Newton's 2nd law is the same: the y'-component just tells us the size of the normal force, and the x'-component tells us that

$$Mg\sin(\theta) - F_f = Ma_{x'}. \tag{5.38}$$

- The rotational version of Newton's 2nd law is a little bit different, though. Now, the ball *will* have both orbital and spin angular momentum. In particular, $L = RMV_{CM} + I\omega$ so that

$$\tau_{net} = RMA_{CM} + I\alpha = RMa_{x'} + I\alpha. \tag{5.39}$$

What is the net torque? Well, now the friction force produces no torque (because it is directed right back toward the origin). The normal force and the y'-component of the weight force both produce torque, but these cancel one another. So it is only the x'-component of the weight force that produces torque, and this is: $\tau_{W_{x'}} = RMg\sin(\theta)$. Thus, the rotational version of Newton's 2nd law reads:

$$RMg\sin(\theta) = RMa_{x'} + \frac{I}{R}a_{x'}. \tag{5.40}$$

where we have used the fact that the same rolling-without-slipping condition as before still applies: $a_{x'} = \alpha R$.

Notice especially that the rotational version of Newton's 2nd law is completely different than it was before. (The expression for L is different, because now there is orbital angular momentum in addition to spin angular momentum, and the expression for the net torque is also completely different.) But solving Equation (5.40) for the acceleration gives the same result as before, Equation (5.37) – as, of course, it must.

It is interesting to note that, if our goal is just to solve for the translational acceleration of the ball, the algebra is easier when we use the x'-y'-system. The reason for this is that, using this coordinate system, the friction force produces no torque, so the variable F_f does not appear in Equation (5.40). This illustrates an important practical point: although the answer should always be the same at the end of the day no matter where you put the origin of your coordinate system, it is often possible to use the freedom to choose a coordinate system to make the math work out in a simpler way than it otherwise might. Some of the Exercises will illustrate this point further.

§ 5.3 Calculating Moments of Inertia

There is a lot going on in that last section, and we will only have time to dip our toes into the shallow end this week. (A future course on Classical Mechanics will provide an opportunity to go into greater depth for those interested!) But we can make some headway on a few of the ideas. Let's start with the "moment of inertia", which can be thought of as the rotational-motion analog of mass: in the same sense that we can think of mass as "resistance to (translational) acceleration" (i.e., the bigger the mass is, the less acceleration will result from a given applied force), so the "moment of inertia" of a body (about a certain axis) can be understood as its "resistance to angular acceleration" (i.e., the bigger the moment of inertia is, the less angular acceleration will result from a given applied torque).

The formula for the moment of inertia that popped out of our analysis in the last section is:

$$I = \sum_i m_i(r_i^\perp)^2 = \sum_i m_i\rho_i^2 \tag{5.41}$$

where $r_i^\perp = \rho_i$ is the (perpendicular) distance from the particle whose mass is m_i to the rotation axis.

This way of writing the formula for the moment of inertia is appropriate for an object composed of a finite number of particles with (finite) masses m_i. For a continuous object, we can always imagine chopping it up into an infinite number of pieces with infinitesimal masses dm. In that case, the more appropriate way to write the formula is:

$$I = \int \rho^2\,dm. \tag{5.42}$$

Let us practice using this formula to calculate the moments of inertia of some simple objects.

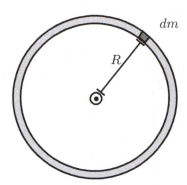

Figure 5.7: To find the moment of inertia of a ring (for rotation about the symmetry axis, in and out of the page, through the black dot in the Figure) we imagine chopping the ring up into (infinitely small) pieces of (infinitely small) mass dm.

5.3.1   First example

As a first example, consider a ring of total mass M and radius R, rotating about its symmetry axis as shown in Figure 5.7.

We should imagine chopping the ring up into an infinite number of infinitely small pieces. One such generic piece is highlighted in the Figure. Its mass is dm and its distance from the rotation axis is just $\rho = R$. We may therefore write that its contribution dI to the moment of inertia of the whole ring is

$$dI = R^2 \, dm. \tag{5.43}$$

The moment of inertia of the entire ring is then the sum of the contributions from all the little pieces, i.e.,

$$I = \int dI = \int R^2 \, dm. \tag{5.44}$$

But every piece of the ring has exactly the same distance from the axis. The R^2, that is, is a *constant*, that can come outside the integral. So we are left with

$$I = R^2 \int dm = MR^2 \tag{5.45}$$

which makes sense, since all of the mass is at the same distance, R, from the rotation axis.

5.3.2   Second example

As a second example, let's calculate the moment of inertia of a thin cylindrical shell of height H, radius R, and total mass M, as shown in Figure 5.8.

As usual, we imagine chopping the object up into an infinite number of infinitely small pieces, and focus our attention, to begin with, on a generic piece like the one highlighted in the Figure. This piece has mass dm and its distance from the rotation axis is just the radius R of the cylinder. Thus,

$$dI = R^2 \, dm. \tag{5.46}$$

To find the moment of inertia of the whole cylinder, we simply add up the contributions from all the pieces:

$$I = \int dI = \int R^2 \, dm = R^2 \int dm = MR^2. \tag{5.47}$$

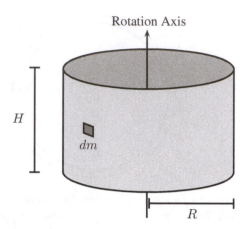

Rotation Axis

H

dm

R

Figure 5.8: To find the moment of inertia of a cylindrical shell (about the symmetry axis shown) we imagine chopping it up into an infinite number of infinitely small pieces of mass dm. Since the ρ in Equation (5.41) means the distance from the rotation axis – rather than the distance from any particular point on the rotation axis, such as the center of mass point – every piece of the cylinder has the same ρ, namely, R.

Just as with the ring, every piece of the cylinder has the same distance from the rotation axis, and so there isn't really a non-trivial integral to actually do. We just need to recognize that $\int dm$ means the sum of the masses of all the little pieces, i.e., the total mass M. And so the formula for the moment of inertia of a thin cylindrical shell, about its symmetry axis, is the same as that for a ring.

The lesson of this second example – the only reason for bothering with it, actually – is just to stress the point that the ρ (aka $r^{\perp}$) in the formula for the moment of inertia, means the distance from (the closest point on!) the rotation axis... *not* the distance from the center of mass point or any other particular point.

5.3.3   Third example

OK, let's consider an example where we actually have to do a non-trivial integral. Suppose there is a thin rectangular slab of total mass M (uniformly distributed over its area), height H, and width W, rotating about the y-axis as shown in Figure 5.9.

If we imagine slicing the rectangle into thin vertical strips, like the one that is highlighted in the Figure, we can see that each strip has a sharply defined distance from the rotation axis (despite being extended in the y-direction), namely: $\rho = x$. The area of the little piece is

$$dA = H\,dx \tag{5.48}$$

and so its mass is

$$dm = \sigma\,dA = \frac{M}{HW}H\,dx = \frac{M}{W}\,dx \tag{5.49}$$

since the mass-per-unit-area σ is just the constant $M/A = M/HW$.

The contribution of the one highlighted slice to the moment of inertia is thus

$$dI = \rho^2\,dm = \frac{M}{W}x^2\,dx. \tag{5.50}$$

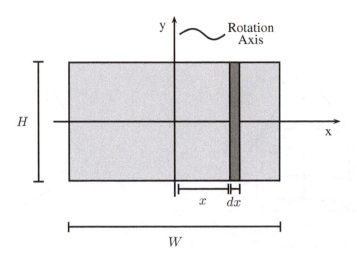

Figure 5.9: Diagram for finding the moment of inertia of a rectangular slab. The highlighted piece has a sharply-defined distance from the rotation axis (namely, $\rho = x$) despite being extended in the y-direction.

All that is left to do is to add up the contributions from all the little pieces:

$$I = \int dI = \frac{M}{W} \int_{x=-W/2}^{x=W/2} x^2 \, dx = \frac{1}{12} MW^2. \tag{5.51}$$

Does this result make sense? The units (mass times the square of a length) are appropriate. And what about the pre-factor, 1/12? Well, if all the mass were concentrated at the left and right edges, whose distance from the rotation axis is $W/2$, the moment of inertia would evidently be $I = M(W/2)^2 = \frac{1}{4}MW^2$. So it makes sense that we get a pre-factor smaller than 1/4 since, in fact, the mass of our rectangle is not concentrated out at the edges: some of the mass is even closer to the axis than that, and so the pre-factor should be even smaller than 1/4.

5.3.4   Fourth example

Let's do one last example that requires us to think about a 3D object. Suppose there is a solid cone of mass M (uniformly distributed throughout its volume), whose height is H and whose base is a circle of radius R. The cone is rotating about its symmetry axis, which we can take to be the z axis as shown in Figure 5.10.

There are, of course, many ways to slice a cone. But, to use the formula

$$I = \int \rho^2 \, dm \tag{5.52}$$

we have to slice the cone up into little pieces, of mass dm, each of which has a sharply-defined distance, ρ, from the rotation axis. There is one and only one way to do that (at least, if we are going to perform only a single integral at the end of the day!), and it's a little hard to visualize: we have to slice the cone into a bunch of nested, co-axial cylindrical shells. One such shell, of radius r and (infinitely small) thickness dr, is highlighted, from two different perspectives, in the Figure. The height of the highlighted shell can be written

$$z = H - \frac{H}{R}r \tag{5.53}$$

Side View: Top View:

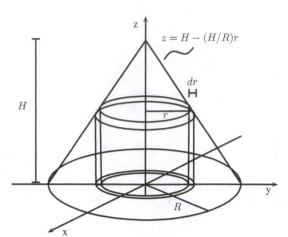
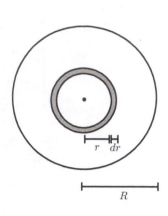

Figure 5.10: A 3D object like this cone must be sliced into nested, concentric cylindrical shells to calculate the moment of inertia, because a cylindrical shell is the only 3D shape all of whose parts are the same distance from a rotation axis.

which is basically just the equation for the cone's top surface. (If you replace r with y, this is just the equation for the line, slanting down and to the right in the Figure, from the apex of the cone at the top to the point where the edge of the base intersects the y-axis. But of course the cone is rotationally symmetric about the z-axis, so it is appropriate to write this in terms of $r = \sqrt{x^2 + y^2}$ instead of just y.)

The volume of our little piece can be written

$$dV = 2\pi r\, z\, dr = 2\pi r \left(H - \frac{H}{R}r \right) dr \tag{5.54}$$

since we can think of it as a rectangular slab of thickness dr, height $z = \left(H - \frac{H}{R}r \right)$, and width equal to the circumference $2\pi r$. (Picture it as a sheet of paper that has been rolled into a cylinder; if we unroll it and lay it flat, it's just a rectangular slab.)

This allows us to write the mass of our little piece (in terms of its density D) as

$$dm = D\, dV = \frac{M}{\pi R^2 H/3}\, dV = \frac{6M}{HR^2} \left(H - \frac{H}{R}r \right) r\, dr \tag{5.55}$$

where we have used the formula for the volume of a cone: one-third the area of the base times the height.

And of course the radius of our little piece – the distance of its sub-pieces from the rotation axis – is just $\rho = r$. So the contribution of this one piece to the moment of inertia is

$$dI = \rho^2\, dm = \frac{6M}{HR^2} \left(H - \frac{H}{R}r \right) r^3\, dr. \tag{5.56}$$

Finally, all that remains to do is to add up the contributions from all the cylindrical shells that jointly compose the cone:

$$I = \int dI = \frac{6M}{HR^2} \int_{r=0}^{r=R} \left(H - \frac{H}{R}r \right) r^3\, dr = \frac{3}{10} MR^2. \tag{5.57}$$

Does this answer make sense? The units are OK. How about the pre-factor, 3/10? It also seems reasonable. Most of the mass of the cone is significantly closer to the rotation axis than $r = R$, so it makes sense that we'd get a pre-factor significantly less than 1.

Hopefully that gives you a sense of how to setup these problems. In the Exercises, you'll have a chance to try a few examples of your own.

§ 5.4 The Cavendish Experiment

Let's return to our historical narrative and see how some of the ideas we've been developing – torque, angular momentum, and moment of inertia – can help us understand how, about a century after Newton, Henry Cavendish first accurately measured what we have been calling "Newton's constant", G, using a "torsional pendulum" – and also why Cavendish described his experiment not as determining G but, rather, as "weighing the Earth".

But first, a little motivational context-setting...

Suppose an object of mass m_1 gravitationally orbits around another object of (much larger) mass M_1, with an approximately circular orbit of radius R_1 and period T_1. The centripetal acceleration of the orbiting object will then be given by

$$a_1 = \frac{v_1^2}{R_1} = \frac{4\pi^2 R_1}{T_1^2}. \tag{5.58}$$

The gravitational force between the objects will have magnitude

$$F = \frac{GM_1 m_1}{R_1^2}. \tag{5.59}$$

Relating the acceleration and force using Newton's second law ($F = ma$), we arrive at a relation between the orbital radius and period and the mass of the central body:

$$\frac{GM_1}{R_1^2} = \frac{4\pi^2 R_1}{T_1^2} \tag{5.60}$$

or equivalently

$$M_1 = \frac{4\pi^2}{G} \frac{R_1^3}{T_1^2} \tag{5.61}$$

which you probably recognize as a statement containing Kepler's third law.

Now suppose there is *another* such orbital system, with an object of mass m_2 orbiting around a (much heavier) object of mass M_2 with orbital radius R_2 and period T_2. Then, by the same argument as above, we should have

$$M_2 = \frac{4\pi^2}{G} \frac{R_2^3}{T_2^2}. \tag{5.62}$$

Now let's divide the last two equations. The proportionality constants (involving Newton's constant G) cancel out, leaving

$$\frac{M_1}{M_2} = \frac{R_1^3/T_1^2}{R_2^3/T_2^2}. \tag{5.63}$$

Thus, if one can empirically determine the orbital characteristics (radius and period) of two orbiting bodies, one can work out the relative masses of the two bodies the orbiting bodies are orbiting around. For example, suppose the first system is the Earth orbiting around the Sun, and the second is the Moon orbiting around the Earth. Then $R_1 = 1\,\mathrm{AU}$, $T_1 = 1\,\mathrm{year}$, $R_2 = 60\,R_{earth}$, and $T_2 = 27.3\,\mathrm{days}$. Plugging in numbers and reducing the units gives

$$\frac{M_1}{M_2} = \frac{M_{sun}}{M_{earth}} = 330,000 \tag{5.64}$$

i.e., the Sun is about three hundred and thirty thousand times more massive than the Earth.

Using the same methods, one can also relate the Earth's mass to that of Jupiter and Saturn, which also have moons that were known to Newton. Based on just this line of reasoning, Newton reports that the mass of Jupiter is about 1/1,000 that of the Sun (or about 330 times that of Earth), with Saturn being about a third that heavy.

It is remarkable that the relative masses of the planets can be so determined. But – perhaps you are already wondering – why not just compute the *masses*? If it is remarkable to know that the Sun is 330,000 times as massive as the Earth, how much more remarkable to know the mass of the Sun *in kilograms*? This would clearly be possible if the mass of the Earth (in kilograms) were known. And one can see from Equation 5.61 that (e.g.) the Sun's mass could also be calculated from the orbital characteristics of (e.g.) the Earth, if only the value of Newton's constant G were known.

But – perhaps surprisingly – neither of these quantities was known to Newton at the time the *Principia* was written. In fact, the value of G was first measured – by directly measuring the strength of the gravitational force between lead balls in a laboratory – about a *century* after the *Principia* was published, by the amateur physicist Henry Cavendish, in 1798, using an apparatus called the "torsional pendulum" which had been invented by Coulomb to measure small electrostatic forces.

Because of the equivalence, noted above, of knowing the value of G and knowing the mass of the Earth, the Cavendish experiment is sometimes referred to as "weighing the Earth." We will describe the experiment in some detail here, with the hope that you will have the opportunity to reproduce it in your class.

To begin with, here is Cavendish's description of the experiment:

> "The apparatus is very simple; it consists of a wooden arm, 6 feet long, made so as to unite great strength with little weight. This arm is suspended in an horizontal position, by a slender wire 40 inches long, and to each extremity is hung a leaden ball, about 2 inches in diameter; and the whole is inclosed in a narrow wooden case, to defend it from the wind. As no more force is required to make this arm turn round on its center, than what is necessary to twist the suspending wire, it is plain, that if the wire is sufficiently slender, the most minute force, such as the attraction of a leaden weight a few inches in diameter, will be sufficient to draw the arm sensibly aside. .... One of these [weights] was to be placed on one side of the case, opposite to one of the balls, and as near it as could conveniently be done, and the other on the other side, opposite to the other ball, so that the attraction of both these weights would conspire in drawing the arm aside; and, when its position, as affected by these weights, was ascertained, the weights were to be removed to the other side of the case, so as to draw the arm the contrary way, and the position of the arm was to be again determined; and, consequently, half the difference of these positions would show how much the arm was drawn aside by the attraction of the weights. .... I resolved to place the apparatus in a room which should remain constantly shut, and to observe the motion of the arm from without, by means of a telescope; and to suspend the leaden weights in such manner, that I could move them without entering into the room."

A sketch of the torsional pendulum apparatus is shown in Figure 5.11.

Analyzing the experiment in more detail will allow us to apply and concretize some of the concepts – in particular "torque" and "angular momentum" – introduced earlier in this chapter. To begin with,

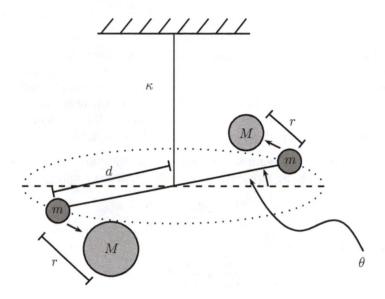

Figure 5.11: Schematic diagram of the Cavendish torsional pendulum apparatus. A "barbell" (made of two masses m connected by a thin rod of length 2d) hangs from a thin fiber. The barbell is thus free to rotate in the horizontal plane. In the absense of external forces, the barbell will orient itself along some equilibrium position (represented by the dashed line). In the Cavendish experiment, two additional balls of mass M are brought in, to a distance r from the respective ends of the barbell. The larger masses then exert (tiny!) gravitational forces on the smaller masses, which cause the barbell to rotate by a (tiny!) angle θ relative to the equilibrium position. Knowing the torsional character of the fiber (represented by the "torsion constant" κ) allows one to infer the absolute magnitude of the gravitational force between the balls, and hence, knowing their masses and separations, the value of Newton's constant G.

note that the motion of the two masses constituting the hanging barbell is in a horizontal plane. Let us pick the center of the barbell (which should never move) as the origin of our coordinate system, and the x-y-plane to be horizontal. Then, because of the vector cross products in the definitions of these quantities, all the relevant quantities (torques and angular momenta) will be purely in the z-direction. So we can represent the vector character of these quantities with a simple plus or minus sign, indicating that the quantity is either "up" or "down" along the z-axis.

Let us first consider the hanging barbell in the absense of the larger masses. If the barbell is rotated slightly with respect to its equilibrium orientation, the twist in the fiber will produce a very tiny torque that will tend to turn the barbell back toward its equilibrium configuration. (This is really the definition of "equilibrium.") For small angular displacements, the resulting torque will be *proportional* to the displacement, i.e.,

$$\tau = -\kappa\,\theta \tag{5.65}$$

where the angle θ is measured from the equilibrium position. Note that a positive angular displacement θ (counterclockwise as seen from above, as shown in the figure) produces a torque that tends to turn the barbell back toward $\theta = 0$ – i.e., a *negative* torque. We thus put an explicit minus sign in the above equation so that the torsional constant of the fiber, κ, is a positive quantity. One should think of κ as characterizing the "rotational stiffness" of the fiber: for example, a thicker fiber will produce a larger restoring torque given the same angular displacement, i.e., will have a larger κ.

Since we have picked the center of the symmetrical barbell (i.e., its center of mass) as the origin about which to calculate torques and angular momenta, the barbell will have only spin angular momentum: $L = I\omega$ and therefore

$$\tau = \frac{dL}{dt} = I\frac{d\omega}{dt} = I\frac{d^2\theta}{dt^2}. \tag{5.66}$$

Combining with Equation (5.65) gives

$$\frac{d^2\theta}{dt^2} = -\frac{\kappa}{I}\,\theta \tag{5.67}$$

where $I = 2md^2$ is the moment of inertia of our barbell.

You might recognize this differential equation as mathematically-equivalent to the equation one arrives at by applying $F = ma$ to a mass m on a spring of spring constant k, namely:

$$\frac{d^2x}{dt^2} = -\frac{k}{m}\,x. \tag{5.68}$$

Such a mass on a spring will undergo simple harmonic motion (meaning that x is a sinusoidal function of time) with period

$$T = 2\pi\sqrt{\frac{m}{k}}. \tag{5.69}$$

But, to quote Feynman, "the same equations have the same solutions". So the torsional pendulum must also undergo (angular) simple harmonic motion with period

$$T = 2\pi\sqrt{\frac{2md^2}{\kappa}}. \tag{5.70}$$

The point of all this is that it allows a way to actually *measure* the torsional constant κ of the fiber. Rearranging, we have

$$\kappa = \frac{8\pi^2 md^2}{T^2}. \tag{5.71}$$

So, given the mass of the balls m and their distance d from the center, the torsional constant of the fiber can be determined by simply setting the torsional pendulum in (rotational) oscillation, and measuring the period.

And that is actually the hard part. The rest of the analysis of the Cavendish experiment will be relatively straightforward. The basic idea is now to bring in two additional balls (mass M) at distances r from the smaller balls, as shown in the Figure. Each of the large balls will exert, on the nearby smaller ball, a force F and hence a torque of magnitude

$$\tau = d\,F = \frac{GMm\,d}{r^2}. \tag{5.72}$$

Since there are two such torques on the barbell, the total torque produced by the gravitational influence between the balls has a magnitude

$$\tau = \frac{2GMm\,d}{r^2}. \tag{5.73}$$

When the heavy balls are put in place, this gravitational torque will cause the barbell to turn slightly and subsequently oscillate about a *new equilibrium* orientation for which the gravitational torque is just cancelled by the restoring torque produced by the (now slightly twisted) fiber. Using Equation (5.65) for the latter, we get an expression for the angle θ of the new equilibrium orientation

$$\kappa\theta = \frac{2GMmd}{r^2} \tag{5.74}$$

which shows that, by *measuring* the angle θ, one can compute Newton's constant G:

$$G = \frac{\kappa\theta r^2}{2Mmd} = \frac{4\pi^2\theta r^2 d}{MT^2}. \tag{5.75}$$

There are a number of subtle issues that arise when one actually performs the experiment. Here we will mention only one, which is that typically one measures not the angle θ between the "original" and "new" equilibrium orientations of the barbell, but rather the angle $\Delta\phi = 2\theta$ between the two "new" equilibrium positions one gets by placing the heavy masses in the two possible ways (where they pull the barbell counterclockwise and clockwise). The resulting expression for Newton's constant G in terms of directly-measureable quantities is then:

$$G = \frac{\kappa\,\Delta\phi\,r^2}{4Mmd} = \frac{2\pi^2\,\Delta\phi\,r^2\,d}{MT^2}. \tag{5.76}$$

We include this here just to make life simpler should you be asked to perform this experiment yourself in class and analyze the results.

For what it is worth, Cavendish's own 1798 experiment implied a final result for G that was within 1% of the best contemporary value:

$$G = 6.67 \times 10^{-11}\ \frac{\mathrm{N\,m^2}}{\mathrm{kg^2}}. \tag{5.77}$$

As mentioned, though, Cavendish himself wasn't really interested in measuring what we now call Newton's constant, but was instead interested in "weighing the earth." What he actually reports as his final conclusion is that the average density of the Earth is 5.48 times that of water, i.e.,

$$\rho = \frac{M_{\mathrm{earth}}}{\frac{4}{3}\pi R_{\mathrm{earth}}^3} = 5.48\,\mathrm{g/cm^3}. \tag{5.78}$$

Plugging in the known radius of the Earth

$$R_{\text{earth}} = 6.37 \times 10^6 \, \text{m} \tag{5.79}$$

and solving for the Earth's mass gives

$$M_{\text{earth}} = 6 \times 10^{24} \, \text{kg} \tag{5.80}$$

from which one can, if desired, compute the mass of the Sun (and other planets) explicitly, by using the relations that began this section. It is amazing and beautiful that Newton's theory of gravitation allows us to infer the masses of distant astronomical bodies from the results of an experiment involving lead balls and string in a terrestrial laboratory!

§ 5.5 Precession

The torsional pendulum that Cavendish used to weigh the Earth (or, equivalently, to determine Newton's constant G) is a nice example of a system for which the rotation axis is fixed. But the general formalism we developed earlier applies in the more general case, too. Let us therefore briefly explore the motion of the familiar toy, the top, to illustrate the power and generality of the results developed above. The amazing thing about a top, of course, is that once it is set spinning, it – as if by magic – refuses to tip over, but instead rotates around with a motion called "precession." To understand how this comes about, let's start by thinking about what happens to a non-spinning object (a ruler, say) if it is placed at an angle and then released.

Figure 5.12 shows the initial situation (in solid lines) and then the situation at some later moment (in dotted lines). We can understand the down-and-to-the-left acceleration of the center of mass point just by considering the down-and-to-the-left net force which is produced jointly by the weight, normal, and friction forces (shown in the figure). But in addition to *translating*, the ruler also *rotates*. This can be understood by considering the torque acting on the ruler, and how this causes its angular momentum to change. If we pick an origin at the point of contact between the ruler and the table, the two contact forces which act at that point will produce no torque. The net torque is then just the torque produced by the weight force. This torque is nonzero in magnitude, and out-of-the-page (counter-clockwise) in direction.

Equation (5.33) tells us that the net torque equals the rate of change of the total angular momentum. Hence, during some short period of time Δt, the angular momentum $\vec{L}$ will change by $\vec{\tau}\Delta t$. Since the ruler is initially at rest, it has zero initial angular momentum, and so its total angular momentum after Δt will be just $\vec{\tau}\Delta t$ – which of course points out-of-the-page. Since the ruler is pivoting about a fixed point, it will have both some orbital and spin angular momentum. The pivoting ruler is similar to the rolling ball, and there is some equation similar to the earlier "rolling without slipping" condition that relates the orbital and spin motions. All that matters here, though, is that the orbital and spin contributions to the total angular momentum will both be in the same direction, hence both of them are out-of-the-page. This means a counter-clockwise orbital motion of the center of mass about the pivot point, and also a counter-clockwise rotational motion of the ruler about its center of mass. Which is just a fancy way of describing the motion we already know will happen as the ruler begins to fall.

The point of discussing the ruler in that way is just to set up a contrast for the following discussion of a top. What is the difference between a (spinning) top and a ruler? Only that the top is spinning – and so *already has considerable angular momentum* when it is let go. Let us run through the same sort of analysis, and see how this changes things.

To begin with, the forces and torques acting on the top are the same as they were with the ruler. With the origin of the coordinate system taken at the contact point between the top and the floor, only the

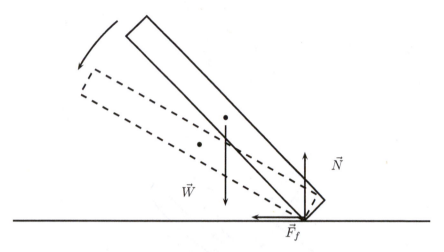

Figure 5.12: A ruler resting on a table, released from rest at an angle, begins to fall. The motion of the center of mass point can be understood on the basis of the three forces shown: a weight force, a normal force, and a friction force. These produce a net force in the "bottom-left" direction. But we can understand this – plus the purely rotational aspect of the falling motion – on the basis of torque and angular momentum, too. Taking the contact point between the ruler and the table as the origin, the normal and friction forces will produce no torque. There is therefore a net torque equal to the torque produced by the weight force. This torque is counter-clockwise or, as a vector, out-of-the-page. The initially stationary ruler has zero angular momentum, and so, at the end of some short period of time Δt, Equation (5.33) requires that the ruler have new total angular momentum $\vec{L} = \vec{\tau}\Delta t$, which will (like $\vec{\tau}$) point out-of-the-page. Since the ruler is pivoting about the fixed point of contact with the table, its total angular momentum will have both an orbital and a spin contribution, and these will be related by something like the "rolling without slipping" condition we've used in previous examples. The details aren't important. What matters is just that the basic dynamical equation for rotational motion requires that the ruler have, after Δt, both some out-of-the-page (counter-clockwise) orbital motion and some out-of-the-page (counter-clockwise) rotational motion – precisely as shown in the Figure and expected from common experience.

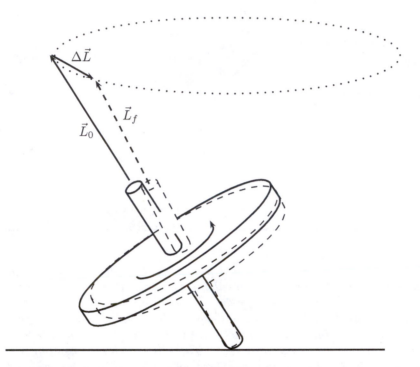

Figure 5.13: A spinning top, released in the same way as the ruler considered previously. Unlike the ruler, the top has some initial (spin) angular momentum. So although the forces and torques acting on the top are the same as the corresponding ones acting on the ruler, the new angular momentum after some short time Δt is quite different. It is, instead of zero plus $\Delta \vec{L} = \vec{\tau}\Delta t$, the large initial spin angular momentum vector – $\vec{L}_0$ in the Figure – plus $\Delta \vec{L}$. The new angular momentum vector – $\vec{L}_f$ in the Figure – is therefore (remember, Δt is small) of the same magnitude as the original one; it just points in a slightly different direction. The angular momentum vector will be again rotated in the next short time interval Δt, and the next, and the next – the net result being that the angular momentum vector (hence also the spin axis of the top) sweeps out a cone as suggested by the dotted-line circle in the figure. This motion – which you can see in real life in class – is called "precession."

weight force contributes to the net torque, and this is out-of-the-page. The key thing is that, where the ruler had *zero* angular momentum initially, the (spinning) top has a *lot* of (spin) angular momentum – as shown by the vector $\vec{L}_0$ in Figure 5.13. It is then required by the basic dynamical equation for rotation, that the angular momentum vector after some short period of time Δt be

$$\vec{L}_f = \vec{L}_0 + \Delta\vec{L} \tag{5.81}$$

where, just as for the ruler, $\Delta\vec{L} = \vec{\tau}\Delta t$ is small and in the out-of-the-page direction.

It is then clear from the properties of vector addition that the new angular momentum vector $\vec{L}_f$ will have (in the limit of small Δt) the same magnitude as $\vec{L}_0$, but will point in a slightly different direction (a little bit toward us, out of the page). What does such a final angular momentum vector tell us about the new state of motion of the top?

You might think the top should start tipping over, just like the ruler did. But this isn't right. Here's why. Suppose it did tip over like the ruler. The falling motion would have associated with it some little bit of angular momentum out-of-the-page, just as it did for the ruler. But in order to have fallen, the spin axis of the spinning top would have to have changed, i.e., the spin angular momentum vector would have to now point slightly more toward the horizontal than it did initially. This would imply a (contribution to) $\Delta\vec{L}$ in the down-and-to-the-left direction. But such a change in the angular momentum would require a *torque* in that same direction. And, simply put, there *is* no torque in that direction! So (in a perhaps not fully satisfying way!) that is the proof that the top, unlike the ruler, cannot just tip over.

We can now start to see what it must actually do. The final angular momentum vector has the same magnitude as the initial one, and is just turned a little bit in direction. This will be achieved if the top doesn't fall at all, but instead just turns its orientation, as shown in the Figure. And then the situation is the same as it was initially (but just now oriented in a slightly different direction) so the same thing happens again in the next short period of time Δt. And so on. The result is a continuous re-orientation of the spin axis around a cone, as suggested by the dotted line path in the Figure. This motion is called *precession*.

As already mentioned, the fact that a spinning top doesn't fall, but instead precesses, seems somehow magical (or at least very counterintuitive), and the above sort of formal analysis somehow leaves one not fully satisfied. There should be no shame in acknowledging this. Actually, what the above "formal analysis" shows is really just that the precessional motion described is a consistent steady state solution. The non-obvious aspect that is so counter-intuitive is: how does the top get into this precessional steady state in the first place?

The answer to this is subtle and complicated, but gesturing vaguely in the direction of some of the subtleties can at least give one a sense that there is nothing magical happening. To begin with, when the top is first released, it *does* actually fall just a tiny bit, at least for a split second. We argued above that this is impossible, but actually it isn't. We neglected to mention it before, but the precessional motion of the top actually implies that the top posseses not only *spin* angular momentum, but also some *orbital* angular momentum. For the orientation of the top shown in the Figure, the center of mass velocity associated with the precessional motion will be out-of-the-page, which implies (by the right hand rule) an orbital angular momentum that is up-and-to-the-right. So actually there is no question of finding some torque to account for the down-and-to-the-left $\Delta\vec{L}$ associated with a tiny bit of initial falling. What actually happens is that, while falling just a bit, the top *converts* a little bit of its *spin* angular momentum into *orbital* angular momentum. The *total* angular momentum is actually *constant*, so no external torques are required to explain a change in $\vec{L}$. There is only some relatively complicated story about internal forces rearranging the overall distribution of total angular momentum.

But that isn't precisely true either! It's true that, when first released, the top falls just a bit, trading some of its spin angular momentum for orbital angular momentum associated with the precessional motion. But the top actually falls a little bit *too far* and overshoots the orientation at which it could stably precess. Something like the same story then happens in reverse. The result is that the spin axis of the top exhibits *another*, secondary sort of precession – it precesses in a (smaller) cone centered on the (moving!) precession axis already discussed. This secondary wobbling motion is called "nutation" and can easily be observed with a real top. Of course, for a real top there will always be a little bit of friction associated with the contact "point" – this works to damp out the nutational motion, helping the top achieve the smooth, steady state precessional motion we discussed at the beginning.

Of course, in time, the same sorts of frictional effects will reduce the magnitude of the spin angular momentum and change the character of the orbital (precessional) angular momentum – until, for example, the spinning part of the top hits the table, substantial new forces are introduced, and the magical motion turns more mundane. It is definitely worth spending some time with a top or gyroscope to observe – and contemplate – some of these effects.

Let us finally calculate the *rate* of the precessional motion (assuming it has reached a constant, steady-state rate). We have already argued that in a short time period Δt, the angular momentum of the top will change by an amount

$$\Delta \vec{L} = \vec{\tau} \Delta t. \tag{5.82}$$

The component of its (spin) angular momentum that is in the horizontal plane will be

$$L_{horizontal} = I\omega \sin \phi \tag{5.83}$$

where ϕ is the angle the top's spin axis makes with the vertical. The precessing top thus sweeps through an angle

$$\Delta \theta = \frac{\Delta L}{L_{horizontal}} = \frac{\tau \Delta t}{I\omega \sin \phi} \tag{5.84}$$

in time Δt. Which means that its precessional angular velocity $\Omega = \Delta \theta / \Delta t$ is

$$\Omega = \frac{\tau}{I\omega \sin \phi}. \tag{5.85}$$

This implies a precessional *period*

$$T_{prec} = \frac{2\pi}{\Omega} = \frac{2\pi I\omega \sin(\phi)}{\tau}. \tag{5.86}$$

For a top, the magnitude of the torque τ is given by

$$\tau = MgR \sin(\phi) \tag{5.87}$$

where R is the distance from the pivot point to the center of mass. This gives

$$\Omega = \frac{MgR}{I\omega}. \tag{5.88}$$

or a precessional period of

$$T_{prec} = \frac{2\pi I\omega}{MgR} \tag{5.89}$$

Note that for a given ω, the precession rate (or period) is independent of the tilt angle ϕ. And – more interestingly – the precession rate is inversely proportional to the spin rate ω. So, for example, as friction works to steadily decrease the spin rate ω, the precession rate Ω will *increase*. Or in terms of the period, as ω decreases (due to friction, say), the top will take less and less time to precess. This feature too is readily observable with a real top.

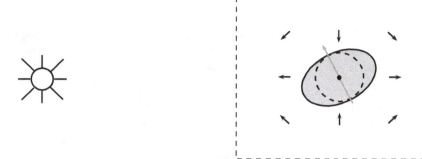

Figure 5.14: The gray blob represents the Earth (with its equatorial bulge greatly exaggerated). The Sun – not, obviously, to scale – is shown on the left for reference. Note that the moment pictured has the Earth's rotation axis (represented by the gray arrow) tilted directly toward the Sun, i.e., this is the Summer Solstice. The arrows in the dashed box represent the tidal force associated with the Sun, i.e., the residual force after the average attraction toward the Sun of the Earth as a whole is subtracted off. The important point is that the equatorial bulge on the bottom left experieces a tidal force which pulls it up-and-to-the-left, while the equatorial bulge on the upper right experiences a tidal force which pulls it down-and-to-the-right – i.e., there is a net clockwise torque on the Earth, due to the interaction of its non-spherical shape with the tidal force from the Sun. This torque causes the rotation axis of the Earth to precess – just like a top – giving rise to the "Precession of the Equinoxes" that we encountered as a mysterious observational phenomenon in earlier weeks.

Now, why are we bothering to worry about how the precession of a top works? Partly because it's a fascinating example of a surprising kind of rotational motion and thus gives us a chance to apply the full version of the rotational-motion analog of Newton's 2nd law. But there is also an important astrophysical application of this idea, which we will just sketch here briefly and then analyze in more depth in one of the Projects.

The Earth, as it turns out, is not perfectly spherical. It "bulges" a bit around the "waist" – i.e., the "radius" from the center out to a point on the equator is a little bigger than the "radius" from the center out to one of the poles. (The technical name for this shape is an "oblate spheroid".) The *reason* for this bulging – oblate – shape is the fact that the Earth is *rotating*. It is like a lump of pizza dough, flung up into the air and rapidly spinning: the rotation stretches it out somewhat in the plane of the rotation.

This not-quite-perfectly spherical shape can then interact with the tidal force from the Sun, as indicated in Figure 5.14. In particular, as shown in the Figure and explained in more detail in the caption, during certain parts of the Earth's orbit around the Sun, the tidal forces (i.e., the residual forces that remain when the average force on the Earth as a whole is subtracted off) exert a *torque* on the Earth's equatorial bulge. That is, in addition to accelerating translationally toward the Sun, the Earth also experiences a "twist" from the Sun due to its (the Earth's) not-quite-perfectly-spherical shape.

But the earth – just like a spinning top – has a large spin angular momentum due to its daily rotation. So instead of turning the Earth clockwise (in the Figure), the clockwise torque causes the spin axis to *precess* with a period given by Equation (5.86).

This, then, is the Newtonian explanation for the "precession of the equinoxes" phenomenon that we encountered previously. You'll have the opportunity in one of the Projects to try to estimate the relevant quantities and convince yourself that the period is indeed in decent agreement with the 26,000 year period that was inferred, quite early, from observation.

Questions:

Q1. One of the Exercises in Chapter 3 involved a puck, connected by a spring to a peg on a frictionless tabletop, "orbiting" around. There we were interested in the analog, for this puck-spring system, of Kepler's Third Law. What is the analog, for this puck-spring system, of Kepler's Second Law?

Q2. The two versions of Newton's second law, $F = ma$ and $F = dp/dt$, are only equivalent if the mass m is constant. In the (relatively uncommon, but still possible) case where the mass m of an object is not constant, it turns out that $F = dp/dt$ is true and $F = ma$ is wrong. (In fact, $F = ma + v\frac{dm}{dt}$.) Based on this, which do you think is always true, $\tau = I\alpha$, or $\tau = dL/dt$? Give an example of a situation in which the equation that is not always right, would be wrong. (This should help you appreciate why we focused most of our attention on "rigid bodies" here.)

Q3. When a spinning ice-skater pulls his arms in close to his body, his angular velocity ω increases. Explain why in terms of torque, angular momentum, and moment of inertia.

Q4. Imagine a child standing near the edge of a merry-go-round which is spinning clockwise (as seen from above). Treat the child as a particle. Using the center of the merry-go-round as an origin, what is the direction of the child's angular momentum vector? How should one think about what this direction means? If the angular velocity of the merry-go-round is decreasing (say, due to friction), what is the direction of the torque acting on the child? What force produces this torque?

Q5. In the example of section 2.3, explain qualitatively why it makes sense that the ball has no orbital angular momentum with respect to the x-y coordinate system, but does have orbital angular momentum with respect ot the x'-y' coordinate system.

Q6. Can you think of a situation (other than the one mentioned in the reading) in which an object's spin angular momentum is (at least partially) converted into orbital angular momentum? How about vice versa?

Q7. A man on a motorcycle is riding a "wheelie." Can the wheelie be maintained if he moves with constant velocity? Explain why or why not.

Q8. When you slam on the brakes in your car, the front end dips down a bit toward the ground and the back end rises up. (A more extreme example of the same phenomenon is braking while riding a bike – if you brake too hard, you can spill forward right over the handlebars.) Relatedly, when you step hard on the gas, the opposite happens: the front end lifts up a little and the back end dips down. Explain this effect using the concepts of torque and angular momentum.

Q9. Figure 14 shows how the tidal forces from the Sun exert a torque on the equatorially-bulging Earth. In the perspective of the Figure, that torque is clockwise. Which way does this torque cause the Earth to precess? Does the tip of the Earth's spin angular momentum vector (or, equivalently, does the North Pole) move toward us, or away from us?

Exercises:

E1. It was stated/implied in section 2, but not explained too clearly, that the net torque on a rigid body will come only from the external forces. That is, just as the internal forces (the forces that are exerted by one piece of the object on another piece) cancel out, because of Newton's third law, when one adds everything up to calculate the net force on the object as a whole, so the "internal torques" (the torques that are exerted by one piece of the object on another piece) cancel out when one adds everything up to calculate the net torque on the object as a whole. This, however, does not follow merely from Newton's third law alone. Another mathematical condition is required on the forces. What is this condition? (Hint: consider the simple barbell case to start. Suppose the two balls exert forces on each other that are equal and opposite, but that the forces are not directed along the line connecting the two balls. For example, in Figure 5.3, suppose that the force exerted by ball 1 on ball 2 is directed to the southeast, while the force exerted by ball 2 on ball 1 is directed to the northwest. Would the torques associated with these two forces cancel one another?)

E2. A kid makes a primitive yo-yo by wrapping a string around a solid disk. He releases it and lets it unwind downward. Apply the translational and rotational versions of Newton's 2nd law, and the appropriate relationship between the translational and angular accelerations (similar to the rolling-without-slipping conditions), to find a formula for the downward translational acceleration of the disk. Note that this is similar to the ball-rolling-down-a-ramp example from the text, in that it can be done in a couple of interestingly-different ways using coordinate systems with different origins.

E3. A pool ball is hit so that it is initially skidding across the tabletop with speed v_0, without any rotation. There is some kinetic friction with the table surface, however, which simultaneously slows the ball down and causes it to begin rotating. At some point, the translational velocity v has decreased (from v_0) and the angular velocity ω has increased (from zero) such that $v = \omega R$. At this point the ball is rolling without slipping, so there is no longer any kinetic friction and the ball just rolls with constant velocity (until it runs into something). Finding the final velocity is rather difficult using conventional methods (Newtons' second law, formulas for kinetic friction, 1-D translational and rotational kinematics), but becomes almost trivially easy if one anayzes it the right way using the rotational version of Newtons' 2nd law. See if you can work this out. Hint: if you choose a coordinate system whose origin is on the table surface, the net torque (produced by the normal, gravitational, and friction forces) on the ball is *zero*.

E4. Consider a rigid body made of N particles, all with equal masses, m. (So the total mass is $M = Nm$.) If the body is near the surface of the Earth, so that the gravitational force on each particle is just $F_{grav} = -mg\hat{k}$, it is obvious that the total gravitational force acting on the whole body is the same as the gravitational force that would act on an "equivalent particle" of mass M. The corresponding thing for torque, however, is a little more subtle. Show that the total gravitational torque acting on the body (i.e., the sum of the torques produced by the gravitational forces on each component particle) is the same as the gravitational torque that would act on an "equivalent particle" of mass M – if the "equivalent particle" is located at the center of mass point of the body. (Note that this provides the justification for something you are often encouraged to "pretend" in introductory mechanics classes, namely, that the gravitational force is "concentrated" at the center of mass point. And note also that this is only valid where the gravitational force is independent of position! For example, as we discussed in the final section, the gravitational torque exerted by the Sun on the Earth is not the same as it would be if the Earth were a point at its center of mass!)

E5. Calculate the moment of inertia of a solid (uniform density) disk of mass M, radius R, and thickness d.

E6. Calculate the moment of inertia of a solid (uniform density) sphere of mass M and radius R.

E7. A painter is standing on a piece of scaffolding near the side of a building. The platform is a long board, whose mass is 100 pounds, attached to the scaffolding at each end. Suppose the man weighs 200 pounds. How big are the forces exerted on the board by the scaffolding (on both ends) if the man stands right in the center of the board? What if he stands 3/4 of the way toward one side?

E8. A ladder is propped up against the side of a house. It makes a 30° angle with the vertical. The top end of the ladder is slippery, so the side of the house exerts a (horizontal) normal force on it, but no (vertical) friction force. The ground exerts both a (vertical) normal and a (horizontal) friction force on the bottom end of the ladder. What is the minimum coefficient of static friction between the ladder and the ground that will allow the ladder to lean stably, without sliding out? Now qualitatively, what happens when somebody gets on the ladder? Is the ladder more likely to slip out with a person near the bottom or near the top?

E9. An object called a "neutron star" can be formed when a large star runs out of nuclear fuel, loses internal pressure as it cools down, and finds itself unable to hold up the outer layers against gravitational collapse toward the center. So the core of the star gets crushed down to enormous density. A typical neutron star might have a mass comparable to the mass of our Sun, but a radius of only about 10 kilometers! That is already pretty amazing, but it is also quite interesting that some neutron stars have been observed to be rotating more than one thousand times per second! Can we make sense of this incredibly fast rotation rate purely on the basis of angular momentum conservation during the collapse process? Suppose the progenitor star has the mass, radius, and rotation period (about a month) of our Sun.

E10. Consider a particle moving inertially (i.e., in a straight line with constant speed). Draw its trajectory and pick a (random) origin point that is not on the trajectory. Draw several $\Delta \vec{r}$ vectors representing the displacement of the particle during (equal) finite durations Δt at several different points along the trajectory. Now consider the (triangular) areas "swept out" during each of these Δt periods. Does the inertially moving particle sweep out equal areas in equal times? Relate this to the angular momentum and torque concepts introduced in the text, and explain the overall connection to Kepler's second law.

E11. Now consider the case depicted in Figure 5.15 in which the particle moves inertially, but receives a "kick" toward the sun at an intermediate time. Prove that the areas of the two triangles in the Figure are equal, so that, still, equal areas are being swept out in equal times.

Projects:

P1. Reproduce the Cavendish experiment in your classroom and come up with a range of values for Newton's constant, G.

P2. Make an order-of-magnitude estimate of the torque exerted on the (equatorially bulging) Earth by the Sun. You may find it helpful to first derive a simple approximate formula for the tidal force exerted on a particle of mass m at position(x, y) (where $x = 0$, $y = 0$ is the center of the Earth), and to then model the equatorially-bulging Earth as a barbell. Use this, along with any other needed estimates and known values, to calculate the rate of the precession of the equinoxes. You should find a result that is at least in the ballpark of the actual 26,000 year precession period.

P3. In our previous work with simple pendulums, we always kept the amplitude small. Let's now explore how the period depends on the amplitude when the amplitude is not small. Start by setting up a pendulum and measuring its period as it swings with several different amplitudes, say, $10°$, $20°$, and $30°$. (Note that it gets increasingly difficult to measure the period accurately for larger amplitudes, because air drag decreases the amplitude quite rapidly when the amplitude is large! So be careful: don't just assume that the pendulum keeps swinging with the same amplitude you started it at.) Can you see a difference between the periods? Iterate/improve your experimental procedure until you do. Then write a computer program to numerically solve the equation of motion for this sytem. (Hint: if θ is the angle of the pendulum at some moment, the torque produced by the gravitational force, about the pivot point at the top of the string, is $\tau = -mgL \sin(\theta)$, and the moment of inertia of the pendulum about this same point is $I = mL^2$.) Compare your measured and calculated periods.

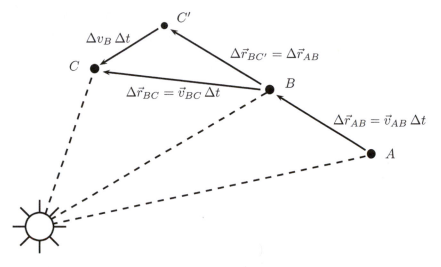

Figure 5.15: A planet receives intermittent regular impulsive forces from a central Sun. Its trajectory is thus not a smooth curve, but a polygon. The positions A, B, and C are occupied by the planet at three successive times separated by the same constant Δt. It is helpful also to consider the position C' that the planet would occupy if no force acted at point B. The displacement vector $\Delta \vec{r}_{BC'}$ is then equal to $\Delta \vec{r}_{AB}$, and so the displacement vector pointing from C' to C is proportional to the velocity change at B: $\Delta \vec{r}_{C'C} = \Delta \vec{r}_{BC} - \Delta \vec{r}_{BC'} = \Delta \vec{r}_{BC} - \Delta \vec{r}_{AB} = (\vec{v}_{BC} - \vec{v}_{AB}) \Delta t \sim \vec{a}_B$ where a_B is the acceleration of the particle at or around point B. (We simply avoid the question of whether a_B represents an average or instantaneous acceleration by claiming only a proportionality at the end of that string of equalities.) The point is that this acceleration at point B is toward the Sun. Hence, the line connecting C and C' is parallel to the line connecting the Sun to point B. This is the key insight that then allows a relatively simple geometric argument for the claim that the triangular areas Sun-A-B and Sun-B-C are equal.

Part II

Electricity and Magnetism

CHAPTER 6

Electric and Magnetic Forces

I N the first five Chapters, we have been focusing on Newtonian mechanics, including especially Newton's theory of universal gravitation. We spent a couple of weeks tracing the pre-history of Newton's
discoveries, going all the way back to Ancient Greek astronomical models, and then a couple of weeks
exploring some interesting applications, some of which brought us almost up to the present day.

For the next few weeks, we're going to turn our attention away from the heavens and toward a more
terrestrially-focused, parallel thread: the development of the scientific study of electrical and magnetic
phenomena. The most interesting and important aspects of this development occur (not coincidentally)
in the centuries following Newton, so although our treatment is by no means strictly chronological, it
does make some sense to place this here. But, as with Newtonian mechanics, there is a pre-history to the
subjects of electricity and magnetism which stretches all the way back to the ancient Greeks. So let us
start with a brief look backwards.

§ 6.1 Early Observations

In one of his dialogues (the *Timaeus*), the 4th-century-BCE Greek philosopher Plato mentions "the
wonderful attracting power of amber and the Heraclean stone." Amber, a gemstone which we now know
to be fossilized tree resin, and which the Greeks called "elektron", was known to develop this "wonderful
attracting power" when it was rubbed with fur or cloth. Subsequent exploration across the following
centuries, carried out by many individuals in many parts of the world, revealed that many other materials
besides amber also developed similar attracting powers when rubbed and these, collectively, came to be
called "electrics".

Plato's "Heraclean stone" referred to a type of iron ore (also known today as "loadstone") which displayed
a similar attractive ability. A particular sample of this type of material which had been studied by Thales
– one of Plato's early Greek philosopher predecessors – had come from the island of Magnesia. This
explains how such objects came to be called "magnets".

Despite sharing the seemingly-magical power to attract other objects without any apparent physical
contact, there were some readily observable differences between magnets and electrics. Magnets, for
example, attracted only objects made of (or containing) *iron*, whereas the attractive power of amber and
other electrics seemed restricted to dry, light objects such as feathers and dried leaves or grass. It was
also recognized early on that a small magnet, left free to rotate (e.g., by mounting it on a cork floating
in water), would spontaneously align itself so that one end pointed North and one South. (This of course
turned out to be a tremendously valuable fact for navigation.) Rubbed amber, on the other hand, showed
no such propensity to align itself in a particular direction. Still, despite these differences, the phenomena

of electric and magnetic attraction were sufficiently similar – and sufficiently myserious – that they were not cleanly distinguished, as distinct phenomena, until something like the 16th century, i.e., around the time of Copernicus.

For example, in his 1550 treatise *On subtlety*, the Italian mathematician and physician Jerome Cardan collected a number of pieces of evidence that electric and magnetic phenomena were indeed distinct:

(i) Amber draws everything that is light; the magnet, iron only.

(ii) Amber does not move chaff toward itself when something is interposed; the attraction of the magnet for iron is not similarly hindered.

(iii) Amber is not attracted by the chaff; the magnet is drawn by the iron.

(iv) Amber does not attract at the end; the magnet attracts iron sometimes at the North, sometimes at the South.

(v) The attraction of amber is increased by friction [i.e., rubbing] and heat; that of the magnet, by cleaning the attracting part.

Having studied Newton's laws (which of course were more than a century in the future in 1550), we might be suspicious of point (iii) in particular. And there are aspects of the other points that also perhaps turn out to be not quite right. But the point is just that this is the period when electricity and magnetism were first being openly and explicitly distinguished as separate phenomena.

We will follow Cardan's lead and discuss, separately, the development of theoretical models for electric and magnetic phenomena.

§ 6.2 Development of a Model of Electricity

The Englishman William Gilbert (1544-1603) dedicated one chapter of his book, *On the magnet*, to discussing electrics and distinguishing them from magnets. His experimental research had revealed that

"...it is not only amber and jet [another mineral which displayed electric properties] that attract small bodies when rubbed. The same is true of diamond, saphhire, carbuncle, iris gem, opal, amethyst, ... glass, ... artificial gems made of glass or rock-crystal, antimony glass, many kinds of flourspar from the mines, and belemnites. Sulfur also attracts, as does mastic and hard sealing wax. Rather hard resin and orpiment also attract, but less strongly. Feeble power of attraction is also possessed under a suitable dry sky by rock salt, mica, and rock alum. This one may observe when in midwinter the atmosphere is sharp and clear and rare..."

You should have the opportunity in class to play around with some hands-on examples of electrical attraction and you will probably also notice that the effectiveness correlates with the weather and especially the humidity.

Several observed facts suggested to Gilbert and others that the electrical attraction might be associated with and explained by an outward flow of invisible material. One such fact was just that a rubbed electric would lose its attractive potency over time (especially in humid weather). This suggested that whatever was done to or given to the amber by rubbing was dissipated and would eventually run out so that further rubbing was required to restore the attractive potency. (This would later be described as "re-charging" the electric material, on the basis of the analogy with a gun which needed to be re-charged with powder before firing again.) Another fact was that the attraction (of, say, a piece of straw) toward the charged amber diminished with increasing distance, just as one might expect if the attraction resulted from an outflow or "effluvium" whose intensity naturally decreased with distance from the source as it spread

Figure 6.1: William Gilbert (1544-1603)

out over a greater and greater area. Similarly, as Cardan had noted, the electrical attraction could be blocked by an interposed object such as a thin slice of wood.

The idea of an aethereal effluvium, as the physical basis for electrical attraction, was also appealing to Gilbert on more philosophical grounds. Raising an issue to which we will return in more detail next week, Gilbert insisted that

> "... no action can be performed by matter save by contact[. And yet] these electric bodies are not seen to touch, [so] something of necessity is sent from the one to the other, something that may touch closely and be the beginning of that incitement..."

This early theoretical model, however, ran into difficulties with some observations that were made in the century following Gilbert's work. For example, by hanging rubbed amber and other (uncharged) objects on threads near one another, it was seen that, contrary to Cardan's point (iii), the electrical attraction was *mutual*. For example, when a charged piece of amber was hung next to an uncharged wooden ball, not only did the amber attract the wooden ball (i.e., not only did the ball move toward the amber with the ball's thread deviating noticeably from vertical), but, in addition, the amber was attracted toward the ball (i.e., the amber's thread, too, deviated noticeably from vertical). This suggested that the electrical phenomenon was a genuine *interaction*, of a more symmetrical character than could be explained on the basis of the charged amber unilaterally emitting effluvium. And it is, of course, in accordance with what would later be formalized as Newton's third law.

Another important 17th century discovery was that electrical interactions can involve not just attraction, but also *repulsion*. For example, the Italian scholar Niccolo Cabeo noticed that an uncharged object that is initially attracted toward a rubbed electric, may fly violently away from it immediately after touching it. Cabeo speculated that this behavior might be explained by the surrounding air which, after initially being pulled in toward the electric by the moving object, could bounce off the electric and pull the object out with it. But the newly-invented air pump allowed this hypothesis to be refuted rather conclusively by Robert Boyle, who showed that the effect occured in vacuum just the same as it does in air.

The possibility of not only attractive, but also repulsive, electrical forces provided important new input to theoretical model builders. In France, for example, Charles Dufay wrote in 1734 about an experiment

"... in which a [tube] of sulfur rendered electrical repels a down-feather[;] I perceived that the same effects were produced not only by the [rubbed tube] but by all electrified bodies whatsoever; and I discovered a very simple principle that accounts for a great part of the irregularities and, if I may use the term, caprices which seem to accompany most of the experiments on electricity. This principle is that an electrified body attracts all those that are not themselves electrified, and repels them as soon as they become electrified by [contact with] the electrified body. Thus gold leaf is first attracted by the tube. Nor is it re-attracted while it retains its electrical quality. But if ... the gold leaf chance to light on some other body, it straightaway loses its electricity and consequently is reattracted by the tube, which, after having given it a new electricity, repels it a second time.... Upon applying this principle to the various experiments on electricity, one will be surprised at the number of obscure and puzzling facts it clears up..."

Others would later interpret Dufay's idea in terms of a sort of invisible fluid which charged objects somehow acquired through rubbing. A charged object will attract an uncharged one. But contact between a charged and uncharged object will result in the charged object transfering some of its electrical fluid to the initially uncharged one. The two objects, which now both possess an excess of this electrical fluid, will repel one another.

This hypothesis, however, raises an interesting empirical question: do any two objects, each independently charged by rubbing, repel one another, as they should if both now contain an excess of this same type of electrical fluid? Dufay writes that an "examination of this matter has led me to a discovery which I should never have foreseen, and of which I believe no one hitherto has had the least idea." His experimental investigations had led Dufay to the realization that

"... there are two distinct electricities, very different from each other: one of these I call *vitreous electricity*; the other, *resinous electricity*. The first is that of [rubbed] glass, rock crystal, precious stones, hair of animals, wool, and many other bodies. The second is that of [rubbed] amber, copal, gum lac, silk, thread, paper, and a vast number of other substances.

The characteristic of these two electricities is that a body of, say, the *vitreous electricity* repels all such as are of the same electricity; and on the contrary, attracts all those of the *resinous electricity*.

Thus, if electric charge is associated with some sort of electrical fluid, there must be *two fluids* – that is, two distinct types of electric charge. Two objects containing excesses of the same type (both vitreous or both resinous) will repel one another, while a pair of differently-charged objects (one vitreous and one resinous) will attract. And, of course, an object containing an excess of either type of fluid – an object charged in either way – can attract an uncharged object.

You might wonder why the hypothesized material substrates for the electrical phenomena were assumed to be *fluids*. The reason is largely that, somewhat prior to Dufay's discoveries, it had become apparent that electrical charge can *flow* through certain materials – largely metals, but also other kinds of things, especially when moist – that came to be called "electrical conductors". For example, the electrical attractive power of a charged piece of amber – evidenced by the ability of the amber to attract small scraps of paper, say – could transfer along a metal rod, or even an extremely long metal wire, and manifest as the ability of the far end of the rod or wire to similarly attract scraps of paper.

The ability of metals to conduct electricity also explained why metals had for so long appeared to be impossible to charge (in the way that initially amber, but eventually most other materials, had been demonstrated to be). Evidently, whatever charge one did manage to produce by rubbing a piece of metal would flow freely through the metal and into whatever it was touching, eventually finding its way into the ground and disappearing. The discovery that metals can conduct electricity – and the coincident

Figure 6.2: Benjamin Franklin (1706-1790)

recognition that other materials, dubbed "electrical insulators", do not – thus made electricity into a truly universal phenomenon and led to great practical advances in efficiently producing, transmitting, and storing electrical charge.

One important result of such practical advances was the suggestion, by the American Benjamin Franklin, in 1747, that a single electrical fluid would suffice to explain all of the known electrical phenomena. Discussing a set of demonstration experiments in which several people acquire electrical charges from a friction-charging-machine involving a glass sphere connected to a hand-crank by a metal tube, Franklin describes the situation in which two people acquire charge by touching, respectively, the two parts of the machine. This is the kind of situation that proponents of the two-fluid theory would describe as one person acquiring a "vitreous", and the other a "resinous", charge. Franklin acknowledges that

> "they will both of them (provided they do not stand so as to touch each other) appear to be electrized to a person standing on the floor; that is, he will perceive a spark on approaching each of them with his knuckle."

However,

> "If they touch each other after exciting the tube and drawing the [electrical] fire as aforesaid, there will be a stronger spark between them than was between either of them and the person on the floor. [And a]fter such strong spark, neither of them discovers any electricity [i.e., both people are no longer charged].

> "These appearances we attempt to account for thus. We suppose, as aforesaid, that electrical fire [i.e., electrical fluid] is a common element, of which every one of the three persons afore-mentioned has his equal share before any operation is begun with the tube. A, who stands on wax [an insulator, so the acquired charge can remain on his body] and rubs the tube, collects the electrical fire from himself into the glass; and his communication with the common stock being cut off by the wax, his body is not again immediately supply'd. B (who stands on wax likewise), passing his knuckle along near the tube, receives the fire which was collected by the glass from A: and his communication with the common stock being likewise cut off, he retains

the additional quantity received. To C, standing on the floor, both appear to be electrized: for C, having only the middle quantity of electrical fire, receives a spark upon approaching B, who has an over-quantity, but gives one to A, who has an under-quantity.

"If A and B approach to touch each other, the spark is stronger, because the difference between them is greater; after such touch there is no spark between either of them and C, because the electrical fire in all is reduced to the original equality. If A and B touch while electrizing, the equality is never destroy'd, the fire only circulating.

"Hence have arisen some new terms among us. We say B (and bodies like circumstanced) is electrized *positively*, A, *negatively*. Or rather, B is electrized *plus*; A, *minus*. And we daily in our experiments electrize objects *plus* or *minus*, as we think proper. To electrize *plus* or *minus*, no more needs to be known than this: that the parts of the tube or sphere which are rubbed do, in the instant of the friction, attract the electrical fire, and therefore take it from the thing rubbing; the same parts immediately, as the friction up on them ceases, are disposed to give the fire they have received to any body that has less."

Several comments about this are in order.

First, note that the fundamental idea here is that the so-called "vitreous" and "resinous" types of charge are in some sense *opposites*: when allowed to inter-mingle (as when persons A and B touch each other) the two types of charge cancel one another out, resulting in bodies that are no longer electrified. This indeed strongly suggests that instead of two different fluids, we are dealing with, respectively, the *excess* and *deficit* of a single fluid, with what we described as un-charged objects possessing some normal baseline "middle quantity" of the fluid.

Second, note that Franklin's single-fluid model implies the principle that would later be formalized as the conservation of electrical charge: when (say) the rotating glass sphere is rubbed, the rubbing does not simultaneously call into existence, out of nothing, two distinct fluids. Instead, the rubbing (for reasons that remain unclear) results in a *transfer* of fluid between the rubber and the rubbed. One object acquires an excess (beyond the normal default quantity of "uncharged" objects) at the expense of the other, which is left with a deficit.

Finally, note that Franklin's introduction of the less qualitative and more mathematical notions of "plus" and "minus" (or "positive" and "negative") suggests that electric charge is a genuine physical *quantity* which can be measured in a precise way. We will see the payoff of this in the following section, where we discuss the experiments which led to the first quantitative description of how the force between two electrical charges depends on the magnitudes of the charges and their separation, i.e., the electrical analog of Newton's gravitational inverse square law.

Before turning to that, however, let us step out of the historical perspective and discuss the controversy between the "two fluid" and "single fluid" models of electricity from a more contemporary point of view. Everybody now "knows" – because they are told in kindergarten – that matter is made of atoms; and that atoms have a positively-charged and relatively heavy nucleus surrounded by a sea of very light, negatively-charged, orbiting electrons; and that the electrons are relatively mobile and easy to strip away. So, for example, when we rub a glass rod with rabbit's fur, what is happening at the micro-scale is that some electrons are getting transferred from one material onto the other, leaving one material with an excess of electrons and the other with a deficit.

There is thus a sense in which the sea of electrons corresponds to Franklin's single electrical fluid or "fire". But there are two things worth pointing out about this. One is that it is not at all obvious whether, for example, in rubbing glass with rabbit's fur, the electrical fluid is transferring *from* the fur *to* the glass, or vice versa. Franklin in effect flipped a coin and guessed, arbitrarily, that this is what was happening. And this was the basis for his (thus arbitrary) convention to call the charge of the glass rod "positive" (because, he guessed, it had an excess of the electrical fluid), while the rabbit's fur was left (he guessed) with a deficit and was hence "negative".

It was more than 100 years after Franklin established this convention (around the beginning of the 20th century) that the structure of the atom was worked out. And note that it is precisely Franklin's convention which implied that the electric charge of the atomic nucleus (which was of the sort that would *repel* the glass rod that had been rubbed with rabbit's fur) was called "positive" and the electric charge of the electrons (which would *attract* the glass rod that had been rubbed with rabbit's fur) was called "negative". Thus, an object with an excess of electrons has a *negative* charge, and an object with a deficit of electrons has a *positive* charge.

There are two equivalent ways to put the implications of this. One is that, after all, Franklin's single electrical fluid – the "electrical fire" – does not correspond to electrons at all, but, in some weird sense, to their lack. In this case, it would be slightly odd to say that Franklin's single fluid model of electricity turned out to be right. But one could also equivalently say that Franklin was just a bit unlucky. The single-fluid model was basically correct, but it was impossible to tell which of the two objects involved in charging-by-rubbing was acquiring "electrical fire" from the other. Had Franklin simply guessed the other way – had he decided to assume that the rabbit's fur acquired electrical fluid from the glass rod, so that the glass rod's charge was called "negative" and the fur's charge was called "positive" – the model would have worked just as well as a qualitative account in the 1700s and 1800s, and we would now say that atoms have a negatively-charged nucleus surrounded by a sea of positively-charged electrons.

As we will see again later, the fact that Franklin was unlucky in this sense, has the annoying consequence that we still speak of the flow of electrical charge, in what is called an electrical current, as if it corresponded to Franklin's "electrical fire". That is, an electrical current flowing left-to-right through a wire corresponds, in fact, to negatively-charged electrons flowing right-to-left. The so-called "conventional current" direction is *opposite* the direction in which the actually-flowing charged particles are moving (because those charged particles, namely electrons, turn out to have a negative charge according to Franklin's unlucky convention). This is just one of those things one has to learn to live with.

Perhaps it is worth pointing out as well that its being in some sense "backwards" (from the point of view of the more advanced, 20th century, context) is not the only problem with Franklin's single-fluid model. If there is just a single electrical fluid, and if the repulsion of (for example) two positively-charged objects (i.e., two objects which both possess an excess of that fluid) is explained by the fact that every bit of the fluid exerts a repulsive force on every other bit of the fluid, then shouldn't ordinary, uncharged, "electrically neutral" objects have a tendency to explode due to the mutual repulsion of their parts? For, on the single-fluid model, those parts all still possess some of the fluid – just an ordinary or baseline amount.

So, a case can be made that actually Dufay's two-fluid model – with one fluid corresponding to all the electrons, and the other fluid corresponding to all the atomic nuclei – comes nearer to capturing what turned out to be the real situation. It's just that, in most ordinary situations, it is only the electrons that are *mobile*; the atomic nuclei are fixed in place, unable to move. So it is as if only one of the two distinct electrical fluids is actually a fluid – the other is frozen, like a solid.

Anyway, we raise this here only to encourage you to ponder how scientific development works. It is not always the case, in a dispute between rival factions, that one side is simply right and the other side is simply wrong. Sometimes both/all of the competing models capture some important element of the truth but also miss some aspects, and the truth involves some way of combining or unifying or just moving beyond the different models, in a way that would have been difficult to anticipate at the time. But progress is always made by those who try.

§ 6.3 The Electric Force Law

Newton's *Principia*, in which he put forward the theory of universal gravitation and in particular the inverse-square force law, was published in 1687. By the mid-to-late 1700s, the theory was widely known and highly respected by scientists. So it is not surprising that, during this period, many people had begun to wonder if there was some analogous law for the force of electricity. Indeed, the qualitative observation that electrical attraction decreased with increasing distance from the charged object suggested that electrical forces, too, might fall off with distance in the same inverse-square way that gravitational forces had been shown to do by Newton.

An interesting early hint in this direction was provided by Joseph Priestley in 1767. He had been following up on an interesting observation, originally due to Ben Franklin: if a metal cup is given an electrical charge, an uncharged cork ball, held on an insulating string outside of the cup, is, of course, attracted to it. But if the cork ball is instead dangled on the interior of the cup, it seems to experience no electrical attraction to the cup. Of course, if the ball is in the exact center of the cup, it is perhaps not too surprising, just on grounds of symmetry, that it is not preferentially attracted to any particular side. But even if the cork ball is held off-center, nearer to one side of the cup than the other, there is no apparent attraction.

This curiousity reminded Priestley of Newton's proof that a massive particle, on the *interior* of a spherical shell of mass, would experience no net gravitational force, the inverse-square-law character of the force law implying a perfect cancellation between the attractive forces toward the small-but-nearer side of the sphere and the larger-but-farther-away side. Priestley points out that the analogy to the gravitational case suggests that electrical forces, too, might obey an inverse square law:

> "May we not infer from this experiment that the attraction of electricity is subject to the same laws with that of gravitation, and is therefore according to the [inverse] squares of the distances; since it is easily demonstrated that, were the earth in the form of a shell, a body in the inside of it would not be attracted to one side more than another."

The question was addressed in a direct, experimental way just a few years later in France by Charles Coulomb. Coulomb had invented the "torsion balance" as a way of measuring small forces. As we saw last week, Henry Cavendish would borrow the design just a few years later to directly measure gravitational forces and thereby "weigh the earth". But the first important application of the device was to measure how the *electrical force* between two balls (carefully insulated so that their charges remain constant) varies with the distance between them.

A schematic version of Coulomb's apparatus is shown in Figure 6.3. A "barbell" – which, for simplicity, we assume here to consist of two identical balls of mass m fixed, by a massless rod, at distance d from the center – hangs by a thread and is free to rotate in the horizontal plane. When rotated by angle θ from its equilibrium orientation, the twisted thread produces a restoring torque proportional to the angular displacement:

$$\tau = -\kappa\theta. \tag{6.1}$$

The rotational-motion analog of Newton's second law thus implies that

$$-\kappa\theta = I\frac{d^2\theta}{dt^2} \tag{6.2}$$

where $I = 2md^2$ is the moment of inertia of the hanging barbell. Just as we saw in the case of the Cavendish experiment, this implies that the barbell can oscillate, rotationally, in the horizontal plane, with period

$$T = 2\pi\sqrt{\frac{2md^2}{\kappa}} \tag{6.3}$$

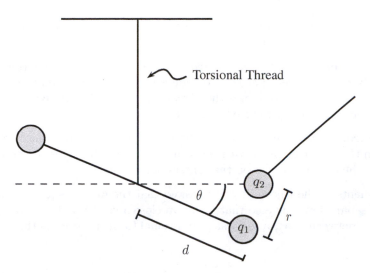

Figure 6.3: Schematic diagram of the torsional balance Coulomb used to measure the distance-dependence of electrical forces. A "barbell" consisting of two balls of mass m hangs from a torsional thread whose torsion constant κ can be determined from the period of angular oscillations. Then if one of the balls is given a charge q_1 and another ball of charge q_2 is introduced, the force between them produces a torque that makes the barbell twist through some angle θ which can be measured. The magnitude of the force can then be determined from θ.

so the torsional constant κ of the thread can be determined.

Once the torsional constant of the thread is known, we can apply an additional force and infer its magnitude from the observable rotation of the barbell. For example, in the situation shown in the Figure, if one of the barbell balls is given an electric charge q_1, and an additional charge q_2 is introduced, the (say, repulsive) force F between q_1 and q_2 will cause the barbell to rotate into a new equilibrium orientation making some angle θ with respect to the original equilibrium orientation (when no additional forces were present). The equilibrium condition that the net torque on the barbell should vanish then implies

$$Fd = \kappa\theta \tag{6.4}$$

where we have assumed that the charged balls are arranged so that the repulsive electrical force between them is in the horizontal plane and the force on q_1 is perpendicular to the barbell rod. (Note also that, in the actual apparatus, the top end of the torsion thread – where it is attached to the support structure from which it is hanging – can also be manually turned using a rotational micrometer. So the total angular twist of the thread does not necessarily coincide with the angular separation θ between the charged balls.) Anyway, the magnitude of the force F can thus easily be determined:

$$F = \frac{\kappa\theta}{d}. \tag{6.5}$$

Let us briefly quote Coulomb's 1785 report about the results of the experiment:

"We found in trial (1) ... that the balls were separated by 36 degrees; at the same time a torsional force equivalent to 36 degrees was produced....

"In trial (2) the distance between the balls was 18 degrees. But, as the [top end of the torsion thread was manually] turned through 126 degrees, it results that, for a distance of 18 degrees, the respulsive force was equivalent to 144 degrees. So at half the first distance the respulsive force between the balls is quadrupled.

"In trial (3) the suspension wire was twisted through 567 degrees, and the two balls were separated by only 8 1/2 degrees. The total torsion was consequently equivalent to 576 degrees, four times that of the second trial, and the distance between the two balls in this third trial lacked only 1/2 degree of being reduced to half of what it was in the second trial.

"It results then from these three trials that the repulsive force which the two balls exert on each other when they are electrified with the same kind of electricity is inversely proportional to the square of the distance [between the centers of the balls]."

More detailed experiments in the following years showed that the inverse square dependence was obeyed over a great range of separations and also that, again in close analogy with Newton's gravitational force law, the electrical force between two charges was proportional to the product of their charges, as expressed in "Coulomb's Law":

$$F_{elec} = \frac{k\, q_1\, q_2}{r^2}.$$

(6.6)

Note that the forces exerted by one charged particle on another are, as in the case of gravity, directed along the line connecting the two particles. But, unlike the case of gravity, the electrical force can be either attractive (for one positive and one negative charge) or repulsive (for two positives or two negatives).

The contemporary unit of measurement for electric charge is, appropriately enough, the "Coulomb", about which we will say more later. But one Coulomb, as it turns out, is an extremely large amount of charge. Rubbing a glass rod with rabbit's fur, or even using a charge-production machine like the one used by Ben Franklin or the similar van de Graaff generators that we can play with in class, one is likely to produce only a few nano-Coulombs, or at best a few micro-Coulombs, of charge.

In contemporary units, the proportionality constant k appearing in Coulomb's Law takes the value

$$k = 9 \times 10^9 \ \text{Nm}^2/\text{C}^2.$$

(6.7)

So, for example, two 1.0 nano-Coulomb charges, separated by a distance of exactly one meter, will repel one another with forces of magnitude 9.0 nano-Newtons.

§ 6.4 Electrical Forces on Non-Point Charges

Let's again step out of the historical narrative and practice the important skill of applying Coulomb's Law to find the electrical force that would be exerted on, or by, extended objects of various shapes and sizes, by treating the extended objects as collections of point charges and summing up (by integration) the forces from all the point charges.

6.4.1   Polarization

As a first warm-up example, though, which doesn't actually involve any integration, let's return to a qualitative point from the first section that may remain puzzling. We have seen that two positive or two negative charges will repel each other, and two opposite charges (one positive and one negative) will attract, in accordance with Equation (6.6). But doesn't Coulomb's law imply that the electrical force will be *zero* between two objects, one of which has a nonzero (positive or negative) electrical charge, but the other of which is uncharged, electrically neutral? In short, if either q in Equation (6.6) is zero, won't F_{elec} be zero? But then, how does (for example) a charged piece of amber or glass attract an uncharged scrap of paper?

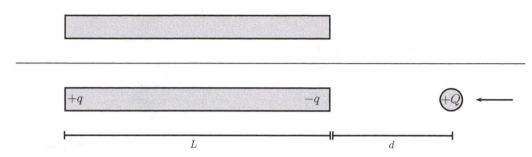

Figure 6.4: (Above) When no external charges are nearby, the electrically neutral metal rod has zero electric charge density everywhere. (Below) But if a (say, positive) point charge $+Q$ is brought near it, the metal rod develops a polarization: *the rod as a whole remains electrically neutral, but some electric charge moves around so that the end of the rod nearest the outside positive charge ends up with a negative charge $-q$, while the far end of the rod ends up with a positive charge $+q$. The net force exerted on the rod by the charge is then attractive, because of the inverse-square character of the force: the attractive force between the $+Q$ and the $-q$ is stronger than the repulsive force between the $+Q$ and the $+q$.*

The answer is that the "electric fire" – or, we would now say, the electrons – in the uncharged object are, to at least some extent, able to *move*, and will do so when another charge is nearby. Consider to begin with the case of a (say) positive point charge near a metal rod, as shown in Figure 6.4. Franklin would say that the "electric fire" in the rod is *repelled* by the nearby positive charge and hence moves, to the left in the Figure, leaving a deficit of "electric fire" (i.e., a negative charge) on the right end of the rod (up close to the outside positive charge), and an excess of "electric fire" (i.e., a positive charge) on the left end of the rod (farthest from the outside positive charge). We would now say, instead, that the negatively-charged electrons, that are free to move in the conducting metal rod, are *attracted* to the outside positive charge, so there is a deficit of them (and hence a positive charge) on the left end of the rod, and an excess of them (and hence a negative charge) on the right end of the rod. Note, though, that the distribution of charge on the rod – how much positive or negative charge you say ends up in each location – is the same, regardless of which model you use. This, of course, is why Franklin's "conventional" model became entrenched, with successful application over the course of more than a century, before it was realized that the convention involved (according to which electrons came out as negatively charged) was unfortunate.

With the distribution of charge as shown in the Figure, we can understand how the charge $+Q$ can attract the rod, despite the rod being overall electrically neutral: the negative charge $-q$ on the right end of the rod is *attracted* to $+Q$, and the positive charge $+q$ on the left end of the rod is *repelled* by $+Q$. But since the force has an inverse-square-law character, the attraction is *stronger* than the repulsion. In particular, the net (attractive) force, that the point charge $+Q$ exerts on the rod, will be

$$F = \frac{kQq}{d^2} - \frac{kQq}{(L+d)^2} = kQq\left[\frac{1}{d^2} - \frac{1}{(L+d)^2}\right] \tag{6.8}$$

which is clearly greater than zero.

This discussion might suggest that although a charge can exert an attractive force on an electrically neutral *conductor* – because the "electrical fire" / electrons in the conductor are, by definition, free to move around in response to outside electrical forces – a charge should not be able to exert an attractive force on an *insulator*, whose internal charged particles are (by definition) not free to roam around. But this is not the case. Even in an insulator, at the level of individual atoms and molecules, the negative and positive elementary particles can move with respect to each other *a little bit*. That is, down at the microscopic scale, each individual molecule in an insulator can become slightly polarized.

Figure 6.5: Even in an insulator, the microscopic charges are free to move a little bit within their home molecules, so each individual molecule can become polarized just like the macroscopic conductor considered previously. In the middle of the insulator, the charge density (averaged over a region containing at least several molecules) is zero despite the polarization. But on the right hand end of the material (closest to the outside positive charge), there is no next-molecule-over to the right, so there is a thin layer of "bare", uncancelled negative charge. And similarly, on the left hand end of the material (farthest from the outside positive charge), there is no next-molecule-over to the left, so there is a thin layer of "bare", uncancelled positive charge.

The situation is sketched in Figure 6.5. With each molecule slightly polarized, the net effect is that the overall charge density (averaged over a suitably large region containing many molecules) is zero in the interior of the rod: at a given location, there are some molecules just to the right whose left-hand-ends have a slight positive charge, but also some molecules just to the left whose right-hand-ends have a slight negative charge. But at the *ends* of the rod, there are no next molecules over on one side or the other, hence nothing to cancel the small charge from the ends of the molecules that are there on one side. So the overall charge density is just like that shown in the previous Figure: neutral in the middle, negative on the right, and positive on the left. The *amount* of positive and negative charge on the ends may not be the same, of course, but qualitatively the situation is the same, and the same overall attractive force results.

6.4.2   Long Charged Wire

Let us consider a different type of example. Suppose that a long, straight wire is electrically charged and, in particular, carries a uniform charge-per-unit-length λ. What force will the charged wire exert on a point charge Q located, say, a perpendicular distance r away from the center of the wire?

The setup needed to calculate this is shown in Figure 6.6. As usual we imagine chopping up the extended object – here, the charged wire – into an infinite number of infinitely small pieces, and then focus our attention on one generic piece like that highlighted in the Figure. The charge on the little piece can be found by multiplying the charge-per-unit-length of the wire (of which the piece is a piece) by the length of the piece:

$$dq = \lambda\,dx. \tag{6.9}$$

The force $d\vec{F}$ that this one piece exerts on the charge Q has a magnitude given by Coulomb's law:

$$dF = \frac{k\,Q\,dq}{s^2} = \frac{k\,Q\,\lambda\,dx}{x^2 + r^2} \tag{6.10}$$

and the direction shown in the Figure. But it is more convenient to work with the components of the vector (rather than its magnitude and direction). So we may write

$$dF_x = -dF\,\frac{x}{s} = -\frac{k\,Q\,\lambda\,x\,dx}{(x^2 + r^2)^{3/2}} \tag{6.11}$$

and

$$dF_y = dF\,\frac{r}{s} = \frac{k\,Q\,\lambda\,r\,dx}{(x^2 + r^2)^{3/2}} \tag{6.12}$$

where we have used the fact that the two triangles in the Figure are similar.

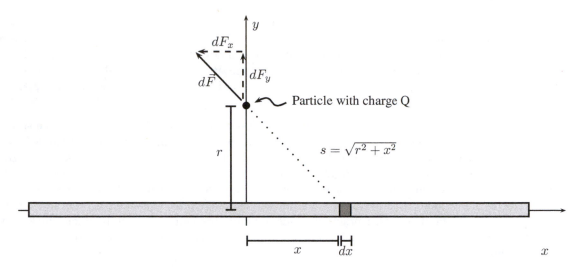

Figure 6.6:   A particle with charge Q is located a distance r away from a long straight wire with charge per unit length λ. To find the force exerted on the particle by the charged wire, we consider the wire to be composed of an infinite number of small, pointlike pieces of width dx; work out the force exerted on the particle by one such piece; and then finally add up the contributions from all the pieces.

We may then calculate the x- and y-components of the total force, exerted not by just one piece of the wire, but by all the pieces which compose it, by summing up the contributions from each piece. For the x-component, we have

$$F_x = \int dF_x = -kQ\lambda \int_{-\infty}^{\infty} \frac{x\,dx}{(x^2 + r^2)^{3/2}} = 0 \tag{6.13}$$

since the integrand is an odd function of x. (Note that we are integrating here from $-\infty$ to ∞ – i.e., we are treating the "long" wire as *infinitely long*! If the length were instead some finite value L, we would, of course, instead integrate from $-L/2$ to $L/2$. The x-component of the force would still come out zero for the reasons described here, though the expression for the y-component, calculated below, would be more complicated.)

The fact that the x-component of the total force (exerted on the particle by the wire) vanishes should not be surprising. For each little piece of wire a certain distance to the right (of the "middle point" just below where the particle is located), there is an identical little piece of wire that same distance to the left. The x-components of the forces associated with these two pieces cancel, so there is perfect "pairwise cancellation" of the x-components associated with *all* of the little pieces.

Here is another, slightly more formal way to see the the x-component of the total force must vanish. This is a "proof by contradiction" in which we assume the opposite of the thing we suspect is true, and then establish that it is true by showing that its opposite implies a logical contradiction. So, let's assume that the total force exerted on the particle by the charged wire *does* have a nonzero x-component. Suppose, to be more concrete, that the total force has a positive x-component, i.e., points at least a little bit to the right. But then here is an interesting fact about the charged wire: a 180-degree rotation, about the y-axis in the Figure, is a "symmetry operation". That is, if we rotate the charged wire by 180 degrees about the y-axis, we do not change the distribution of charges at all: there is exactly the same amount of charge, at every place, after the rotation, as was there before.

But then we can arrive at a contradiction in the following way. On the one hand, if performing this symmetry operation leaves the distribution of charge unaffected, then surely the force on the particle (which after all depends on the distribution of charge in the wire) should also be unaffected, and so if the force pointed a little bit to the right before rotating the wire, it should still point to the right after rotating the wire. But on the other hand, if the force depends on how the charge in the wire is distributed, then the force on the particle should rotate around with the wire if the wire is rotated. And in particular, if the force is initially to the right, and then the wire is rotated through 180 degrees, the force should also rotate through 180 degrees and end up pointing to the left.

To summarize, if the force points to the right then, after rotating the wire 180 degrees, the force should, on the one hand, still point to the right, but should also, on the other hand, now point to the left. But that's a contradiction: the force cannot both point to the right and point to the left! And so the original supposition – that the force points to the right – must be false. It should be obvious that the only way to escape such a contradiction is for the force to have no rightward (or leftward) component, i.e., F_x must be zero. Of course, here, this symmetry argument was just telling us something we had already shown, by calculation. But it is worth practicing this style of reasoning; often it is much quicker and easier and more illuminating to use the symmetry that is inherent in a problem to save trouble and avoid unnecessary calculations.

Returning to the other half of our problem, it should be clear that there will be a nonzero y-component to our force, since every piece of the wire contributes a positive dF_y. To find the y-component of the total force, we just add up the contributions from each little piece like this:

$$F_y = \int dF_y = kQ\lambda r \int_{-\infty}^{\infty} \frac{dx}{(x^2 + y^2)^{3/2}}. \tag{6.14}$$

This integral can be done with a trig substitution ($\sin(\theta) = x/\sqrt{x^2 + r^2}$) with the result that

$$F_y = \frac{2kQ\lambda}{r}. \tag{6.15}$$

Does the result make sense?

Well, the expression does have appropriate units since λ is a charge-per-unit-length: $Q\lambda/r$ thus has the same units as a product of charges divided by a distance squared, as appears in Coulomb's law.

And it makes some sense that the force falls off with increasing distance r more slowly than the inverse-square r-dependence characteristic of the force between point charges. When r is small, only the central part of the wire contributes $d\vec{F}$ with appreciable dF_y: the parts of the wire that are far from the center contribute $d\vec{F}$s that are nearly horizontal. If, then, r is (say) doubled, we are twice as far away from the part of the wire that was previously making appreciable, non-cancelling contributions to the total force, and, all other things being equal, we would expect the force to be only about 1/4 as big. But when we are twice as far away, the part of the wire that contributes appreciable dF_y is also roughly twice as big. And twice as much charge, about twice the distance away, implies that the force should be only about *half* as big. In short, the force going as $1/r$ seems quite plausible.

6.4.3   A Charged Plane

One more example: let's try to find the force that a large planar "sheet" of charge, with uniform charge-per-unit-area σ, would exert on a particle of charge Q a distance r away. See Figure 6.7.

A finite-size sheet of total charge Q and area L^2 would have charge-per-unit-area $\sigma = Q/L^2$. If we slice that sheet up into narrow strips of width dx (and length L), the charge of each strip would be $dQ = \sigma \, dA = \sigma L \, dx$. Each of the length-$L$ strips can then be though of as a charged wire, with charge-per-unit length $d\lambda = dQ/L = \sigma \, dx$. This relationship should hold even if Q and L are both infinite (such that the charge-per-unit-area σ stays finite).

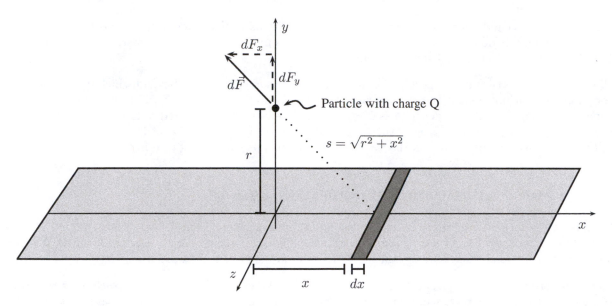

Figure 6.7:   The force exerted on a charged particle by an infinite planar sheet with uniform charge density can be calculated by thinking of the sheet as an infinite number of parallel wires.

Thus we can think of the generic strip highlighted in the Figure as a charged wire with charge-per-unit-length

$$d\lambda = \sigma \, dx. \tag{6.16}$$

We can thus re-use the formula we developed in the previous example, according to which (noting that the distance r from before now corresponds to the distance s!)

$$dF = \frac{2 \, k \, Q \, d\lambda}{s}. \tag{6.17}$$

The x-components will cancel, as before, so we needn't bother worrying about them. The y-component of the force exerted by our highlighted generic strip will be

$$dF_y = dF \, \frac{r}{s} = \frac{2 \, k \, Q \, r \, \sigma \, dx}{r^2 + x^2}. \tag{6.18}$$

The y-component of the total force, exerted by all the strips which compose the sheet, can be found by adding up the y-components of the forces contributed by each piece:

$$F_y = \int dF_y = 2kQr\sigma \int_{-\infty}^{\infty} \frac{dx}{r^2 + x^2} = 2\pi kQ\sigma. \tag{6.19}$$

The units of this result do make sense: σ is a charge per unit area, so $kQ\sigma$ has the same units as kQq/r^2, which we know is a force.

It is rather surprising, though, that our answer doesn't depend on the distance r at all! Does this make sense? Actually it does, for a reason rather like what we saw in the previous example. The sheet has charge distributed over a two-dimensional area. When the particle is close to the sheet, only the charge in a small roughly-circular area contributes appreciably to the non-cancelling y-component of the total force. If the distance r is now doubled, that same part of the sheet, being twice as far away, would exert only 1/4 the force it did previously. But if the particle is twice as far away, the region of the sheet that contributes appreciably is roughly four times bigger – twice as wide in each of the two dimensions of the sheet! So when r is doubled, the overall force should be something like $1/4$ of something that is roughly 4 times as big... i.e., it should stay the same!

Of course, for a real – finite – sheet of charge, there will be some distance r beyond which it is no longer a good approximation to treat the sheet as infinitely big. So for a real sheet, F will eventually decrease as r increases (and, indeed, will decrease as $1/r^2$ when r becomes large compared to the size L of the sheet). But as long as $r \ll L$, i.e., as long as the particle is so close to the sheet that it is a good approximation to treat the sheet as infinitely big, the force will be independent of its distance from the sheet.

§ 6.5 Forces Between Permanent Magnets

We have been focusing on the concept of electric charge and the electrical forces that charged objects exert on one another. Let's back up and discuss magnets and magnetic forces in a similar way. We will, however, be a little quicker, leaving many details to be worked out next week.

As noted earlier, a magnet that is left free to rotate in a horizontal plane (e.g., by attaching it to a cork and letting it float on water) will spontaneously orient itself roughly parallel to the local longitude lines. Such a device is called a "compass" and the magnet is usually referred to as the "compass needle". The end of the needle that points toward the geographic North is called the "north pole" and likewise, the "south pole" end of the compass needle is the end that points toward the geographical South.

Having established that naming convention, let's discuss the forces that two magnets exert on each other. Consider first the probably-familiar case of two standard "bar"-shaped permanent magnets. As shown in the top frame of Figure 6.8, the magnets *repel* one another if their North pole ends are facing each other. (They also repel if the South pole ends face one another.) On the other hand, the magnets will *attract* if one of the magnets is reversed, as shown in the bottom frame of the Figure, so that now the North pole end of one magnet faces the South pole end of the other magnet.

What if the magnets are arranged side-by-side instead of end-to-end? As shown in Figure 6.9, the magnets will repel one another if they are side-by-side in the same orientation (i.e., N next to N and S next to S), but will attract one another if one of them is flipped (so that the N of one is next to the S of the other, and vice versa).

Note that all of the observations described so far would be exactly the same if the North and South poles in all of the scenarios were replaced by positive and negative electric charges. The point here is

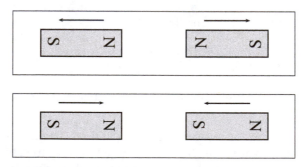

Figure 6.8: Top frame: two magnets with their North pole ends facing each other (or, not shown, with their South pole ends facing each other) repel one another. Bottom frame: two magnets with their North and South pole ends facing each other attract one another.

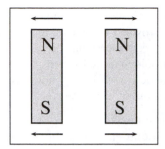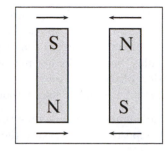

Figure 6.9: Left frame: two magnets aligned, side-by-side, with the same orientation (so that the North pole end of one is next to the North pole end of the other, and likewise for the South pole ends) will repel one another. Right frame: if one of the magnets is flipped, so that opposite poles are next to each other, the magnets now attract one another.

emphatically *not* that a North magnetic pole *is* a positive charge. It's not! For example, a positively-charged cork is not repelled by the North pole end of a magnet. Indeed, it interacts with the magnet exactly as it would a similarly-shaped non-magnetic hunk of metal – which is to say that the electric charge doesn't interact with the North and South magnetic poles *at all*. The point is rather that North and South magnetic poles seem to interact with other North and South magnetic poles in exactly the same way that positive and negative charges interact with other positive and negative charges. Namely: Ns repel other Ns, and Ss repel other Ss, but Ns and Ss attract one another.

This analogy can help us predict and understand some slightly more complicated observations. For example, suppose we place several small compasses at various points in the vicinity of a larger bar magnet. The idea that the N and S ends of the compass needles are respectively attracted to the S and N ends of the bar magnet (and also respectively repelled from the N and S ends of the bar magnet!) can help us make sense of the directions that the compass needles end up pointing at different locations.

See, for example, Figure 6.10. The compass on the left, along the axis of the bar magnet and near the South pole end, orients with its N pole end to the right, because the N pole end of the compass is strongly attracted to the nearby S pole of the bar magnet and the S pole end of the compass is strongly repelled by the nearby S pole of the bar magnet. (Of course, the N and S poles of the compass needle are also, respectively, repelled and attracted by the more distant N pole end of the bar magnet; but since this pole is further away, the forces are smaller, so the forces described in the previous sentence dominate and determine the orientation.)

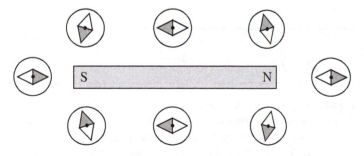

Figure 6.10: Several compasses are placed around a larger bar magnet. (The compass needles are diamond-shaped, with the shaded end being the North pole.) The observed orientations of the compasses can be understood using the model that Ns repel Ns and attract Ss.

Note that, for example, the compass that is directly above the labeled S pole of the bar magnet orients so that its N pole end points roughly – but not exactly – toward the bar magnet's S pole. The reason is that the more distant N pole end of the bar magnet also exerts forces on the poles of the compass needle; these are, as before, smaller, because the bar magnet's N pole is more distant, but here their contribution does affect the orientation of the compass needle. You should convince yourself that the orientations of the other compasses shown here can be understood in a similar way.

And you should have the chance to spend some time in class exploring this analogy further, e.g., by performing a Coulomb-like experiment to measure the separation-dependence of the forces between magnets.

But there are some ways in which the analogy between magnetic poles and electric charges breaks down. For example, suppose we build an electric analog to a bar magnet – a "bar electric" – say, by placing positive and negative charges on the two ends of a wooden slab. And suppose we build an "electric compass" by placing smaller positive and negative charges at the two ends of a small needle which is free to rotate. If a small cavity is drilled in the middle of the "bar electric" and the "electric compass" is inserted, the compass needle will orient itself in the way you would probably expect: the positive end of the "compass" will point toward the negative end of the "bar electric", as shown in the upper frame of Figure 6.11.

But, as shown in the lower frame of the Figure, something unexpected happens in the magnetic case: the N end of the compass needle (the shaded end of the diamond in the Figure) points toward the *North* end of the bar magnet! It is as if, *inside* of the bar magnet, the forces are reversed – the N end of the compass needle attracts, instead of repelling, the N end of the bar magnet, and so on.

Here is a similarly puzzling fact. If the "bar electric" is broken in half, one half is simply now an isolated positive charge and the other is simply an isolated negative charge. We could provide observational evidence for this by, for example, removing one of the halves to the next room, and surrounding the remaining half with several "electric compasses". We would observe that the "electric compasses" would all align themselves radially with respect to the charged half "bar electric" that remains in the center. (If, for example, the remaining charge is positive, the "electric compasses" would align so that their positive ends point radially outward, away from the central charge.)

But, again, what happens in the analogous magnetic case is entirely different. Breaking a bar magnet

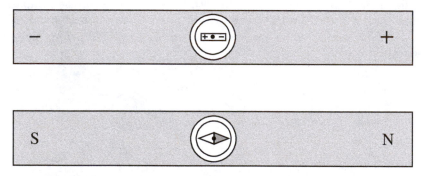

Figure 6.11:   An interesting element of disanalogy between electric and magnetic forces. An "electric compass" (i.e., a needle with a positive charge on one end and a negative charge on the other end), if placed in a hollow cavity in a "bar electric" (i.e., a slab with a positive charge on one end and a negative charge on the other end) will orient itself in the expected way: the positive end of the needle is attracted to the negative charge on the left, and the negative end of the needle is attracted to the positive charge on the right. But in the magnetic version, in which a regular (magnetic) compass is placed in the hollow cavity of a bar magnet, the compass needle will point in the opposite of the (probably) expected direction.

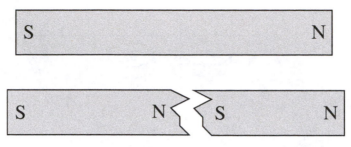

Figure 6.12: If a bar magnet is broken in half, the halves do not become isolated magnetic poles (so-called "magnetic monopoles"). Instead, each half becomes a full-fledged bar magnet of its own, with both a North and a South pole. This illustrates the seemingly-fundamental principle that magnetic monopoles do not exist.

in half does *not* result in two pieces, one of which contains a now-isolated North pole and the other of which contains a now-isolated South pole. Instead, two new poles seem to be created, as indicated in Figure 6.12. That is, each half of the broken bar magnet becomes a full bar magnet, replete with both a South pole and a North pole. Indeed, far more extensive and systematic exploration has never turned up an isolated magnetic pole – i.e., a so-called "magnetic monopole". So it seems that this is another fundamental difference between the electric charges and magnetic poles: isolated (positive or negative) electric charges certainly exist, but isolated (North or South) magnetic poles do not.

§ 6.6 Magnetism and Electric Current

The fact that North and South magnetic poles do not, at the end of the day, really exist, obviously shows that the analogy between magnetism and electricity (in which North and South magnetic poles interact with one another just like positive and negative electric charges do) is not a viable fundamental model. What, then, *is* magnetism, after all?

The answer began to emerge from an unexpected observation made by the Italian scientist Luigi Galvani in 1791. Galvani had previously discovered that the leg muscles of dissected frogs could be made to contract by allowing the electric charge stored in a Leyden jar (an early version of what we now call a "capacitor") to spark through them. The novel observation was that the muscles were similarly caused to contract – but without any deliberately-applied or apparent electric charge – if they were simply touched simultaneously by two different types of metals.

Galvani suspected that the phenomenon was revealing something about a special (and evidently electrical) "force" associated uniquely with living organisms – a kind of "life force" or "animal electricity" as he described it. But as Galvani's fellow Italian physicist Alessandro Volta showed in the following years, there was nothing uniquely biological about the phenomenon: a separation of charge could be produced by simply placing strips of distinct metals (copper and zinc, say) in a bowl of salt water. Subsequent refinements involved amplifying the effect by, for example, stacking disks of alternating metals (copper, then zinc, then copper, then zinc, etc.) with salt-water-soaked cardboard between them. Positive charge would build up on one end of the "Voltaic pile", negative charge on the other, and the charge was continuously replenished as it was drawn off. This was the precursor to modern chemical batteries and provided a source of continuously flowing electrical charge, i.e., an electrical *current*.

The ability to produce and study continuous electric current led to a momentous discovery just a few years later when the Danish physicist Hans Christian Oersted noticed – during a lecture! – that a compass needle swung into a new direction when electrical current was made to flow through a nearby wire. More

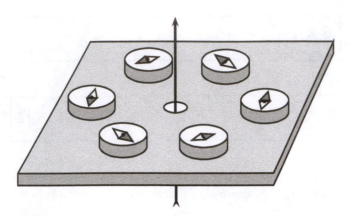

Figure 6.13:   When (conventional) electrical current is flowing upward through the wire (i.e., when negatively-charged electrons are flowing downward through the wire), the needles on the surrounding magnetic compasses arrange themselves as shown. If the direction of current is reversed (so the conventional current flows down instead of up), the compass needles reverse their orientations as well.

careful follow-up investigations revealed that a magnetic compass needle will orient itself in the plane perpendicular to the direction of current flow in a long straight wire and, in particular, perpendicular to the radial line from the wire out to the compass. The needles of a set of compasses, arranged in a circle around a current-carrying wire, will orient themselves in the continuous circular pattern indicated in Figure 6.13. As a little contemplation will make clear, there is no arrangement of magnetic North and South poles (distributed, somehow, in or around the wire) which could produce this pattern of compass needle orientations. So this observation (which we can reproduce in class) is further evidence that viewing magnetism as simply analogous to electricity is inappropriate. But more importantly, it is a massive clue that magnetism is, fundamentally, a result of *moving* electric charges.

Further evidence for this suggestion is provided by the fact that it is possible to mimic the magnetic effects of a bar magnet – including the appearance of a North pole and a South pole end – with electrical current. One need simply wrap the current-carrying wire around and around in a cylindrical shape. This produces a so-called "solenoid" whose magnetic effects – as probed with compasses at various locations – are identical to a bar magnet of the same shape and size. See Figure 6.14.

Note that this suggests something that would not be understood in any detail until well into the 20th century, but which we will mention now anyway: actually, the magnetic effects of a standard bar magnet *are*, in fact, produced by a distribution of flowing electric charge (i.e., a distribution of electric current) identical to that involved in a solenoid. That of course sounds fantastic because there is certainly no apparent electric current flowing around the sides of a bar magnet. You can't get an electric shock, for example, by picking it up.

But, as it turns out, the electrically-charged sub-atomic particles – the electrons – in the magnet can *spin*. It is somewhat misleading, but also helpful in this context, to picture them as little balls of charge that *rotate*, so that the charge is actually flowing around and around in circles. In a permanent magnet, the electron spin axes tend to be *aligned*. Then, for reasons that are analogous to the point conveyed in Figure 6.5 but are illustrated for this case explicitly in Figure 6.15, the effective amount of electrical current (averaged over a suitable region) in the interior of the magnet is zero, but there is an effective "surface current" flowing around and around the edges of the bar magnet in exactly the same pattern as the regular current flowing in a solenoid.

Finally, note that in the same way standard permanent magnets can both exert magnetic forces on

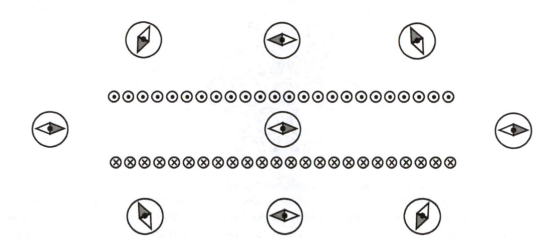

Figure 6.14: A current-carrying wire, looped repeatedly around a cylinder, makes a so-called "solenoid", shown here in cross-section: the smaller circles with dots and Xs are cross-sections through the wire at different points, with the dots representing current flowing out of the page toward us (the dot is meant to suggest the tip of an arrow that is pointing toward us) and the Xs representing current flowing into the page away from us (the X suggests the feathers you'd see at the tail of an arrow moving away). Several compasses are shown to indicate the magnetic forces generated by the solenoid. For locations outside the solenoid, the pattern is identical to that shown, for a standard bar magnet, in Figure 6.10, and the "surprising" orientation of the compass on the interior, shown earlier in Figure 6.11, is reproduced as well.

other magnets, and also have magnetic forces exerted on them, the same applies (not surprisingly) to electrical current. That is, a current-carrying wire can not only generate magnetic effects (as revealed, for example, by compasses, as we've been discussing), but can also have magnetic forces exerted on it by other sources of magnetism... including another current-carrying wire. So, in some sense, the simplest and most fundamental example of a magnetic interaction – first demonstrated in 1820 by the French physicist André-Marie Ampère – is that between two parallel current-carrying wires.

Ampère showed that two parallel wires with currents flowing in the *same* direction *attract* one another, whereas the wires *repel* if the currents flow in opposite directions. More detailed investigations showed that the magnitude of the force on each wire was proportional to its length and to the currents flowing in each wire, and also inversely proportional to the separation between them. Thus, for two parallel wires, 1 and 2, separated by distance d and carrying currents I_1 and I_2, respectively, the force exerted on wire 1 per unit length can be written

$$\frac{F_1}{L_1} = \frac{\mu_0}{2\pi}\frac{I_1 I_2}{d}. \tag{6.20}$$

As you can see, we've – perhaps puzzlingly – written the proportionality constant as $\mu_0/2\pi$. The reason for this is that certain other equations pertaining to magnetism, that we will encounter in the coming weeks, look simpler if the seemingly-pointless extra factor of 2π is inserted here.

In class, we will demonstrate the phenomenon and perhaps attempt to estimate the value of μ_0, which can be regarded as the fundamental constant pertaining to magnetic phenomena, analogous to the "k" from Coulomb's Law for electrical phenomena. It is worth noting, however, that in fact the value of μ_0 (which, unfortunately, goes by the cryptic name "the vacuum permeability constant") is not actually something we get from experiment. Instead, the constant is *stipulated* to have the value

$$\mu_0 = 4\pi \times 10^{-7} \text{ Newtons/Ampere}^2. \tag{6.21}$$

Figure 6.15: A crude model of how the aligned spins of the electrons in a permanent magnet give rise to an effective current equivalent to that present in a solenoid. The picture is meant to show a cross-section through a bar magnet whose North pole end is out of the page toward us and whose South pole end is behind the page away from us. The shaded circles are meant to represent individual spinning electrons (not to scale, obviously) with their spin axes all perfectly aligned. In the interior, the electrical current associated with the spin of one electron is effectively cancelled by the oppositely-directed current associated with the spin of the adjacent electron. But at the edges, there is no adjacent electron, so there is an effective surface current that, as one can see, runs counter-clockwise around the edge of the magnet.

This implies that the magnetic force exerted on a one-meter length of a long straight wire that is separated from another long straight wire by one meter will be exactly 2×10^{-7} Newtons, if each wire carries a current of one Ampère. Or, in other words, we *define* the Ampère to be the amount of electric current that makes the magnetic force, on each meter of two long parallel wires separated by one meter, be exactly 2×10^{-7} Newtons.

The Coulomb – the unit of electric charge that we encountered earlier – is then defined as one Ampère-second, i.e., the amount of charge that passes a given point in a wire in one second when a current of one Ampère is flowing. With the Coulomb defined in this way, the question of (for example) the magnitude of the electric force between two one Coulomb charges separated by one meter, can only be answered by experiment. So although the value of the fundamental magnetic constant, μ_0, is stipulated, the value of the corresponding electric constant k from Coulomb's law is genuinely empirical.

We will continue to explore electric and magnetic interactions next week, after introducing the crucially important "field" concept which makes calculations of (especially) magnetic forces much simpler and also renders electric and magnetic interactions more causally comprehensible.

Questions:

Q1. Which model do you think corresponds more directly to the eventually-discovered truth as you understand it – Dufay's two-fluid model, or Franklin's single-fluid model? Explain.

Q2. If the positive point charge outside the electrically neutral rod in Figure 6.4 were replaced by a negative point charge, how would the charge distribution on the rod change? Would the net force (exerted on the rod by the charge) be attractive or repulsive in this case?

Q3. Do *all* of the mobile electrons (or "electric fire") in the electrically neutral rod of Figure 6.4 move to one end of the rod? Or is it only some fraction of them that end up piling up at the end? Hint: to be sure, all of the electrons in the rod are attracted to the outside positive charge, but don't forget that the electrons also repel one another!

Q4. Explain in your own words why it makes sense that the electric force exerted on a point charge Q

by an infinite rod with uniform charge-per-unit-length λ a distance r away, should be proportional to $1/r$.

Q5. Explain in your own words why it makes sense that the electric force exerted on a point charge Q by an infinite sheet with uniform charge-per-unit-area σ a distance r away, should be independent of r.

Q6. William Gilbert was the first to recognize that magnetic compass needles align themselves North/South because the Earth itself is a huge magnet. Given that the end of the compass needle which points toward the geographical North is called the "North pole", is the (geographical) North pole a magnetic North pole or a magnetic South pole? Explain.

Q7. A single circular loop of current-carrying wire would act, magnetically, like a permanent magnet with what shape?

Exercises:

E1. Play around with some magnets and electrics. Can you identify any similarities or differences between them, beyond those mentioned in our brief historical sketch?

E2. A rod of length L extends from $x = -L/2$ to $x = L/2$ and carries a non-uniform charge-per-unit-length $\lambda = c|x|$ where c is a constant. What force would the rod exert on a point charge Q located a distance r away from the rod's center (along a direction perpendicular to the rod)?

E3. In the reading we derived Equation (6.19) by treating the sheet as a collection of long wires and using the previous result. Here, you will derive this same result in an alternate way. First, calculate the force exerted on a point charge Q a perpendicular distance z away from the center of a circular ring of charge q and radius r. Second, treat the sheet as a collection of concentric rings and add up their contributions to find the total force exerted by the sheet.

E4. A "surface current" K flows in the +y-direction in the x-y plane. (This means that an infinitely long but narrow strip of width dx, parallel to the y-axis, can be treated as a long straight wire carring current $K\,dx$.) If another infinitely long wire, parallel to the y-axis and carrying current I in the negative y-direction, is placed a distance z above the x-y-plane, what is the magnetic force per unit length exerted by the surface current on the wire? Setup an integral to calculate this and compare your result to a related electrical situation.

Projects:

P1. Reproduce Coulomb's torsion balance experiment, or some other modern equivalent, to see how the electric force between two charges varies with their separation.

P2. Conduct an experiment to measure the magnitude of the force between two magnets as the distance between them is varied. There are a couple of different interesting cases (e.g., finding the force between two disk magnets arranged co-axially, or finding the force between the North poles of two long bar magnets with the South poles swung out in opposite directions to get them as far out of the way as possible). Is your data consistent with an inverse-square (or some other simple mathematical) relationship between F and r? Perhaps different groups can explore different cases and share their findings.

P3. Setup and examine a reproduction of Ampère's demonstration that parallel wires attract if current flows through them in the same direction, and repel if current flows in opposite directions. Try to infer, from your observations, an estimate for the force exerted and use this (along with the known magnitude of the electric currents and the separation between the wires) to compute a value for the magnetic constant μ_0.

CHAPTER 7

Fields

S o far we have encountered three fundamental categories of forces: gravitational, electric, and magnetic. One thing these forces all have in common – the very thing that made electrical and magnetic forces seem somewhat magical already to their earliest observers – is that they allow for interactions between *separated* objects. This is particularly dramatic in the case of gravity: the Sun, for example, evidently exerts a force on the Earth, even though the Earth is separated from the Sun by something like a hundred million miles of apparently empty space.

This sort of apparent action-at-a-distance bristled the intuitions and expectations of many thinkers. Critics argued, for example, that Newton's theory of the solar system was incomprehensible and absurd. How can an object like the Sun act where it is not? Indeed, Newton himself agreed with the critics' intuitions. He wrote, in a famous 1693 letter to Richard Bentley:

> "It is inconceivable that inanimate brute matter should, without the mediation of something else which is not material, operate upon and affect other matter without mutual contact.... That gravity should be innate, inherent, and essential to matter, so that one body may act upon another at a distance through a vacuum, without the mediation of anything else, by and through which their action and force may be conveyed from one to another, is to me so great an absurdity that I believe no man who has in philosophical matters a competent faculty of thinking can ever fall into it."

It is not entirely clear how Newton did or would reconcile this philosophical position with his theory of universal gravitation.

Similar concerns had been expressed about electric and magnetic forces as well. For example, William Gilbert suggested that electrical interaction must involve some invisible material outflow – the electrical *effluvium* we encountered before on the grounds that

> "... no action can be performed by matter save by contact[. And yet] these electric bodies are not seen to touch, [so] something of necessity is sent from the one to the other, something that may touch closely and be the beginning of that incitement..."

And Ampère, who had systematically explored the magnetic effects of electrical current, similarly condemned "the supposition of an action between bodies that do not touch each other."

Our goal this week is to appreciate the concept of "field" – which developed in the 19th century studies of electricity and, especially, magnetism – and which has become the physicist's main tool for rendering apparent action-at-a-distance comprehensible.

Figure 7.1: Michael Faraday (1791-1867)

§ 7.1 Faraday

Michael Faraday was born on the outskirts of London in 1791. His family was relatively poor and Michael mostly educated himself by reading books he encountered during an apprenticeship with a local bookbinder. He developed an interest in chemistry and electricity and, after attending a lecture by the eminent English chemist Humphrey Davy, wrote to Davy and soon found himself working as Davy's assistant.

Faraday would go on to be one of the most accomplished scientists of all time, which is particularly interesting and impressive because he never developed the mathematical expertise that many people erroneously think of as a requirement for (or even the meaning of) doing good physics. Faraday once wrote to his mathematically sophisticated French colleague Ampère:

> "I am unfortunate in a want of mathematical knowledge and the power of entering with facility any abstract reasoning. I am obliged to feel my way by facts closely placed together."

For Faraday, "feeling his way by facts" meant developing experimental demonstrations to clarify theoretical proposals and exhibit novel phenomena. For example, early in his career, Faraday took the idea of magnetism arising from electric current, that we discussed last week, and used it to design the first primitive electric motors in which a permanent magnet would rotate continuously around a current-carrying wire, or vice versa. He also made important discoveries, at the boundary of physics and chemistry, involving electrolysis and was the first to recognize that the polarization direction of light can be rotated when the light propagates through a magnetized material.

One of Faraday's most important contributions, however, was not directly experimental, though it was certainly grounded in and motivated by his experimental "facts closely placed together". The contribution was instead of a theoretical nature, but more conceptual than mathematical. In particular, Faraday became convinced, due to the magnetic influences he observed on light and material substances, that magnets and electrical currents produced a kind of invisible magnetic "power" in the surrounding region.

Last week we used magnetic compass needles to visualize the magnetic effects at different places in the vicinity of a magnet or current-carrying wire. We saw, for example, that the needles of a set of compasses surrounding a long straight wire will orient themselves in a circle around the wire. We can represent this fact mathematically by saying that there are circular "lines of magnetic force" surrounding the wire, with the (curved!) "line of force" just representing the direction of the force on a magnetic North pole at each location. These "lines of force" could also be visualized by sprinkling iron filings in the vicinity of a magnet. Each tiny piece of iron would act like a small compass needle and would therefore line itself up along the "line of force" through its location. But with hundreds or thousands of such pieces, the result

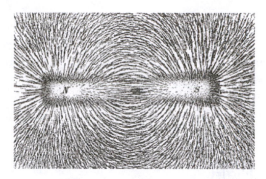

Figure 7.2:   Iron filings sprinkled near a magnet align themselves with and reveal the structure of the "magnetic field" surrounding the magnet.

was a visually compelling "map" of the magnetic influences surrounding the magnet itself. See Figure 7.2.

Faraday's innovation was to take such "lines of force" seriously, as representing the detailed structure present in a physically real (but intangible and invisible) "stuff", that surrounded the magnet or wire whether compass needles or iron filings were present or not. Faraday, that is, interpreted the "lines of force" not just as telling us about the force that a magnetic pole *would* feel, *hypothetically*, *if* it were placed at a certain location in the vicinity of the magnet or current-carrying wire. Instead, he saw the compass needles and iron filings as ways to *reveal* the magnetic power that was already present in the (apparently) empty space surrounding the magnet/wire.

As Faraday would explain this idea:

> "External to the magnet these concentrations which are named poles may be considered as connected by what are called magnetic curves, or lines of magnetic force, existing in the space around. These phrases have a high meaning, and represent the ideality of magnetism. They imply not merely the directions of force, which are made manifest when a little magnet, or a crystal or other subject of magnetic action is placed amongst them, but these lines of power which connect and sustain the polarities, and exist as much when there is no magnetic needle or crystal than as when there is; having an independent existence analogous to (though very different in nature from) a ray of light or heat, which, though it be present in a given space, and even occupies time in its transition, is absolutely insensible to us by any means whilst it remains a ray, and is only made known through its effects where it ceases to exist."

As we now say, following terminology that Faraday introduced elsewhere, the permanent magnet or current-carrying wire sets up a magnetic *field* in its vicinity. The presence and detailed structure of this field can be revealed by its effects on material objects (or light rays), but the field "exists as much" even when no such objects are present.

The "field" idea is conceptually relevant to the problem of action-at-a-distance with which we began. If what happens, for example, when a current-carrying wire magnetically attracts another nearby current-carrying wire is that the first wire produces a magnetic field in the space surrounding it, and then the part of this field which is present at the location of the other wire exerts the force on the other wire, there is no longer any mystery about how the first wire could "act where it is not". The force on the other wire is not (directly) exerted by the distant first wire, but is rather exerted by the magnetic field at the location of that other wire. And the magnetic field at that location evidently acquired its structure as a result of perfectly-comprehensible, contiguous causality: the electrical current in the first wire directly influences the field in its immediate surroundings, which in turn influences the field in its immediate surroundings, and so on.

It is relevant here that Faraday allows that the field "occupies time in its transition". The idea, evidently, is that, for example, the magnetic field at some distance from a wire does not respond immediately when current begins to flow. There is rather some temporal delay, while the effect of the current on the surrounding field propagates, at some finite speed, outward.

We will learn more about the finite propagation speed of such field-based influences in a couple of weeks. For now, we turn to developing the concept of "field" in a more systematic (and, yes, mathematical) way.

§ 7.2 The Electric Field

Faraday himself was most interested in – and most adamant about the reality of – the *magnetic* field. But the same idea can also be developed for electricity and even gravity. Let us start with the case of the electric field, which is probably a little more familiar than the gravitational field and definitely simpler to treat mathematically than the magnetic field.

Recall that the force $\vec{F}_{elec}$ on a particle with charge q, located at a distance r from another particle with charge Q located at the origin, is given by

$$\vec{F}_{elec} = \frac{k\,Q\,q}{r^2}\,\hat{r} \tag{7.1}$$

where $\hat{r}$ is a unit vector in the radially-outward direction. This is the force that the charge q would feel if it were located at a given location, and the formula for the force, not surprisingly, involves its charge, q. The force is, after all, a kind of interaction between the two charges, and depends on the properties of both of them.

But if we simply divide by q, we are left with a quantity that differs from one location to another, and depends *only* on the charge Q of the fixed particle at the origin:

$$\vec{E} = \frac{\vec{F}_{elec}}{q} = \frac{k\,Q}{r^2}\,\hat{r} \tag{7.2}$$

This quantity, that is, can be regarded as a description of the "electric power" – the *electric field* – that the charge Q sets up in its vicinity. This power can be *revealed* by placing a second "test charge" q at a certain location and observing the force exerted on it. But the field can be regarded as existing in the space surrounding Q whether any test charge is present or not.

Figure 7.3 shows, and its caption explains, two different ways of visualizing the electric field – the $\vec{E}$-field – in the vicinity of a single point charge. Both the "field vector" and "field line" representations capture, in a visually comprehensible way, what is expressed mathematically in Equation (7.2): the $\vec{E}$ field around an isolated positive point charge has a direction that is everywhere radially outward, away from the charge, and a magnitude that falls off as $1/r^2$ as one moves away from the charge.

Note that the $\vec{E}$ field in the vicinity of an isolated *negative* point charge also has a magnitude that falls off as $1/r^2$. But its direction is radially *inward* – *toward* the negative charge – instead of radially outward. In terms of the field line representation, we can say that whereas the field lines begin (and then spread out from) positive charges, the field lines are directed in towards (and terminate at) negative charges.

7.2.1   Superposition

Last week, we worked extensively with the idea that the total electric force on a charged particle can be computed by simply adding up the contributions from each point-like piece of a given charge distribution.

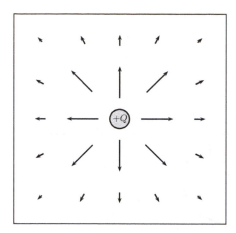 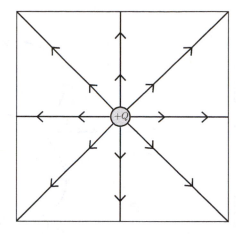

Figure 7.3: Two representations of the electric field surrounding an isolated (positive) point charge. On the left, the "field vector representation" shows the $\vec{E}$ field, represented by an arrow whose length is proportional to the magnitude of $\vec{E}$ and whose direction matches the direction of $\vec{E}$, at a sampling of points. The field at other points (in between the arrows) is not zero, of course, but can be estimated by interpolating using the nearby arrows. On the right, the "field line representation" shows continuous (directed) "lines of force" or "field lines". The direction of the $\vec{E}$ field at a given point is just the direction of the "field line" passing through that point. (For points not on a line, some interpolation is again needed.) The magnitude of $\vec{E}$ at a given point is related to the spacing of nearby lines: where the lines are close together (e.g., here, near the central fixed charge) the magnitude is large, and where the lines are more widely separated (e.g., here, farther from the fixed central charge) the magnitude is small. Note finally that these images just show the $\vec{E}$ field in the vicinity of a positive point charge, in two-dimensional cross-section. The field is really three-dimensional. It is worth thinking briefly about how, in a fully three-dimensional field line diagram for this situation, the density of lines would fall off as $1/r^2$ just like the magnitude of the actual $\vec{E}$.

For example, if a charge q is in the vicinity of N other point charges, the force on q is

$$\vec{F}_{elec} = \sum_{i=1}^{N} \vec{F}_{elec}^{i} \tag{7.3}$$

where $\vec{F}_{elec}^{i}$ is the force exerted on q by particle i. This individual force can be written in terms of Coulomb's law and will be proportional to q and Q_i.

The electric field $\vec{E}$ at a given location is, of course, the electric force $\vec{F}_{elec}$ that a test charge q would feel at that location, divided by q. That is:

$$\vec{E} = \frac{\vec{F}_{elec}}{q} = \sum_{i=1}^{N} \frac{\vec{F}_{elec}^{i}}{q}. \tag{7.4}$$

But $\vec{F}_{elec}^{i}/q$ is just the electric field that would exist, at that location, if only the charge i were in the vicinity. If we call that $\vec{E}_i$ we can therefore write

$$\vec{E} = \sum_{i=1}^{N} \vec{E}_i. \tag{7.5}$$

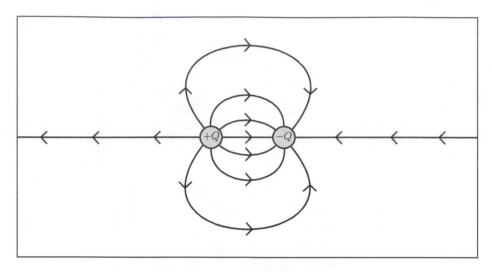

Figure 7.4: Field line representation of the electric field in the vicinity of an electric dipole, i.e., two separated charges $+Q$ and $-Q$. Note that, up close to either of the charges, the field lines look like they would look in the vicinity of that charge if the other one weren't present. This is because the $\vec{E}$ field at each point is, by "superposition", the sum of the contributions from each charge. If we are much closer to one than the other, the contribution of the other is negligible (due to the $1/r^2$ character of the field from a given point charge). Note also that, as you move out away from the dipole (say, along the vertical line that is equidistant from the two charges) the magnitude of the field – as represented by the spacing between the field lines – decreases.

The electric field, that is, obeys the "principle of superposition": the (total or actual) electric field at a point is just the sum of the contributions from all the nearby charges. (The "contribution" from a given charge is just the $\vec{E}$ field that would exist, at that point, if only that one charge were nearby.)

And of course the same idea holds if, instead of a finite number of finite charges, we have a continuous charge distribution which we can think of as an infinite number of infinitely small charges. Thus, the continuum version of Equation (7.5) also applies:

$$\vec{E} = \int d\vec{E}. \tag{7.6}$$

In short, we can calculate electric fields using the same techniques we practiced previously for calculating electric forces. Indeed, the calculations are almost identical – just a factor of q is missing!

We will practice this in class, but for now we illustrate the principle of superposition with just one fairly simple example. Figure 7.4 shows the field line representation for a slightly more complicated situation: a so-called "electric dipole", consisting of a postive and negative charge with some separation. You should convince yourself that, at a couple of randomly-chosen locations, the magnitude and direction of the field that are implied by the field lines, match up with what you would get by using the principle of superposition, i.e., adding up the (here, just two) contributions from the (here, just two) charges.

In class we'll play with some nice computer simulation tools that allow you to visualize the fields for more complicated distributions of charges.

7.2.2   Gauss' Law

The German mathematical physicist, Carl Friedrich Gauss, announced what amounts to a re-formulation of Coulomb's law, in terms of the electric field, in 1813. In words, Gauss' law states that the surface integral of the $\vec{E}$ field, over any closed surface (i.e., any surface that completely closes off a region that is "interior" to the surface), is proportional to the total amount of electric charge in that interior region.

It is easy to see that this is true for the simplest possible surface (a sphere) surrounding the simplest possible charge distribution: a single point charge. In particular, consider a spherical surface of radius R surrounding an isolated point charge Q. The electric field is given by Equation (7.2), and the surface integral of the field over the spherical surface (call that surface "S") – also known as the "flux of $\vec{E}$" through the surface – is

$$\Phi_{elec} = \oint_S \vec{E} \cdot d\vec{A}. \tag{7.7}$$

This means the following. Break the surface S up into an infinite number of infinitely small patches. For each patch, we define an associated "area vector" $d\vec{A}$ whose magnitude is just the area of the patch and whose direction is perpendicular to the patch (in the outward direction). For each patch, compute the dot product of the $d\vec{A}$ vector with the electric field $\vec{E}$ at that patch's location. Then add that quantity up for all the patches.

For the spherical surface, we can write $d\vec{A} = dA\,\hat{r}$. Plugging Equation (7.2), evaluated on the surface ($r = R$), into Equation (7.7) gives

$$\Phi_{elec} = \oint_S \frac{kQ}{R^2} \hat{r} \cdot \hat{r}\, dA. \tag{7.8}$$

But $\hat{r} \cdot \hat{r} = 1$ and kQ/R^2 is just a constant that can come outside. We are left with

$$\Phi_{elec} = \frac{kQ}{R^2} \oint_S dA = \frac{kQ}{R^2} 4\pi R^2 \tag{7.9}$$

since $\oint_S dA$ is just the sum of the areas of all the patches composing S, i.e., the total area $4\pi R^2$ of the spherical surface. Cancelling the R^2 in the numerator and denominator, we are left with:

$$\Phi_{elec} = 4\pi kQ. \tag{7.10}$$

This is Gauss' Law: the flux (i.e., surface integral) of $\vec{E}$ through a closed surface is proportional to the total amount of charge – here Q – enclosed within the surface.

Note that Coulomb's Law is sometimes written with the constant k replaced with $1/4\pi\epsilon_0$ where ϵ_0 is called the "permittivity constant". This makes Coulomb's law look a little more complicated, but replacing k with $1/4\pi\epsilon_0$ makes Gauss' Law look a little simpler:

$$\Phi_{elec} = Q_{enclosed}/\epsilon_0. \tag{7.11}$$

It is crucial to appreciate that although we have illustrated Gauss' Law here with the simplest possible example, it is true in general – not just for a spherical surface surrounding a point charge, but for any arbitrary shaped surface surrounding a point charge... or any other arbitrarily complicated charge distribution.

Let's think about why this is true in such generality. First, note that there was already a little bit of generality in our example: we showed that the flux of $\vec{E}$ through a spherical surface surrounding an isolated point charge Q is Q/ϵ_0 for a spherical surface of *any radius* R. To see that this remains true even for a non-spherical surface, it is helpful to consider a surface that consists of two different-radius hemispheres plus the annulus that connects them. See Figure 7.5.

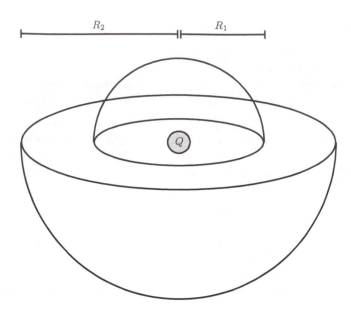

Figure 7.5:   A point charge Q surrounded by a closed surface with three parts: a small hemisphere of radius R_1, a larger hemisphere of radius R_2, and the annulus that connects them (i.e., the circle of radius R_2 with a hole of radius R_1). As explained in the text, it is easy to see that Gauss' Law holds even for this non-spherical surface: $\Phi_{elec} = Q/\epsilon_0$.

Since the electric field produced by the charge Q is perpendicular to the surface everywhere on the smaller upper hemispherical part of the surface, the flux of $\vec{E}$ through that sub-surface is just the magnitude of the field on that sub-surface (kQ/R_1^2) times the area of the sub-surface ($2\pi R_1^2$), i.e., $2\pi kQ$. Similarly, the flux through the larger lower hemispherical part of the surface will be kQ/R_2^2 times $2\pi R_2^2$, i.e., $2\pi kQ$. And since the electric field is everywhere radially outward, away from the charge, there is no flux through the annular sub-surface: $\vec{E} \cdot d\vec{A} = 0$ because $\vec{E}$ and $d\vec{A}$ are perpendicular. So, adding up the three contributions from the three sub-surfaces, the total flux is $\Phi_{elec} = 4\pi kQ = Q/\epsilon_0$.

That example is a helpful stepping-stone, because any arbitrary surface can be approximated, with arbitrary accuracy, as a bunch of little pieces, each of whose $d\vec{A}$ vectors are either radial (proportional to $\hat{r}$) or perpendicular to $\hat{r}$. The individual flux contributions through the radial pieces will add to Q/ϵ_0, for the same reasons we saw in the simpler example, and there will be no further flux contributions from the pieces whose $d\vec{A}$s are perpendicular to $\hat{r}$. Thus, the surface need not be spherical for $\Phi = Q_{encl}/\epsilon_0$ to be true. (A more rigorous proof can of course be given using vector calculus, but the math is a little beyond the level of this course, so it's enough to just see qualitatively that the result does hold.)

Note also that Gauss' Law holds, in the vicinity of an isolated point charge, whether the point charge is enclosed within the surface or not. For example, consider the surface depicted in Figure 7.6. This surface again consists of three sub-surfaces: a small hemisphere, a big hemisphere, and an annulus. The electric flux through the annulus is again zero for the same reason as before. On the small hemispherical sub-surface, the electric field is given by $\vec{E} = \dfrac{kQ}{R_1^2}\hat{r}$ and the area vectors can be written $d\vec{A} = -dA\,\hat{r}$. Note in particular the minus sign: on this sub-surface, the "outward" direction – away from the interior region that is enclosed within the surface – is radially inward, toward the charge Q. This makes the contribution of this sub-surface to the flux *negative*: $\Phi_1 = -2\pi kQ$. The contribution of the bigger hemispherical sub-surface is positive, just as before: $2\pi kQ$. So in this case the total flux through the closed surface is *zero* – just as one would expect from Gauss' Law since no electric charge is enclosed within the surface!

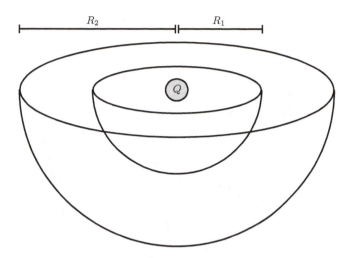

Figure 7.6: A point charge Q is located near – but is not enclosed by – a closed surface with three parts: a small hemisphere of radius R_1, a larger hemisphere of radius R_2, and the annulus that connects them (i.e., the circle of radius R_2 with a hole of radius R_1). As explained in the text, it is easy to see that the total flux through the surface is zero, so that Gauss' Law holds (since $Q_{encl} = 0$).

OK, so far we have been arguing that Gauss' Law is true, for the electric field set up by a single isolated point charge, no matter the size and shape of the surface and no matter whether the surface encloses, or fails to enclose, the charge.

It is then relatively easy to see that it must be true in full generality, because of the principle of superposition obeyed by electric fields. Suppose, for example, there are N point charges distributed around in some way. The electric field at each point in space can be written

$$\vec{E} = \vec{E}_1 + \vec{E}_2 + \cdots + \vec{E}_N \qquad (7.12)$$

where $\vec{E}_i$ is the electric field that would be there if only the ith charge were present. But then the flux of $\vec{E}$ through an arbitrary closed surface S is

$$\Phi_{elec} = \oint_S \vec{E} \cdot d\vec{A} = \oint_S \vec{E}_1 \cdot d\vec{A} + \oint_S \vec{E}_2 \cdot d\vec{A} + \cdots + \oint_S \vec{E}_N \cdot d\vec{A}. \qquad (7.13)$$

The ith term on the right will be either Q_i/ϵ_0, or zero, depending on whether the ith charge is enclosed by, or not enclosed by, the surface S. Hence

$$\Phi_{elec} = Q_{encl}/\epsilon_0 \qquad (7.14)$$

in full generality.

One other helpful point about Gauss' Law is that it can be understood in an intuitive way in terms of the "field line" representation of electric fields. The basic point is just that the flux of electric field through a surface is closely related to the number of field lines which pierce through the surface. (And note that these piercings are "directed": a field line that pierces through the surface from inside to out contributes a positive flux, whereas a field line that pierces through from out to in contributes a negative flux.) Of course, how many field lines you *draw* on a given diagram depends on various subjective factors. But as long as your field line diagram is consistent – meaning that the number of field lines which emerge from each unit of positive charge and terminate on each unit of negative charge is the same for all charges and

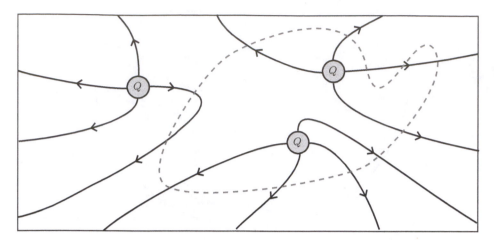

Figure 7.7: A (two-dimensional) field line diagram representing the electric field produced by three equal magnitude positive charges. The dashed gray curve can be thought of as (a two-dimensional cross-section through) a Gaussian surface. Each field line that originates on one of the charges inside the surface will contribute exactly one net outward piercing of the surface. (Most lines just pierce through the surface once. But, as illustrated by one line near the upper right corner, the lines may also pierce through the surface once, then pierce back in, and then pierce back out again. But this, too, represents exactly one net signed piercing.) Field lines which originate on charges outside the surface either never pierce it, or pierce into it, and then back out, which counts as zero net signed piercings. In short, all – and only – the field lines that originate inside the surface will at some point pierce their way out of it. So assuming the number of lines originating on each charge is proportional to the amount of its charge, the net number of outward piercings will be proportional to the total enclosed charge. This is a helpful way of understanding Gauss' Law in terms of the electric field line representation of fields.

the field lines are appropriately spaced so as to accurately reflect the relative magnitude of the field at different locations – then the flux through a given surface will be *proportional to* the number of (directed) piercings.

For example, consider a random surface in the vicinity of, say, three point charges, two of which are inside the surface and one of which is outside. See Figure 7.7. It is clear that every field line which originates on a charge *inside* of the surface must eventually pierce through the surface in order to escape to infinity. (It may pierce the surface just once, or it may pierce its way out, then back in again, and then back out again, etc. But either way there will be exactly one net signed piercing.) And it is clear that every field line which originates on a charge *outside* the surface will either not pierce the surface at all, or will pierce into it and then back out (perhaps some number of times!), but either way the net number of signed piercings will be zero. And so, since the number of field lines emerging from a given charge should be (in a proper diagram) proportional to its amount of charge, the total number of net signed piercings of the surface will be proportional to the total amount of electric charge enclosed within the surface.

So, thinking about the electric flux Φ_{elec} in terms of electric field lines provides a nice, alternative, visual way to understand why Gauss' Law is true and what it means.

To summarize, Gauss' Law can be derived from Coulomb's Law and is equivalent to it. It is important, though, because it is formulated in terms of the electric field (produced by a given distribution of charges), instead of the forces that charges do (or would) exert on one another. Indeed, Gauss' Law is the first of the four fundamental field equations for the electric and magnetic fields – equations that would be

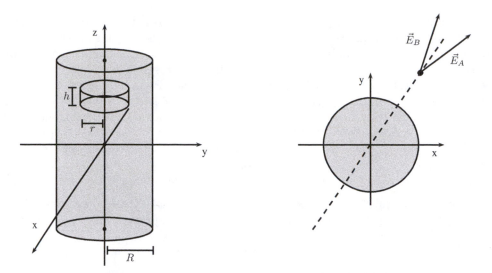

Figure 7.8: An infinitely long cylindrical tube with uniform charge density is shown, on the left, in side view, and, on the right, in an end-on view.

reformulated, collected, and fixed up by James Clerk Maxwell in the 1860s – that we will be slowly developing in the coming weeks.

7.2.3   Using Gauss' Law and Symmetry to Calculate Fields

Before turning to a discussion of magnetic and gravitational fields, let's see how Gauss' Law can be used to *calculate* electric fields for certain symmetrical charge distributions. We will just illustrate here with one example and then practice with more examples in class.

So, suppose there is an infinitely long cylinder of radius R with a uniform three-dimensional charge density (i.e., a charge-per-unit-volume) ρ. See Figure 7.8. We can take the z-axis to coincide with the axis of the cylinder and use the cylindrical radial coordinate $r = \sqrt{x^2 + y^2}$. Thus, we can describe the cylinder by saying that ρ is different from zero where $r < R$. We will also make use of the associated cylindrical radial vector $\hat{r}$ which points outward, perpendicularly away from the z-axis.

Let's begin by thinking about symmetry. We can prove that the $\vec{E}$ field at any location must be in the radial direction (in the cylindrical coordinates sense) as follows. Suppose it wasn't. Suppose, for example, that the field at the point marked with the black dot in the right frame of the Figure had a nonzero azimuthal (around the z-axis) component, like the $\vec{E}_A$ shown there. Well then, consider the operation of *reflecting* everything across the plane indicated by the dashed line in the Figure. This is a symmetry operation for the charged cylinder: after the reflection, there is exactly the same amount of charge everywhere as was there originally. So, on the one hand, the electric field shouldn't change. But, on the other hand, if everything reflects across this plane, that should include the $\vec{E}$ field as well, so after the reflection the field should look like the vector labeled $\vec{E}_B$. But then we have a contradiction: the field, at this location, cannot be both $\vec{E}_A$ and $\vec{E}_B$ since these are different! The only way to avoid the contradiction is for the azimuthal component of the field to be zero everywhere.

The z-component of the field must also be zero everywhere (though this is only true because the charged cylinder is infinitely long!). See if you can identify a symmetry operation that allows one to establish this with a proof-by-contradiction similar to the one rehearsed in the previous paragraph.

So, just from the symmetry of the charge distribution, we have learned the direction of the electric field everywhere. That's pretty amazing, but we can go farther: we can also establish, just using symmetry, that the *magnitude* of the field can only depend on the distance r from the z-axis. Again, suppose this weren't true. That is, suppose there were two points at the same r – say, one along the x-axis and one along the y-axis – where the magnitude of the field was different. But a 90 degree rotation about the z-axis is another symmetry operation for the cylinder. So, on the one hand, the field at each point should stay the same if we make that rotation. But, on the other hand, a 90 degree rotation should take whatever field was there along the x-axis, onto the other point on the y-axis. So if the fields there were originally different, the field at the point along the y-axis would change when we make the rotation. But it can't change since the rotation is a symmetry operation! So, again, the only way to avoid the contradiction is for the field to have the same magnitude at all points with the same distance r from the z-axis.

A similar argument establishes that the magnitude of the field cannot depend on z, either – but, again, only because this cylinder happens to be infinitely long! So we have proved, just using the symmetry of the given charge distribution, that

$$\vec{E} = E(r)\,\hat{r}. \tag{7.15}$$

Now we are ready to use Gauss' Law to find out exactly what the electric field is, both for points inside the cylinder ($r < R$) and for points outside the cylinder ($r > R$).

Let's start with $r < R$. Consider the little Gaussian surface sketched in the Figure – an imaginary cylinder of radius r and height h that is coaxial with the actually-existing charged cylinder that is producing electric field here. This Gaussian surface consists of three faces: (1) the top circle, (2) the curved sides, and (3) the bottom circle. The area vectors $d\vec{A}$ for (1) and (3) are in the (plus or minus) z-direction. This is perpendicular to the direction of the $\vec{E}$ field, and so there is no flux through these two faces. All of the flux is through the curved sides, (2), where $d\vec{A} \sim \hat{r}$, and each point on the sides is at the same r, so

$$\Phi_{elec} = \oint \vec{E} \cdot d\vec{A} = \int_{(2)} E(r)\, dA = 2\pi r h E(r) \tag{7.16}$$

where we have used the fact that the total area of the curved side face of the cylinder is its circumference, $2\pi r$, times its height h.

Since the charge density is just constant within our Gaussian surface, we can write the charge enclosed by it as density times volume, i.e.,

$$Q_{encl} = \rho\,\pi r^2 h. \tag{7.17}$$

Plugging the last two equations into Gauss' Law gives

$$2\pi r h E(r) = \rho\,\pi r^2 h/\epsilon_0 \tag{7.18}$$

or, solving for the magnitude of the electric field,

$$E(r) = \frac{\rho\,r}{2\epsilon_0}. \tag{7.19}$$

So, evidently, the electric field is radially outward (in the cylindrical coordinates sense) and its magnitude increases linearly from zero as we move from the z-axis out toward the edge of the cylinder at $r = R$.

What if we are outside the cylinder? If we consider a similarly-shaped – but bigger – Gaussian surface (a cylinder whose radius r is now larger than R), we have again that the flux through the surface can be written

$$\Phi_{elec} = 2\pi r h E(r). \tag{7.20}$$

The charge enclosed by this larger Gaussian surface, though, is not the density ρ times the volume of the Gaussian surface. With $r > R$, some part of the interior of our Gaussian surface is *outside* of the charged cylinder. So the total charge enclosed is just the density ρ multiplied the part of the volume enclosed by the Gaussian surface where the charge density is in fact ρ rather than zero, i.e.,

$$Q_{encl} = \rho \pi R^2 h. \tag{7.21}$$

Putting the pieces together, we find that, for $r > R$,

$$E(r) = \frac{\rho R^2}{2\epsilon_0} \frac{1}{r}. \tag{7.22}$$

Or, putting everything together in one nice package:

$$\vec{E} = \begin{cases} \dfrac{\rho r}{2\epsilon_0} \hat{r} & \text{for} \quad r < R \\[3mm] \dfrac{\rho R^2}{2\epsilon_0} \dfrac{1}{r} \hat{r} & \text{for} \quad r > R \end{cases}. \tag{7.23}$$

Does this result make sense? Well, the units are OK in both expressions: remembering that ρ is a charge per unit volume, we see that both expressions have the same units as a charge divided by ϵ_0 and divided by a distance squared, which is correct for an electric field. It also makes sense that the electric field vanishes along the axis of the cylinder ($r = 0$), which we can see it must do from symmetry, and also that it falls off with distance as $1/r$ when we are outside of the cylinder. (So the field goes to zero as we get very far from the cylinder, but it falls off more slowly than the field from a point charge would. This makes sense for the reasons discussed last week.) One final nice check is that the expressions for $r < R$ and $r > R$ match at $r = R$.

§ 7.3 The Magnetic Field

The magnetic field is a little bit more complicated than the electric field, for the reason we came to grips with last week: the correct fundamental description of magnetism involves *moving charges* (i.e., electrical currents) rather than isolated North and South magnetic poles. The full expression for the magnetic force between two moving charges (or small segments of current) is a little bit complicated; perhaps you noticed that no magnetic analog to Coulomb's law was even presented last week? But the introduction of the magnetic field concept makes this all much easier to understand. So let's jump in with that.

7.3.1   The Biot-Savart Law

The expression for the magnetic field produced by a small element of electrical current was first put forward in 1820 by the French scientists Jean-Babtiste Biot and Felix Savart. In contemporary mathematical notation, the Biot-Savart law reads:

$$d\vec{B} = \frac{\mu_0}{4\pi} \frac{I \, \vec{dl} \times \vec{r}}{r^3} \tag{7.24}$$

where I is the electric current running through an infinitessimal piece of wire with a length and direction given by $\vec{dl}$. The vector $\vec{r}$ is the position vector of the field point (i.e., the location where the magnetic field contributed by the current element is $d\vec{B}$) relative to the current element.

Let's jump straight into an example, which will clarify both how the formula works, mathematically, and also how it relates to what we have seen previously.

So, consider an infinitely long straight wire carrying current I along, say, the z-axis. We will calculate the magnetic field $\vec{B}$ at a location a distance s from the wire. We can choose the field point, without loss

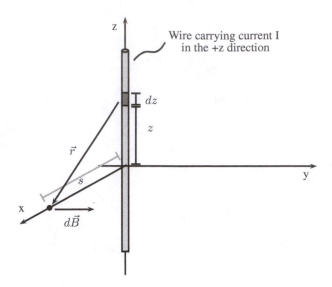

Figure 7.9: *Diagram for using the Biot-Savart law to calculate the magnetic field in the vicinity of an infinitely long straight wire carrying current I.*

of generality, to lie along the x-axis. Figure 7.9 shows the now-standard type of setup, in which we think of the wire as an infinite collection of infinitely small current elements to which Equation (7.24) can be applied.

The dark-shaded piece of the current-carrying wire has a length dz and has current flowing through it in the +z direction. So, for that piece,

$$I\,d\vec{l} = I\,dz\,\hat{k}. \tag{7.25}$$

The position vector $\vec{r}$ can be thought of as the displacement vector from the location of the current element (whose position is $z\hat{k}$), to the field point (whose location is $s\hat{i}$). So

$$\vec{r} = s\,\hat{i} - z\,\hat{k}. \tag{7.26}$$

The cross-product in the numerator of the Biot-Savart law is thus given by

$$I\,d\vec{l} \times \vec{r} = I\,dz\,\hat{k} \times \left(s\,\hat{i} - z\,\hat{k}\right) = I\,dz\,s\,\hat{j} \tag{7.27}$$

where we have used $\hat{k} \times \hat{i} = \hat{j}$ and $\hat{k} \times \hat{k} = 0$. Note that this explains why the contribution $d\vec{B}$ – of the highlighted current element to the magnetic field at the field point – is drawn, in the Figure, in the positive y-direction. Of course, one can also work this direction out directly from the right hand rule.

Putting the various pieces together, we have

$$d\vec{B} = \frac{\mu_0}{4\pi} \frac{I\,s\,\hat{j}\,dz}{(z^2 + s^2)^{3/2}}. \tag{7.28}$$

We can now finally add up the contributions from all of the pieces that compose the wire to find the total magnetic field at the field point:

$$\vec{B} = \int d\vec{B} = \frac{\mu_0\,I\,s\,\hat{j}}{4\pi} \int_{-\infty}^{\infty} \frac{dz}{(z^2 + s^2)^{3/2}} = \frac{\mu_0 I}{2\pi s}\,\hat{j} \tag{7.29}$$

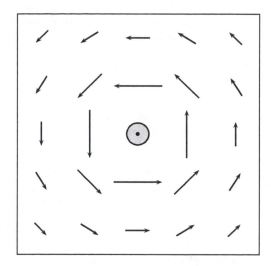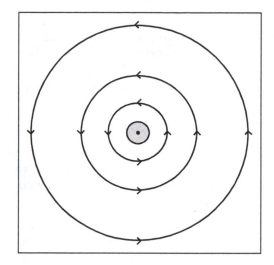

Figure 7.10: 2D cross-sectional view of the magnetic field in the vicinity of a long straight current-carrying wire. Here the wire is represented by the gray circle in the middle. The dot (which recalls the tip of an arrow pointing toward us) is meant to indicate that the current is flowing out of the page, toward us. The left panel shows the surrounding magnetic field in "field vector" representation, while the right panel shows the same magnetic field structure in "field line" – aka "magnetic lines of force" – representation. Note that the direction of the field at each location could be observed, in practice, by placing a compass at that point; the compass needle aligns itself with the local magnetic field, with its North pole end pointing in the direction of the field. Thus, the circularly-swirling magnetic field shown here is closely related to the pattern of compass needle directions, surrounding a current-carrying wire, in Figure 6.13.

So the magnitude of the magnetic field is proportional to the current I (which makes sense) and falls off with distance s from the wire as $1/s$ (which also makes sense). The direction is also interesting. The magnetic field direction is in a plane perpendicular to the wire, but it is not radially inward or outward (in the cylindrical coordinates sense) – instead, it is in the azimuthal direction, perpendicular to the (cylindrical coordinates) radial direction.

The overall field pattern (which follows immediately from the rotational symmetry of the current distribution about the z-axis) is sketched, using both "field vector" and "field line" representations, in Figure 7.10. The magnetic field "swirls" in circles around the wire, with a direction given by a version of the right hand rule due to Ampère: if you point the thumb of your right hand in the direction that current is flowing through the wire, your fingers naturally curl around in the direction of the magnetic field.

Recall that the *electric* field at a certain location can be understood as telling us something about the magnitude and direction of the force that a charged particle would feel, *if* it were placed at that location. The *magnetic* field, similarly, can be understood as telling us something about the magnitude and direction of the force that an isolated magnetic pole would feel, if it were placed at a given location. The only problem is, isolated magnetic poles don't exist.

Still, it will not lead us astray here (and, indeed, it will help us understand why the magnetic field lines are still sometimes referred to as "lines of force") to think of, for example, a magnetized compass needle as containing, at one end, a North magnetic pole which experieces a magnetic force in the direction of the magnetic field – and, at the other end, a South magnetic pole which experiences a magnetic force opposite the direction of the magnetic field. Hence, a compass needle will experience a *torque*, in a magnetic field, which tends to align it with the local field direction (with the North pole end "pointing" in the direction of the field). So we can now see how, for example, the pattern of compass needle orientations, in the vicinity of a current-carrying wire, that we talked about last week, was really telling us something about the structure of the magnetic field around the wire.

7.3.2   Gauss' Law for Magnetism and Ampère's Law

As with Gauss' Law for electric fields, some fundamental facts about magnetic fields and how they relate to their sources (namely, electric currents) can be captured with some abstract statements about certain integrals of the field.

There are, actually, two such statements. Let's start with the simpler one, which is closely related to Gauss' law for electric fields but unfortunately does not have any widely-recognized name. We'll just call it "Gauss' law for magnetism".

The point is simple. Unlike electric field lines, which begin (on positive charges) and end (on negative charges), magnetic field lines always make complete loops with no beginning or end. (Note that the claim that magnetic field lines never begin or end is equivalent to the statement that isolated magnetic monopoles do not exist!) This means that, for any closed surface S, any field line which pierces into the surface (thus contributing some negative magnetic flux) will eventually also pierce its way back out (thus contributing an equal amount of positive magnetic flux). The total magnetic flux, that is, through any closed surface, will necessarily be exactly zero:

$$\Phi_{mag} = \oint_S \vec{B} \cdot d\vec{A} = 0. \tag{7.30}$$

That's all there is to Gauss' law for magnetism. It is basically just a formal way of expressing, in terms of the magnetic field, that magnetic monopoles do not exist.

The other important statement about an integral of the magnetic field is called "Ampère's Law". It is different from Gauss' Law in that it involves a one-dimensional "path integral" of the magnetic field (rather than the two-dimensional "surface integral" that is also called the "flux"). But it is also more similar to Gauss' Law (for electric fields) than Gauss' law for magnetic fields is, in that it summarizes and generalizes the relationship between the field and its source in the simplest possible case. In particular, in the same way that Gauss' Law (for electric fields) is straightforwardly true for a spherical surface surrounding an isolated point charge (but can then be shown to be true in general as well), so Ampère's Law can be understood as an alternative formulation of the result we derived in the previous subsection – the formula for the magnetic field around a long straight current-carrying wire – which turns out to be true more generally.

Let's spell that out in detail. Recall that the magnetic field in the vicinity of a long straight wire, carrying current I in the positive z-direction along the z-axis, can be written:

$$\vec{B} = \frac{\mu_0 I}{2\pi r} \hat{\phi} \tag{7.31}$$

where r is the (cylindrical coordinates) radius, i.e., the perpendicular distance out from the z-axis, and $\hat{\phi}$ is the cylindrical coordinates unit vector in the "azimuthal" direction.

Now consider integrating the magnetic field, along a circular path P of radius R lying in a plane perpendicular to the z-axis. See Figure 7.11. The path integral $\oint_P \vec{B} \cdot d\vec{s}$ can be understood as follows. We break the path up into an infinite number of infinitely small "baby steps". Each "baby step" can be described as an infinitesimal displacement vector, $d\vec{s}$, like the one highlighted and labeled in the Figure. Now, for each "baby step", we compute the dot product of $d\vec{s}$ with the magnetic field $\vec{B}$ at that location. And then we add that up for all of the baby steps which jointly compose the entire path.

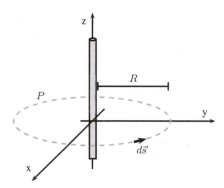

Figure 7.11: The same long straight current-carrying wire as before. To derive Ampère's Law, we consider the path integral of the magnetic field around a circular path P of radius R. The path integral involves adding up, for all of the little "baby steps" that compose the path, the dot product of the magnetic field $\vec{B}$ at the location of that baby step, with the "baby step" vector $d\vec{s}$.

For the circular path, the "baby steps" are all in the azimuthal direction, so we can write

$$d\vec{s} = ds\,\hat{\phi} \tag{7.32}$$

where ds is the magnitude of $d\vec{s}$, i.e., just the (infinitesimal) length of the baby step. Combining this with Equation (7.31) gives

$$\vec{B} \cdot d\vec{s} = \frac{\mu_0 I}{2\pi R}\,ds \tag{7.33}$$

since $\hat{\phi} \cdot \hat{\phi} = 1$. In words, since the "baby step" along the path, and the magnetic field, are in exactly the same direction, their dot product is just the product of their magnitudes.

OK, so, finally, we can compute the path integral around the entire path by adding up the contributions from each baby step:

$$\oint_P \vec{B} \cdot d\vec{s} = \frac{\mu_0 I}{2\pi R} \oint_P ds = \mu_0 I \tag{7.34}$$

since $\oint_P ds$ just means the sum of the lengths of all the baby steps around the entire path, i.e., the circumference of the circle (which is $2\pi R$).

So that is an interesting fact. Just like the surface integral of the electric field (over a spherical surface of radius R surrounding a point charge Q) was proportional to Q but independent of R, so the path integral of the magnetic field (along a circular path of radius R surrounding a wire carrying current I) is proportional to I but independent of R. That is, it turns out that this path integral of $\vec{B}$ is equal to $\mu_0 I$ for *any* circular path around the wire.

And, of course, just like with Gauss' Law for electric fields, this statement turns out be far more general even than this. The path integral of $\vec{B}$ along any path that "loops around" the wire – even if the path is not circular and even if the path doesn't lie in a plane perpendicular to the wire – will give this same result. And again similarly to the situation with Gauss' Law, the path integral of $\vec{B}$ along any path that *doesn't* "loop around" the wire – i.e., any path through the middle of which the electric current does not flow – will be *zero*.

We can summarize all of that by saying that, if there is just a single current-carrying wire, the path integral of $\vec{B}$ around any closed path is $\mu_0 I_{through}$ where $I_{through}$ is either the full current I, or zero, depending on whether the wire goes through the path P, or not.

And then finally, since the magnetic field obeys a principle of superposition – that is, since the magnetic field in the vicinity of an arbitrary collection of current-carrying wires is just the sum of the fields that would exist if each wire alone were present – it will be true in complete generality that

$$\oint_P \vec{B} \cdot d\vec{s} = \mu_0 I_{through}. \tag{7.35}$$

This is Ampère's Law. (Strictly speaking, we've only argued here that this will be true for magnetic fields produced by some combination of infinitely long straight wires. It does turn out to be true for a completely arbitrary distribution of electric current, but this is a little harder to show and so we set the proof aside here.)

In just a minute we will show how Ampère's Law can be used to calculate magnetic fields in situations where the distribution of electric currents is sufficiently symmetric. But let's clarify one mathematical detail first. The path integral in Ampère's Law involves going around the closed path P, with a sequence of "baby steps", *in a certain direction*. For example, in Figure 7.11, the path P is traversed in a counterclockwise direction as seen from above. We picked this direction because this is the same direction that the magnetic field "flows" in around the wire; so $\vec{B} \cdot d\vec{s}$ is positive and everything is a little simpler. But had we instead traversed the path P in the opposite direction, the result would have come out with the same absolute value but the opposite sign. The left hand side of Ampère's Law is therefore ambiguous, up to an overall sign, associated with the direction we traverse the path.

But the right hand side is also similarly ambiguous. If current pierces through the path (more precisely, pierces through some open two-dimensional surface whose edge is the path P) in one direction we should call that a positive $I_{through}$, whereas if it pierces through in the other direction we should call that negative. But which is which?

Here is the convention that we need to use to resolve these ambiguities and make Ampère's Law, as formulated above, true: curl the fingers of your right hand in the direction along which the path is being traversed on the left hand side; then, current that pierces through the path in the direction of your thumb should be called positive, whereas current piercing through in the opposite direction should be called negative. Note that this is really just a new application of Ampère's right hand rule for the direction of the magnetic field loops around a current carrying wire.

7.3.3   Using Ampère's Law to Calculate Magnetic Fields

Just like Gauss' Law (for electric fields), Ampère's Law is not only a fundamental statement about how fields relate to their sources (charges in the one case and currents in the other), but also a practical tool for calculating fields in cases of a high degree of symmetry.

Let's work through an example. Suppose there is an infinitely long and straight but *thick* wire – a cylinder of radius R – carrying a total electric current I along the z-axis. See Figure 7.12. We will assume that the current is uniformly distributed over the cross-sectional area of the wire – i.e., that the current density (current per unit area) is

$$J = \frac{I}{\pi R^2} \tag{7.36}$$

for $r < R$. Notice that the entire situation has a number of symmetries: rotational symmetry about the z-axis, reflection symmetry across any plane that includes the z-axis, and also translational symmetry parallel to the z-axis (because the wire is infinitely long).

If you expect, on the basis of Ampère's right hand rule, that the magnetic field should swirl around the z-axis, i.e., that $\vec{B} \sim \hat{\phi}$, you are correct. But let's see exactly how we can infer this rigorously using symmetry. (There is at least one interesting new subtlety in the magnetic case.)

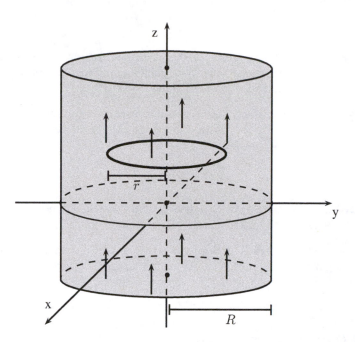

Figure 7.12: A piece of an infinitely long straight "thick" wire of radius R. The black arrows pointing the z-direction are meant to indicate the current flow, which we take to be uniformly distributed across the cross-sectional area of the wire. The bold circle of radius r is an Ampèreian path that we can use to calculate the magnetic field using Ampère's law.

To begin with, the rotational symmetry about the z-axis, and the translational symmetry along the z-axis, imply that the cylindrical coordinates field components cannot depend on the cylindrical coordinates z or ϕ. They can only depend on the cylindrical coordinates radius. That is, if $\vec{B} = B_r\hat{r} + B_\phi\hat{\phi} + B_z\hat{z}$ then $B_r(r, \phi, z) = B_r(r)$, $B_\phi(r, \phi, z) = B_\phi(r)$, and $B_z(r, \phi, z) = B_z(r)$. The only remaining question is: which of these components are nonzero, i.e., what is the *direction* of $\vec{B}$?

Could $\vec{B}$ have a nonzero component in the z-direction, i.e., parallel to the current in the wire? Suppose it did. But then consider a reflection across, say, the $x = 0$ plane. This is a symmetry operation for the current density, i.e., after the reflection there is the same amount of electrical current everywhere as was there before the reflection. And a magnetic field (component) in the z-direction is left unaffected by this reflection. So it may seem that everything is fine – a nonzero z-component to the field is compatible with the symmetry, and we cannot rule this out.

But, actually, that is not correct. There is something else that changes when we reflect across the $x = 0$ plane: right hands turn into left hands! And so, whereas before the reflection the magnetic field is determined according to the Biot-Savarat law (using the *right* hand rule to define the direction of the vector cross product), after the reflection the magnetic field should now be determined by the Biot-Savart law but using the *left* hand rule to define the direction of the vector cross product. And so, after the reflection, the direction of the magnetic field, in relation to the current flow, should reverse! (The formal terminology that expresses this point is as follows: the magnetic field is actually not a vector, but is instead a so-called "pseudo-vector". This means that it acts like a vector under *rotations*, but flips its direction under *reflections*.)

So we have a contradiction. One the one hand, a nonzero z-component is unaffected by a reflection across the $x = 0$ plane. But on the other hand, it should change sign. The only way out is for the z-component of the magnetic field to vanish.

A similar argument establishes that the radial (in the cylindrical coordinates sense) component of the field must also vanish: considering the same reflection, such a component is, on the one hand, unaffected, but, on the other hand, it should change sign. So it must be zero.

That leaves only the expected possibility, namely a non-zero ϕ-component. This is allowed by the symmetry because a magnetic field in the ϕ-direction *does* change direction under a reflection in (for example) the $x = 0$ plane. And so we are left with

$$\vec{B} = B(r)\hat{\phi}. \tag{7.37}$$

Now we are finally ready to use Ampère's law to determine how the magnitude, $B(r)$, varies with r. There are two different regions we need to consider: inside the wire ($r < R$) and outside the wire ($r > R$).

To find the magnitude of the field on the inside of the wire, we can consider a circular Ampèrian loop of radius $r < R$ like the one shown in Figure 7.12. With the magnetic field given by Equation (7.37), we have

$$\oint \vec{B} \cdot d\vec{s} = B(r) \oint ds = 2\pi r B(r). \tag{7.38}$$

On the right hand side of Ampère's Law, we have

$$\mu_0 I_{through} = \mu_0 \pi r^2 J = \mu_0 I \frac{r^2}{R^2}. \tag{7.39}$$

Equating the last two expressions then gives

$$B(r) = \frac{\mu_0 I r}{2\pi R^2} \tag{7.40}$$

so that the magnetic field, for $r < R$, is:

$$\vec{B} = \frac{\mu_0 I r}{2\pi R^2}\,\hat{\phi}. \tag{7.41}$$

To find the field on the *outside* of the wire, where $r > R$, we again consider a circular Ampèreian loop of radius r (but, now, obviously, with $r > R$). The two sides of Ampère's law become

$$2\pi r B(r) = \mu_0 I \tag{7.42}$$

because, now, *all* of the current in the wire pierces through the loop. And so, for $r > R$,

$$\vec{B} = \frac{\mu_0 I}{2\pi r}\,\hat{\phi}. \tag{7.43}$$

To summarize, it is nice to write an equation for the magnetic field *everywhere*, like this:

$$\vec{B} = \begin{cases} \dfrac{\mu_0 I r}{2\pi R^2}\,\hat{\phi} & \text{for} \quad r < R \\[2em] \dfrac{\mu_0 I}{2\pi r}\,\hat{\phi} & \text{for} \quad r > R \end{cases}. \tag{7.44}$$

We will practice with some additional similar examples in class.

§ 7.4 Magnetic Forces and Electro-Magnetic Induction

We have been reviewing the idea of electric and magnetic fields and focusing on how to calculate and visualize field configurations. In the case of electric fields, we understood, from the very beginning, how those fields related to the electric *forces* that we discussed last week. But in the case of magnetic fields, this remains somewhat unclear. Of course, we know that, if we pretend that magnets contain separated North and South magnetic poles, we can understand the magnetic field as telling us about the magnitude and direction of the force that would be exerted on a North magnetic pole at a given location. But, helpful though that pretense is in certain situations, we also know that magnetic poles don't really exist, and that magnetic forces are actually exerted on *moving electrical charges*, i.e., electrical currents.

So what, exactly, is the relationship between the magnetic field $\vec{B}$ at a certain location, and the force that a moving charge or current element at that location would experience?

Let us begin with an observation made by Michael Faraday. In the wake of the discovery that electrical currents acted as a source of magnetic effects – for example, a magnetized compass needle in the vicinity of a current-carrying wire would turn, aligning itself with the direction of the magnetic field produced by the current – Faraday became convinced that the opposite sort of influence should also be possible. If, that is, a current can affect a magnet, then probably a magnet should be able to generate an electrical current.

Faraday searched for such an effect by placing various wires and coils of wire, with various orientations, near strong permanent magnets or solenoids (i.e., electromagnets), and connecting a sensitive electrical current meter called a "galvanometer" to the wires to see if electrical current was flowing. Inevitably, he saw nothing, and remained puzzled for some time. Was there really no "reverse" effect (from magnets back onto currents)? Or was the effect simply too feeble to observe with the available equipment?

As it turned out, though, Faraday hadn't been looking for quite the right thing. He eventually stumbled onto the effect when he noticed that a small "induced" current flows in a coil of wire – but only for a brief moment – when the current in a nearby electromagnet is first turned on, i.e., while the current in that electromagnet is ramping up from zero to its full steady value. Once it reaches that steady value, the galvanometer shows no induced current in the coil. But then, when the current in the electromagnet is turned off and is in the process of ramping back down to zero, there is another brief flow of induced current in the coil, this time in the opposite direction.

Faraday conceptualized this phenomenon in terms of the "lines of magnetic force" that he insisted were physical realities and not mere mathematical abstractions. In particular, when the current through the electromagnet is turned on, the magnetic field lines (which emerge from the North pole end and swirl around to re-enter through the South pole end) increase in number and size, with individual field lines growing outward like an inflating balloon. Faraday inferred that it was the passage – the "cutting" – of these lines of force, across the wires constituting the coil, which caused electrical current to flow in the wire.

Faraday reasoned that the same phenomenon should therefore occur if a permanent magnet (whose field lines are "frozen" around it) was *moved*, rapidly, toward or away from the coil. And this indeed happened: the galvanometer deflected one way (indicating a current flowing in a certain direction through the coil) when a magnet was moved in toward the coil, and then deflected the other way (indicating a current flowing now in the opposite direction through the coil) when the magnet was pulled out away from the coil. While the magnet was just sitting there, next to the coil, there was no induced current.

Induced currents were also observed in other situations in which the wires in the coil were made to "cut through" magnetic field lines: the coil, for example, could be moved toward or away from the magnet, the magnet could be *rotated* in place, etc.

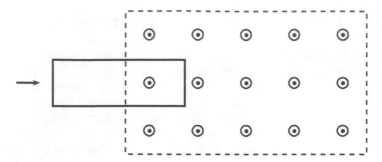

Figure 7.13: A rectangular loop of wire is pushed to the right into a region with a uniform magnetic field (represented by the circles with dots) coming out of the page toward us. As it moves, the front edge of the wire loop (on the right hand side of the rectangle) will cut through magnetic field lines and this, in Faraday's way of thinking about it, gives rise to an electro-motive force that drives a clockwise flow of electrical current in the loop.

The common denominator in all of these cases – which Faraday grasped qualitatively from his experimental demonstrations but which other scientists formulated in mathematically rigorous terms over the following decades – turned out to be the following: an electro-motive force (or "EMF") that can cause electrical current to flow through a coil of wire is induced when the magnetic flux

$$\Phi_{mag} = \int \vec{B} \cdot d\vec{A} \tag{7.45}$$

through the coil *changes*. We will discuss this fundamental principle – Faraday's law of induction – in more detail in subsequent weeks. For now, let's just focus on one simple example of this kind of situation in order to better understand the magnetic forces that magnetic fields exert on moving charges.

So, take the simplified situation depicted in Figure 7.13. A (for simplicity) rectangular loop of wire moves into a (for simplicity) rectangular region in which there is a uniform magnetic field out of the page toward us. It is an experimentally observed fact that in this kind of situation, an electrical current will flow clockwise around the loop. (Recall that this means that hypothetical positive charges that are free to move in the wire will be moving clockwise – or if, as we of course know turns out to be the case, it is instead negatively charged particles which are free to move in the wire, these will be moving counter-clockwise. But we will use the standard and harmless convention of pretending that there are free positive charges in the wire, so the direction of "conventional current flow" can be related to the direction of their motion.)

The "front edge" of the wire loop (i.e., the right hand side of the rectangular loop) is the one that is "cutting across" the magnetic field lines as the wire loop moves to the right, so according to Faraday's conception, that is where the action – which causes current to flow – is happening. In order to generate a clockwise (conventional) current in the wire, positively charged particles in that "front edge" would need to experience a magnetic force that pushes them *down*. Such a downward force would be perpendicular to both the rightward direction in which those charged particles are moving by virtue of the motion of the wire loop as a whole, and also perpendicular to the out-of-the-page direction of the magnetic field where the charges are located. With a velocity $\vec{v}$ to the right and a magnetic field $\vec{B}$ out of the page, one can check that the cross product $\vec{v} \times \vec{B}$ is in the correct, downward direction. Appending a factor of q as in

$$\vec{F}_{mag} = q\,\vec{v} \times \vec{B} \tag{7.46}$$

ensures that, if we are instead talking about a negatively-charged free particle in the wire, the force will be in the opposite direction (namely up), producing a counter-clockwise flow of negatively-charged particles which we would also describe as a clockwise conventional current.

This indeed turns out to be the correct formula for the magnetic force on a single moving charged particle. It is often combined with the related equation for the *electric* force on a charged particle, as in

$$\vec{F} = q\vec{E} + q\vec{v} \times \vec{B}, \tag{7.47}$$

which is usually called the "Lorentz force law" after the Dutch physicist Hendrik Lorentz who first explicitly formulated it in 1895 – well more than half a century after the observations suggesting it were made by Oersted, Ampère, and Faraday.

To help us make sense of this formula, let's make sure it connects up in the proper way with the formula we saw last week, due to Ampère, for the magnetic force that two parallel current-carrying wires exert on one another. So, consider the case, depicted in Figure 7.14 of two parallel wires, separated by a distance r. The wire on the left carries current I_1 away from us (into the page) and the wire on the left carries current I_2 in that same direction. As shown in the Figure, the current I_1 in the first wire produces magnetic field lines that spiral around the wire in a clockwise direction (as seen from this perspective). These field lines pierce down through the second wire, so that the magnetic field (produced by the first wire) at the location of the second wire has magnitude

$$B = \frac{\mu_0 I_1}{2\pi r} \tag{7.48}$$

and a direction that is straight down.

The current I_2 can be imagined to be constituted by positive charges moving away from us in the direction of the conventional current flow. With $q > 0$, $\vec{v}$ away from us along the wire, and $\vec{B}$ down, the magnetic force $\vec{F}_{mag} = q\vec{v} \times \vec{B}$ will be to the left, i.e., back toward the first wire. So it is clear that, qualitatively, our formula for the magnetic force on a moving charged particle can account for the observed fact that parallel electric currents *attract* one another.

Does the formula also get the magnitude of this attractive force correct? Let's model the current I_2 in the second wire as follows: suppose there are many particles, with number density (number per unit volume) n, each with charge q and moving with the same (average "drift") velocity v in the direction of current flow. In a time Δt, each particle will progress a distance $\Delta x = v\,\Delta t$ along the wire. Equivalently, every particle in a length Δx segment of the wire will pass a given point in time Δt. If the cross-sectional area of the wire is A, this is $N = nA\Delta x$ particles whose total charge is $\Delta Q = qN = qnAv\Delta t$. And so the electric current – the rate at which charge is flowing past each point in the wire – can be expressed as

$$I_2 = \frac{\Delta Q}{\Delta t} = qnAv. \tag{7.49}$$

The total magnetic force acting on this collection of N particles, distributed through a segment of the wire of length Δx, has magnitude

$$\Delta F = NqvB = nAqvB\Delta x = I_2 B\Delta x = \frac{\mu_0 I_1 I_2}{2\pi r}\Delta x. \tag{7.50}$$

This is in agreement with the result from last week, that the magnitude of the attractive force, per unit length, on each wire should be

$$\frac{\Delta F}{\Delta x} = \frac{\mu_0}{2\pi} \frac{I_1 I_2}{r}. \tag{7.51}$$

So that is reassuring.

Note finally that one of the intermediate results from that derivation is also convenient. In particular, we saw along the way that the magnetic force on a small piece of current-carrying wire has a magnitude equal

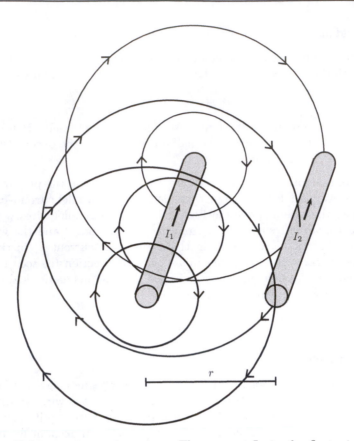

Figure 7.14: *Two parallel current-carrying wires. The current I_1 in the first wire creates the magnetic field shown, which spirals around the wire in a clockwise direction, passing downward through the second wire. Positively charged particles flowing through that second wire, in the direction of the conventional current, will feel a magnetic force to the left. This explains, in fundamental terms, why parallel currents attract each other magnetically.*

to the product of the current flowing in the wire, the length of the piece in question, and the magnitude of the field (when the current and field are perpendicular). This is consistent with the force being the cross product of the "current segment" $I\,d\vec{l}$ with the magnetic field, as in:

$$dF = I\,d\vec{l} \times \vec{B}.\qquad(7.52)$$

This is really just a way of re-writing $\vec{F}_{mag} = q\vec{v} \times \vec{B}$ for current segments.

That is a good place to draw a line. We now have put all the pieces together to understand how electric and magnetic fields are generated (from charge and current distributions, respectively) and how those fields function as intermediaries (avoiding the embarrassment of action-at-a-distance) to generate electric and magnetic forces on other distant charges and currents. We will continue to develop this thread in the coming weeks, turning next to catch up with the concept of "energy" and (especially) its role in electromagnetic phenomena, and then finally turning to Maxwell's completion of electromagnetic theory and his discovery of the fundamental nature of light.

Questions:

Q1. Many people find it intuitively obvious that action-at-a-distance should be impossible. But there is no rigorous proof of this. What do you think? Do you find it comprehensible that, for example, a particle could interact with – could exert a force on – another distant particle, "without the mediation of anything else, by and through which their action and force may be conveyed from one to another"?

Q2. Suppose you want to know the electric field at some distance $r > R$ from the center of a uniformly charged ball of radius R. Explain how the symmetry of the charge distribution allows you to get what you want using Gauss' Law.

Q3. Suppose you want to know the electric field at some distance $r > L/2$ from the center (and along the axis) of a uniformly charged stick of length L. Does the symmetry of the charge distribution allow you to get what you want using Gauss' Law? Explain why or why not.

Q4. In the reading we used Ampère's Law to calculate the magnetic field $\vec{B}$ in the vicinity of a long straight cylindrical wire of radius R and uniform current density. Are there any *different* geometries with enough symmetry to use Ampère's Law to calculate magnetic fields, or can we only use it with electric current distributions that are cylindrical?

Q5. When we used Ampère's Law to calculate the magnetic field in the vicinity of a long straight cylindrical wire of radius R, we assumed the current density in the wire was uniformly distributed, so $J = I/\pi R^2$. Do you think that the current density in a real current-carrying wire is uniformly distributed? Explain why or why not.

Q6. What do you think of Faraday's idea that if current can affect a magnet, a magnet should be able to cause electrical current? What, if anything, is the expectation for this kind of symmetry based on? Also, do you think a more careful formulation would have led Faraday to expect, from the beginning, what he eventually found? (Hint: electrical current produces a magnetic field that can exert a force on a magnet and cause the magnet to *move*. In short, electrical current can cause a magnet to move. Exactly what process would be the "reverse" of that?)

Q7. We discussed the clockwise electric current that runs through the rectangular wire loop in Figure 7.13. Describe the current (if any) that would flow through the loop if it continues moving to the right, through the region with the magnetic field, and eventually out the other side.

Q8. Imagine two permanent magnets shaped like coins, with the "heads" face being the north pole end and the "tails" face being the south pole end. If stacked on top of each other, say with both "heads" faces pointing up, there is an attractive force between the magnets. Explain why by considering the effective current distribution of each magnet and using the ideas developed here about how electrical currents interact magnetically.

Exercises:

E1. Use symmetry and Gauss' Law to find the electric field in the vicinity of an infinite planar sheet with charge-per-unit-area σ. Assess your answer by comparing to our result from last week for the force that such a sheet would exert on a point charge.

E2. Use symmetry and Gauss' Law to find the electric field both inside and outside a sphere of radius R whose charge density is proportional to the distance from the center, $\rho = cr$, where c is a constant, and then zero for $r > R$.

E3. Use symmetry and Ampère's law to find the magnetic field in the vicinity of an infinite planar surface current K flowing in the positive y-direction in the x-y-plane. (You should articulate very slowly and clearly how you can determine the direction of the field from symmetry considerations.)

E4. Use Ampère's law to find an approximate expression for the magnitude of the $\vec{B}$ field in the interior of a solenoid consisting of N circular loops of wire, wound around a cylinder of radius R and length L, carrying current I. (You should understand and explain why the expression you develop is only approximately true.)

E5. Use the Biot-Savart law to develop a formula for the magnitude $B(z)$ of the magnetic field a distance z along the axis from the center of a radius R circular wire loop carrying current I. Could the same formula be derived using Ampère's Law? Explain why or why not.

E6. Use the result of the previous exercise and the principle of superposition to write down a formula for the magnetic field along the axis of a *pair* of circular current loops, separated by distance d. (Assume one is at $z = -d/2$ and one is at $z = d/s$.) Note that, if d is small compared to R, the loops act like one loop (with twice the current) and $B(z)$ will have a single peak at $z = 0$; on the other hand, if d is large compared to R, $B(z)$ will have two distinct peaks separated by distance d. There is some "just right" value of d in the middle, though, that makes $B(z)$ have an extremely broad and extremely flat peak, i.e., a value that makes $B(z)$ as *constant*, in the vicinity of $z = 0$, as it is possible to achieve with two loops. Find this value of d by requiring that $d^2 B(z)/dz^2$ vanish at $z = 0$. What is the magnitude $B(0)$ at the center of a pair of circular coils with this special separation? (This arrangement is described as a pair of "Helmholtz coils" and is widely used to create an approximately-uniform magnetic field.)

E7. What happens if a particle with charge q and mass m is launched with speed v (perpendicular to the field lines) in a region with uniform magnetic field of magnitude B? Develop a formula relating the charge q and mass m of the particle and the magnitude B of the field to the radius R of the resulting orbit.

E8. Combine the Biot-Savart law (which gives the magnetic field produced by a current element) with the expression for the force on a current element, to write down a formula for the magnetic force exerted by one current element (say, with current I_1 and length/direction $\vec{dl_1}$) on another (with current I_2 and length/direction $\vec{dl_2}$). Show that, for two side-by-side current elements (with currents flowing either parallel or anti-parallel), the forces the two elements exert on each other obey Newton's third law and are directed along the line connecting the two elements (i.e., the two elements attract or repel). But is this always the case? Construct an example arrangement in which the magnetic forces the two current elements exert on each other do not obey Newton's third law, and contemplate/discuss the significance of this interesting possibility. (We will not be able to fully resolve the situation now, but it will come up again in a couple of weeks.)

Projects:

P1. Measure the magnetic field at a number of points along the axis of a current-carrying circular loop of wire, and compare your experimental data to the formula you developed in E5.

P2. Measure the charge to mass ratio of the electron by observing an electron beam undergoing cyclotron motion in an approximately uniform magnetic field.

P3. Develop the concept of "gravitational field", which can be defined as the gravitational force a test particle of mass m would feel at a given location (due to other massive objects in the vicinity), divided by m. What are the units of the gravitational field? In an earlier Chapter, you saw a picture of a gravitational field (in "field vector" representation) although it was not labeled as such; it would probably be illuminating to remember what that was and now recognize it for what it was. The close parallel between Coulomb's law (for electric forces) and Newton's law (for gravitational forces) suggests that it might be possible to fomulate a "Gauss' Law for gravity". Do this, by following a gravitational analog of the development we used to derive Gauss' Law (for electric fields) in the Chapter. Finally, generate – and then solve – a problem in which symmetry and Gauss' Law for gravity are used to find the gravitational field $\vec{g}$ in the vicinity of some appropriately-symmetrical distribution of mass.

CHAPTER 8

Energy

I T may seem odd that we have proceeded this far into this book – and this far into our tour of the historical development of physics – without encountering the concept of "energy". But although energy is now regarded as an essential component of the very subject matter of the science of physics, the concept made its appearance surprisingly late.

An early version of the concept of "kinetic energy" – defined as mv^2 (without the now-standard factor of $1/2$) and dubbed *vis viva* ("living force") by Gottfried Wilhelm Leibniz in 1695 – had been recognized by Newton's contemporary Christian Huygens as a quantity that is conserved in certain processes such as (what we now call) "elastic collisions". But there was some confusion between *vis viva* and momentum, and it was also quite clear that *vis viva* was not a conserved quantity in general. There were *in*elastic collisions, and also, more generally, processes (like a book slowing to a stop while sliding across a table) in which, we would now say, the kinetic energy is just obviously not conserved.

The full articulation of the principle of energy conservation – and with it the recognition of the importance of the concept of energy – thus had to wait for the recognition of the additional forms we now call "potential energy" and (especially) "thermal energy". This didn't happen until the mid- to late- 19th century. Important contributions along this road were made by Emilie du Chatelet, Leonhard Euler, Joseph-Louis Lagrange, Hermann Helmholtz, and a number of other people whose work we will encounter in the coming weeks.

For now, we take a largely ahistorical approach, and simply develop some ideas, relating to energy, that can provide deeper insight into the gravitational, electrical, and magnetic phenomena we have been exploring.

§ 8.1 Work and Energy

The technical concept of "work" was introduced as a measure of the net effect of an applied force on a moving object. It can be roughly understood as "force times distance", as in the following sort of case. If you want to lift a 10 pound weight from the floor to a shelf 5 feet off the ground, you have to apply an upward force of magnitude 10 pounds (to prevent the weight from falling, i.e., to keep it moving upward at, say, constant velocity) while the weight moves through a vertical distance of 5 feet. The work you do is therefore the product: 50 foot-pounds.

Or, to change the example and use preferred units, a force of magnitude 10 Newtons, applied (parallel to the direction of motion) to an object as it moves through a distance of 5 meters, does 50 Newton-meters = 50 Joules of Work on the object. (A "Joule" – the standard MKS unit for energy, is equal to a Newton times a meter, and was named after James Prescott Joule, an English physicist who helped establish the principle of energy conservation in the 19th Century.)

In this sort of example, it is important that the force be applied parallel to the direction of motion. A force applied perpendicular to the direction of motion (as, say, when you carry the 10 pound weight from one side of the room to the other) does *no* work, and a force applied *opposite* the direction of motion (as, say, when you lower the weight from the shelf back down to the floor) does *negative* work. All of this can be summarized with the more formal definition of the work done by a force $\vec{F}$ acting on an object as it moves through a displacement $\Delta\vec{r}$:

$$W = \vec{F} \cdot \Delta\vec{r} \tag{8.1}$$

or, for the infinitessimal work done by a force acting on an object as it moves through an infinitessimal displacement,

$$dW = \vec{F} \cdot d\vec{r}. \tag{8.2}$$

This formal definition allows us to prove an important result called the Work - Kinetic Energy theorem. It states that the total work done on an object during some process will equal the change in that object's kinetic energy during the process. To see why this is true, consider first the total work done (by all the different forces acting) during one "baby step" of some bigger process:

$$dW_{total} = \sum_i \vec{F}_i \cdot d\vec{r} = \vec{F}_{net} \cdot d\vec{r}. \tag{8.3}$$

Applying Newton's second law allows us to write

$$dW_{total} = m\vec{a} \cdot d\vec{r}. \tag{8.4}$$

But $\vec{a} = \dfrac{d\vec{v}}{dt}$ and $\vec{v} = \dfrac{d\vec{r}}{dt}$, so $\vec{a} \cdot d\vec{r} = \vec{v} \cdot d\vec{v}$. So

$$dW_{total} = m\vec{v} \cdot d\vec{v}. \tag{8.5}$$

Adding up both sides for all the baby-steps that compose some finite process then gives

$$W_{total} = \int dW_{total} = m \int \vec{v} \cdot d\vec{v} = \frac{1}{2}mv_f^2 - \frac{1}{2}mv_i^2 \tag{8.6}$$

or

$$W_{total} = KE_f - KE_i \tag{8.7}$$

where we define the "kinetic energy" as $KE = \dfrac{1}{2}mv^2$.

8.1.1   Gravitational Potential Energy Near the Surface of the Earth

Let's explore the implications of the Work - KE theorem for the simple case of objects near the surface of the Earth, where the gravitational field is given by

$$\vec{g} = -g\hat{j}. \tag{8.8}$$

(We assume here a coordinate system in which the positive y-direction is up.) The gravitational force on an object of mass m is therefore

$$\vec{F}_{grav} = -mg\hat{j} \tag{8.9}$$

and the work done (by this gravitational force) when the object moves along some path $\vec{r}(t)$ is given by

$$W_{grav} = \int \vec{F}_{grav} \cdot d\vec{r} = -mg \int dy = -mg\Delta y \tag{8.10}$$

where $\Delta y = y_f - y_i$ is the change in the object's y-coordinate – the change in its "altitude" – during the process in question. Note that if Δy is positive (i.e., if the object moves *upward*) the gravitational force does *negative* work, whereas if Δy is negative (i.e., if the object moves *downward*) the gravitational force does *positive* work.

If the gravitational force is the only force acting on the object during the process in question, then we can replace W_{total} in Equation (8.7) with W_{grav}:

$$-mg(y_f - y_i) = \frac{1}{2}mv_f^2 - \frac{1}{2}mv_i^2. \tag{8.11}$$

This is mathematically equivalent to the idea of "energy conservation"

$$E_f = E_i \tag{8.12}$$

if we define the total energy E as

$$E = KE + U_{grav} = \frac{1}{2}mv^2 + mgy \tag{8.13}$$

where $U_{grav} = mgy$ is the so-called "gravitational potential energy".

This is of course a familiar idea. Part of the point of rehearsing it here is to explain the close connection between "energy conservation" and the Work-KE theorem. Really, the statement of "energy conservation" is just the Work-KE theorem, with the terms describing the Work done by the gravitational force shuffled over to the other side of the equation and given a new name: $W_{grav} = \Delta KE$ is obviously equivalent to $\Delta KE - W_{grav} = 0$ which is the same as "energy conservation", $\Delta KE + \Delta U_{grav} = 0$, with $\Delta U_{grav} = -W_{grav}$. The change in the "gravitational potential energy" of some object during some process, that is, is simply minus the work done by the gravitational force during that process.

If a ball is released from height h and falls to the floor, we can say that its kinetic energy (KE) increases from zero to $\frac{1}{2}mv_f^2 = mgh$ either because the gravitational force did positive work mgh on it (and no other forces did any work), or because it started with gravitational potential energy mgh and this was converted into kinetic energy as the ball fell. These are simply two different ways of verbally describing the same mathematical fact that is a consequence of Newton's laws of motion.

Thinking explicitly about the connection to Work, though, does help us remember and understand why $E = KE + U_{grav}$ is sometimes constant and sometimes not constant. If gravity is the only force that does nonzero work on the object, then E should be conserved. This will of course be the case if no forces other than gravity are acting, or if all other forces are so small that their effects are negligible. But it will also be the case if other, non-negligible forces act, but do no work. This will be the case for forces which always act perpendicular to the direction of motion – for example, the tension force exerted by the string on a swinging pendulum bob, or the "normal" force exerted on a block sliding down a frictionless ramp. But if there are additional forces (such as air drag on the swinging pendulum or kinetic friction on the sliding block) which do nonzero amounts of work, then $\Delta E = 0$ no longer follows from $W_{total} = \Delta KE$. Instead, we would have

$$\Delta E = W_{other} \tag{8.14}$$

where W_{other} is the total amount of work done by forces *other than the gravitational force*, whose effects have already been included under the guise of U_{grav} in the definition of E.

One final note. We saw above that the gravitational potential energy U_{grav} can be defined in terms of the work done by the gravitational force as: $\Delta U_{grav} = -W_{grav} = -\int \vec{F}_{grav} \cdot d\vec{r}$. This tells us how to calculate the change in a particle's potential energy, as it moves along a certain path, in terms of the force $\vec{F}_{grav}$ acting on it. But this same relationship can be inverted to give us an expression for the force in terms of the potential energy:

$$\vec{F}_{grav} = -\vec{\nabla} U_{grav}. \tag{8.15}$$

The right hand side here is minus the "gradient" of the gravitational potential energy. The gradient is a vector-valued spatial derivative which can be understood in two equivalent ways. First, $\vec{\nabla} U_{grav}$ can be understood as a vector whose x-component is the partial derivative of U_{grav} with respect to x, whose y-component is the partial derivative of U_{grav} with respect to y, and whose z-component is the partial derivative of U_{grav} with respect to z. Or, second, $\vec{\nabla} U_{grav}$ can be understood as a vector whose direction is the direction along which U_{grav} increases most rapidly, and whose magnitude is the rate of change along that special direction. With $U_{grav} = mgy$, it is clear that both ways of thinking about the gradient reproduce the expected expression for the force: $\vec{F}_{grav} = -mg\hat{j}$.

8.1.2 Long Range Gravitational Potential Energy

We can also define a gravitational potential energy associated with long-range gravitational interactions. Suppose there is a body of mass M (which might be the Sun) which stays fixed at the origin of our coordinate system, and then a second body of mass m which is free to move. The gravitational force exerted by the fixed mass on the mobile mass is

$$\vec{F}_{grav} = -\frac{GMm}{r^2}\,\hat{r}. \tag{8.16}$$

So, in a process in which the mobile mass moves from radial coordinate (i.e., distance from the fixed mass) r_i to r_f, the work done by the gravitational force is

$$W_{grav} = \int \vec{F}_{grav} \cdot d\vec{r} = -GMm \int \frac{1}{r^2} dr = \frac{GMm}{r_f} - \frac{GMm}{r_i}. \tag{8.17}$$

If this gravitational force is the only one acting on the mobile mass during the process in question, the Work-KE theorem tells us that

$$\frac{GMm}{r_f} - \frac{GMm}{r_i} = \frac{1}{2}mv_f^2 - \frac{1}{2}mv_i^2 \tag{8.18}$$

which is equivalent to

$$\frac{1}{2}mv_f^2 - \frac{GMm}{r_f} = \frac{1}{2}mv_i^2 - \frac{GMm}{r_i}. \tag{8.19}$$

We can interpret this as a statement of "energy conservation" with $E = KE + U_{grav}$ as before, but now with the gravitational potential energy of the mobile particle given by:

$$U_{grav} = -\frac{GMm}{r}. \tag{8.20}$$

Note that this formula has U_{grav} being zero when $r = \infty$ and has the gravitational potential energy being negative for finite r. But of course, the gravitational potential energy *increases* as r increases, as one would expect. Indeed, from a given location, the direction along which U_{grav} increases most rapidly is the radial ($\hat{r}$) direction, and the rate of change along that direction is $\frac{\partial}{\partial r} U_{grav} = \frac{GMm}{r^2}$. So, as expected, we have that

$$\vec{F}_{grav} = -\vec{\nabla} U_{grav}. \tag{8.21}$$

8.1.3   Gravitational Potential and Field

If a small "test particle" finds itself in the vicinity of a collection of other particles, it experieces a force that is the sum of the forces it would feel from each of those other particles individually. This is the so-called superposition principle. Last week, we discussed the ways in which it is illuminating to consider such forces as arising from the local action of a *field*, which receives contributions from the collection of other particles.

There is a detailed parallel to all of this having to do with energy. Let us see how it plays out, first in the case of gravitation, and then, in the following section, in the case of electricity.

So: if a small "test particle" finds itself in the vicinity of a collection of other particles, the test particle can be assigned a potential energy that is simply the sum of the potential energies from its pairwise interactions with each of the other particles:

$$U = \sum_i \frac{-GM_i m}{r_i} \tag{8.22}$$

where M_i is the mass of the ith other particle, and r_i is its distance from the test particle of mass m.

But then, in the same way that we define the gravitational field $\vec{g}$ as "the (gravitational) force a particle of mass m would feel at a certain location, divided by m" – we can define a new quantity that is "the (gravitational) potential energy that a particle of mass m would possess at a certain location, divided by m." This quantity is called the "gravitational potential" (not to be confused with "gravitational potential energy") and can be denoted by the letter V. Near the surface of the Earth, where the potential energy of a particle of mass m is just $U = mgy$, the gravitational potential is given by $V = gy$.

Or, for the general case described above, in which the potential energy of a particle of mass m is given by Equation (8.22), the gravitational potential at a given location will be given by

$$V = \frac{U}{m} = \sum_i \frac{-GM_i}{r_i} \tag{8.23}$$

or, for a continuous mass distribution,

$$V = \int \frac{-G\, dM}{r}. \tag{8.24}$$

The gravitational potential V is a little easier to work with, compared to the gravitational field $\vec{g}$, because it is a scalar rather than vector. But the gravitational potential V relates to the field $\vec{g}$ in just the same way that the gravitational potential energy U relates to the gravitational force $\vec{F}$. Indeed, just dividing both sides of

$$\Delta U_{grav} = -\int \vec{F}_{grav} \cdot d\vec{r} \tag{8.25}$$

by m gives

$$\Delta V = -\int \vec{g} \cdot d\vec{r}. \tag{8.26}$$

So the difference in the gravitational potential at two points can be computed by calculating (minus) the path integral of the gravitational field, along a path that connects those two points.

Similarly, dividing both sides of the equivalent inverse relationship,

$$\vec{F}_{grav} = -\vec{\nabla} U_{grav}, \tag{8.27}$$

by m gives

$$\vec{g} = -\vec{\nabla} V. \tag{8.28}$$

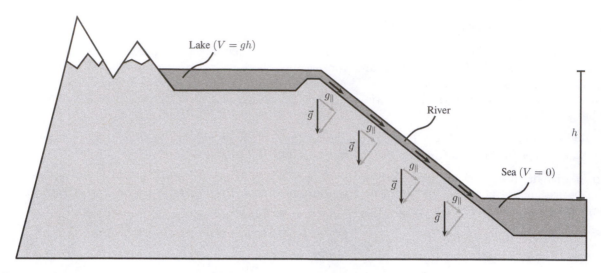

Figure 8.1: Schematic diagram of the flow of water from a region of high gravitational potential (the mountain lake) to a region of lower gravitational potential (the sea).

So the gravitational field $\vec{g}$ can be computed, from the gravitational potential, by simply taking (minus) the gradient. For example, near the surface of the earth, where $V = gy$, we have that

$$\vec{g} = -\vec{\nabla}(gy) = -g\hat{j} \qquad (8.29)$$

which is the familiar expression for the gravitational field near the surface of the Earth.

As a slightly unusual example of the practical applicability of the concept of "gravitational potential", consider a lake in the mountains somewhere. Suppose the lake is at an altitude h above sea-level. Then we can think of the lake as a reservoir of water molecules which are all located in a region with relatively large gravitational potential $V = gh$. This means, of course, that all the mass-m water molecules in the lake have a gravitational potential energy that is higher by $mV = mgh$ than their counterparts in the surface of the sea.

There is a complicated sequence of natural processes that is responsible for these high-altitude water molecules having this unusually large amount of gravitational potential energy. Something like: water down at sea level absorbs energy from the Sun; it warms up and evaporates, eventually condensing into clouds which then eventually rain the water back down into the mountains.

Once existing, though, the reservoir of water molecules in this region of high gravitational potential can serve as a source for a continuous flow of water from the mountain lake down to the ocean. This, of course, we call: a "river"! This is a familiar thing, but a couple of aspects are worth noting.

First, why does the water flow? Sometimes we say that the water that is up in the region of high gravitational potential "wants" to get to a region of lower gravitational potential. (Or, equivalently, the water molecules "want" to reduce their gravitational potential energy.) But that is an abstract and somewhat anthropomorphic way of describing the situation. A more direct and straightforwardly causal explanation is just that there is a path (the riverbed) along which it is possible for the water to move, and the gravitational field has a non-zero component parallel to that path. So, in short, the water molecules are subject to a *gravitational force* that pushes them along the path, giving rise to a steady flow from the region of higher gravitational potential to the region of lower gravitational potential.

But then, second, if there is a gravitational force pushing them along this path, wouldn't the water molecules *accelerate* so that the flow velocity of the water would dramatically increase as the river approached the sea? Actually, if the landscape looked like the sketch in Figure 8.1 – with the river flowing down an extremely steep slope – this probably would be the case. But the water in a real river, with a much more gradual downhill slope, does not dramatically accelerate. It moves with a roughly constant speed, independent of altitude. But this is still mysterious: a block sliding down a frictionless ramp, starting at height h, will be moving at the same speed ($v = \sqrt{2gh}$) when it reaches the bottom, regardless of the steepness of the ramp. The downhill journey will take more time if the slope is very gradual, but it will still be moving fast when it gets to the bottom. Why doesn't the river work the same way?

The point, of course, is that the gravitational force is not the only force acting on the water as it flows. The riverbed offers some *resistance* to the flow. In particular, there is friction between the moving water and the sides and bottom of the riverbed. And in addition there will be friction between a given blob of water and any neighboring blob which happens to be moving with a different instantaneous velocity. And, as it turns out, this happens all the time: the river bed is filled with rocks, trees, fish, bridge pilings, and all kinds of other things that make the water flow rather chaotic and turbulent. A given individual blob of water is constantly running into things, and moving off into some new direction, before it has time to build up appreciable speed from the gravitational force acting on it.

The overall effect of these frictional processes is that the gravitational potential energy, which the water possesses when it is up in the lake, is not converted into macroscopically-obvious kinetic energy of the water as a whole, as the water flows downhill into the region of lower gravitational potential. Instead, the kinetic energy gets churned down into a macroscopically-invisible form, in which each individual water molecule is bouncing around, in random directions, relative to its neighbors, faster than it had been previously (when the water was up at higher altitude). As we will discuss in more detail in a couple of weeks (but as you already know), this random motion down at the microscopic level of individual molecules is called "thermal energy". The gravitational potential energy of the high-altitude water, in short, is converted to *heat* as the water flows toward the sea. If the water were thermally insulated from its environment, it would be noticeably warmer by the time it reached the sea. But, in fact, the thermal energy dissipates out into the surroundings pretty rapidly, so in practice the water temperature doesn't correlate much with altitude.

One final point about the river. If we define the "current" $I = \frac{dM}{dt}$ as the amount of mass that passes a given point in the river per unit time, then the total rate at which gravitational potential energy is being converted into thermal energy (which dissipates out into the surrounding environment) can be expressed in a simple way in terms of the gravitational potential difference $\Delta V = gh$ between the lake and the sea:

$$P = I\,\Delta V. \tag{8.30}$$

Think about the formula this way. In a time dt, each little blob of water moves a short distance downhill along the river bed. But the net effect is equivalent to a blob of mass dM being transported directly from the lake to the sea. The gravitational potential energy of that transported blob would decrease by $\Delta U = dM\,\Delta V = dM\,g\,h$. And so the thermal energy of the river and/or its surroundings should increase by this same amount. And so the *rate* of increase of thermal energy – the rate, that is, at which gravitational potential energy is converted to thermal energy by the flowing river – is $dM\,\Delta V/dt = I\,\Delta V$, as claimed.

The Connecticut river (which flows near Smith College) has an outflow rate, where it empties out into the Long Island Sound, of about 18,000 cubic feet per second. Multiplying by the density of water gives the equivalent current (mass per unit time):

$$I \approx 500{,}000\,\text{kg/sec}. \tag{8.31}$$

The river begins at an altitude of about 800 meters, in northern New Hampshire. Of course, really, the water that eventually flows out into the Long Island Sound did not all start at the same New Hampshire lake. There are uncountably many smaller rivers, streams, and tributaries that all ultimately join together to constitute the total outflow rate. Still, as a rough simplifying model, we can pretend that all the water begins at this same altitude, i.e., at a location where the gravitational potential is

$$\Delta V = gh = 9.8\,\mathrm{m/s^2} \cdot 800\,\mathrm{m} = 7840\,\mathrm{m^2/s^2}. \tag{8.32}$$

Multiplying the current I and gravitational potential drop ΔV gives the river's "power", i.e., the rate at which gravitational potential energy is being converted into thermal energy:

$$P \approx 4 \times 10^9 \text{ Joules/sec} \tag{8.33}$$

i.e., about four GigaWatts. For comparison, the total average energy consumption of New York City is about a GigaWatt, so the river's power is quite considerable!

§ 8.2 Electric Potential and Current

The close mathematical similarity between Newton's law (giving the gravitational force between two masses) and Coulomb's law (giving the electric force between two charges) means that there is a perfect electrical parallel to everything we talked about, in the last section, for gravity. For example, the electrical potential energy associated with a charge q located a distance r from another charge Q is

$$U_{elec} = \frac{kQq}{r}. \tag{8.34}$$

This is the perfect "electrical analog" of Equation (8.20), with charges replacing masses, Coulomb's constant k replacing Newton's constant G, and a minus sign because whereas like masses (which is to say, any two masses, since they're all positive!) attract each other gravitationally, like charges *repel* electrically.

We can define the "electric potential" V as the electric potential energy a "test charge" q would feel at a certain location, divided by q. So for example, at a distance r from a point charge Q,

$$V = \frac{kQ}{r}. \tag{8.35}$$

Note that whereas the gravitational potential can be measured in Joules per kilogram (aka $\mathrm{m^2/s^2}$), the electric potential can be measured in Joules per Coulomb. A Joule per Coulomb, however, is more frequently known as a "Volt" and, indeed, sometimes the electrical potential is simply referred to as the "voltage".

"Volt" and "voltage" are of course named after Allesandro Volta, who (you will recall) invented what would eventually evolve into the modern chemical battery. The chemical reactions inside such a battery are an electrical analog of the "complicated sequence of natural processes" that we described in the last section as giving rise to the mountain lake, i.e., the reservoir of water molecules at relatively high gravitational potential (compared to the potential at sea level). The positive terminal of a battery can similarly be thought of as a reservoir of mobile electrical charges at a location with a relatively high electric potential compared to the electrical potential of the negative terminal – i.e., (positive!) charges at the positive terminal have a higher electrical potential energy than charges at the negative terminal. In practice, a battery does a good (though not perfect) job of maintaining a fixed electrical potential difference between its two terminals – for example, $\Delta V = 1.5$ Volts for many standard modern battery types (AAA, AA, C, and D), but there are also aptly-named "9 Volt" batteries, etc.

If a path is provided between the positive and negative terminals of a battery, along which it is possible for charges to move, a flow of charge will result, just as in the gravitational analog case of the river. For example, connecting the terminals by a wire will allow an electrical current to flow. The current flows – i.e., individual charges move – for the same reasons that the water molecules move in a river. We could say that the charges at the positive terminal of the battery have a large electrical potential energy and "want" to decrease their potential energy by moving toward the negative terminal. Or we could say that charges move because there is an electric field (directed along the wire and pointing from higher potential to lower potential).

Also just as in the case of the river, the mobile charges (whose motion constitutes the flow we describe as an electrical current) do not systematically move faster as they progress along the wire. At least, they do not systematically move faster along the direction of motion. The conduction electrons in a wire move in a way that is far more chaotic and turbulent, even, than the water molecules in a river: each one is bouncing around, this way and that, moving very fast at any given moment, but in a completely new direction from one moment to the next due to frequent collisions with other particles in the wire. The typical instantaneous speed of an electron may be hundreds or thousands of meters per second, but the "drift velocity" – the speed with which the electron makes progress in moving along the wire – is typically a tiny fraction of one meter per second. Just as was the case with the river, the overall energy conversion is thus from (in this case, electrical) potential energy to thermal energy. And in the electrical case, this effect is easily noticeable: if you connect a wire across the terminals of a battery, the wire can get very hot, very fast. Be careful!

We can summarize by saying that the wire offers some *resistance* to the flow of charges. In the 1820s, the German physicist Georg Ohm systematically studied this effect using early batteries and galvanometers. He found that the measured electrical current, flowing through a wire that connected the terminals of a battery with a certain electric potential difference ΔV between them, was inversely proportional to the length of the wire. He also found that, all other things being equal, a larger electrical current would flow through a thicker wire, and that the current flowing through a given wire was proportional to the number of batteries connected in series (i.e., the effective ΔV between the end terminals of the set of batteries).

Today we summarize these findings with "Ohm's Law":

$$\Delta V = IR \tag{8.36}$$

where the "resistance" R (whose units are Volts per Amp – aka, you guessed it, "Ohms") of a wire with length L and cross-sectional area A is

$$R = \frac{\rho L}{A}. \tag{8.37}$$

Here ρ is the so-called "resistivity", which is a characteristic property of a given type of material (and may also depend on the temperature and other factors).

For example, the resistivity of copper at room temperature is $\rho_{copper} = 1.7 \times 10^{-8}\ \Omega m$. So a one meter long copper wire with a 2.0 mm diameter circular cross-sectional shape would have a resistance

$$R = \frac{1.0\,\mathrm{m}\ 1.7 \times 10^{-8}\ \Omega m}{\pi(1.0 \times 10^{-3}\,\mathrm{m})^2} = 0.0054\,\Omega. \tag{8.38}$$

This is a very small resistance, which is not surprising: copper, after all, is an excellent conductor. Even a small potential difference of about one Volt can give rise to a very large current of more than a hundred Amps!

Other materials offer significantly more resistance to the flow of charge. Solid amorphous carbon, for example, has a resistivity of order $10^{-5}\ \Omega\,\mathrm{m}$, silicon has a resistivity of order $1\,\Omega\mathrm{, m}$, and glass has a resistivity of order $10^{10}\ \Omega\,\mathrm{m}$. Modern manufactured resistors (which often consist of a long narrow "wire" of carbon, sitting in a "groove" cut into a ceramic base, all wrapped up in a small cylindrical

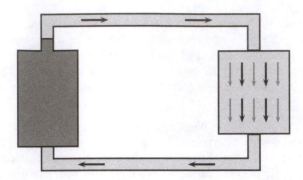

Figure 8.2: Schematic diagram of the flow of electric charge, pushed through a resistor by a battery. The black arrows are meant to represent the electric current, while the gray arrows are meant to represent the electric field which exerts forces on the mobile charges and keeps them flowing. (Note that no electric field vectors are indicated in the, say, copper wires that connect the resistor to the battery. Of course, current flow through the wires requires some electric field, but because the resistivity of copper is so low, a miniscule electric field is sufficient. By comparison, in the more highly resistive material of the resistor, a more significant electric field is required to produce the same overall electrical current.)

tube) are available with nominal resistances between a fraction of an Ohm and many Mega Ohms. Such resistors are used to control electrical potentials and currents in circuits. We will have the opportunity to play with several such applications in class.

Note that although Ohm's Law, Equation (8.36), was an empirical discovery, we can also regard it as providing a definition of the resistance R of a particular circuit element under particular conditions. In practice, though, we only call an object a "resistor" and assign a definite resistance to it, if $R = \Delta V/I$ remains at least approximately constant as the applied voltage ΔV (and any other relevant variable factors such as temperature) change under suitable operating conditions. (Circuit elements with negligibly small resistance, like the copper wire considered just above, are also generally not thought of as resistors.) In class we'll explore how the current I varies with changes in applied voltage ΔV for several different objects, and discuss which can reasonably be classified as "resistors".

One final point about resistors. Since the electrical current is defined as the rate at which electric charge flows past a given point in the circuit,

$$I = \frac{dQ}{dt}, \tag{8.39}$$

we can express the rate at which a resistor converts electric potential energy (provided, say, by the chemical reactions in a battery) into thermal energy in the same way we did for the river:

$$P = \Delta V\, I. \tag{8.40}$$

The reason, again, is that in time dt an amount of charge dQ is transferred from the positive terminal to the negative terminal, which implies that the amount of electrical potential energy that has been converted to thermal energy in the resistor is $dU = \Delta V\, dQ$. So the amount of energy that is converted per unit time is $P = dU/dt = \Delta V\, dQ/dt = \Delta V\, I$.

Combining Equations (8.36) and (8.40) allow us to re-write the formula for a resistor's power – the rate at which it generates thermal energy – as

$$P = \frac{(\Delta V)^2}{R} \tag{8.41}$$

(or $P = I^2 R$). The fact that the power generated by a resistor, connected to a battery that maintains a certain potential difference ΔV, is inversely proportional to R explains why you are in danger of burning

Figure 8.3: An early Leyden Jar.

your fingers if you connect a wire across a battery, but are unlikely to hurt yourself by similarly connecting a resistor with appreciable resistance.

§ 8.3 Capacitors and Electric Energy

Last week, we encountered, in passing, the Leyden Jar, which was a device for storing electrical charge. It consisted of a glass jar, with metal foil wrapped around the outside and also layered along the inside, so the two conducting (metal foil) surfaces were separated by a thin layer of electrically insulating glass. In the example shown in Figure 8.3, a metal rod, connected to the metal foil layer on the inside of the jar, allows for easy electrical access to the inner layer.

If the inner and outer conducting layers are connected, respectively, to the positive and negative terminals of a battery, the battery works to establish its standard electric potential difference between the two foil surfaces. It accomplishes this by pumping positive charge onto the inner foil surface (and simultaneously pulling positive charge out of the outer foil surface, leaving that surface negatively charged). It is important to appreciate that the net charge of the Leyden jar typically remains zero. Although it is frequently said that such devices "store charge", it is more accurate to say that they maintain a *charge separation.*

A "charged" Leyden jar (or more accurately, a Layden jar whose two foil surfaces contain opposite charges) allows the charge (separation) to be maintained, stably, even after disconnecting it from the battery. The reason for this is that the positive charge on one foil surface is strongly attracted to the negative charge on the other foil surface. So, for example, one could touch the metal rod connected to the inner surface and nothing would happen. However, if one simultaneously touched the metal rod connected to the inner surface, and also the outer foil surface, one's body would provide a conducting path through which the positive charge on the one surface could travel to get to the negatively-charged plate to which it is so strongly attracted. One would thus experience an electric shock, as an electric current ran briefly through one's body.

In the early to mid 1800s, Leyden jars were commonly used in electrical experiments and demonstrations, as a way of temporarily storing electrical charge until the moment when – for some scientific or perhaps entertainment purpose – it was released as a spark or other burst of electrical current. In a particularly famous experiment, Benjamin Franklin flew a kite during a thunderstorm and managed to successfully shunt part of a lightning bolt into a Leyden jar. Subsequent investigation showed that the contents of the Leyden jar behaved identically to standard electrical charge (from, e.g., a chemical battery), thus revealing that lightning was, in fact, just an unusually strong electrical phenomenon – a giant spark, in effect, through which electric charge that had built up in the clouds could return to the ground.

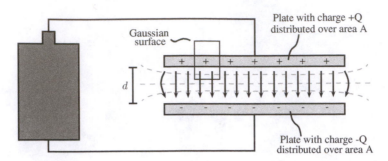

Figure 8.4: A capacitor connected to a battery. The plates acquire opposite charges, $+Q$ and $-Q$, and a strong electric field (indicated by the black arrows) develops between them. The gray dashed curves are equipotential surfaces along which the electric potential has a constant value.

Incidentally, this discovery motivated Franklin to invent what was probably the first practical application of electromagnetic theory: the lightning rod, which provided a safe, electrically-conducting path for the electrical current associated with a lightning strike, between the top of a building and the ground.

8.3.1   Charge and Potential Difference

The modern version of the Leyden jar is called a "capacitor" (or, sometimes, a "condenser"). A capacitor can be modeled as two parallel conducting plates, separated by a layer of insulating material. Suppose the plates each have an area A and are separated (for simplicity) by an air gap of width d. If the capacitor is connected to a battery whose terminals maintain an electric potential difference ΔV, the battery will pump charge until the electrical potential difference between the plates is also ΔV. The situation is depicted in Figure 8.4.

Since the magnitude of the electric field E is the rate of change of the electric potential (along the direction it changes most rapidly), and with the electrical potential changing by ΔV across the inter-plate gap of width d, we have that

$$E = \frac{\Delta V}{d}. \tag{8.42}$$

But the magnitude of the electric field between the plates can also be related to the charge on the plates using Gauss' Law. The flux of $\vec{E}$ through the Gaussian surface shown in the Figure – a cylinder, say, of cross-sectional area a – will simply be Ea. The charge enclosed by the surface can be written σa where $\sigma = Q/A$ is the charge per unit area on the top plate. Plugging into Gauss' Law gives

$$Ea = \frac{Qa}{\epsilon_0 A} \tag{8.43}$$

or, cancelling the as, substituting in Equation (8.42), and solving for Q:

$$Q = C\,\Delta V \tag{8.44}$$

where

$$C = \frac{\epsilon_0 A}{d}. \tag{8.45}$$

Equation (8.44), relating the potential difference between the capacitor plates to the charge on those plates, is the fundamental circuit-analysis law for capacitors, analogous to Ohm's law for resistors.

Equation (8.44) says that the charge (or really we should say the charge separation) stored in the capacitor is proportional to the potential difference between its plates. The proportionality constant, C, is called

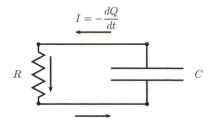

Figure 8.5: A capacitor discharging through a resistor is the simplest RC circuit.

the capacitance; it is proportional to the plate area A and inversely proportional to the separation d. Thus, to store the maximum amount of charge, by connecting to a battery with a fixed ΔV, one wants to make the capacitance as large as possible by using large-area plates that are separated by a very small distance.

The standard contemporary unit for capacitance is called the "Farad" (named after Michael Faraday), with one Farad being the same as one Coulomb per Volt. Hence, if a 1.0 F capacitor is connected to a 1.0 V battery, the capacitor plates will, respectively, develop charges of +1 and -1 Coulomb. One Farad is an extremely large capacitance, however. The early Leyden jars, with (foil) "plates" whose areas were of order $A = .01\,\mathrm{m}^2$, separated by a layer of glass with thickness of order $d = .01\,\mathrm{m}$, had capacitances of order 10^{-11} F, i.e., a small fraction of a nano-Farad. Modern manufactured capacitors achieve much higher capacitance, typically by using long rectangular pieces of metal foil, with extremely thin layers of insulating material between them, rolled into a cylinder and housed in a protective "can". The capacitance of a capacitor can also be increased by using an insulating but polarizable material called a "dielectric" in the space between the plates.

8.3.2   RC Circuits

Suppose a capacitor (with capacitance C) is connected across a battery (with potential difference, i.e., EMF $\mathcal{E}$) so that the capacitor has charge $Q_0 = C\mathcal{E}$. The capacitor will maintain this charge if it is disconnected from the battery. But if the charged capacitor is now connected to a resistor of resistance R, the resistor provides a path through which positive charge from one plate can flow around to neutralize the negative charge on the other plate – i.e., a path through which the capacitor can discharge.

The situation is shown schematically in Figure 8.5. With the arrangement shown in the Figure, the electric potential decreases by

$$\Delta V_R = IR \tag{8.46}$$

from the top to the bottom of the resistor, according to Ohm's Law. Similarly, at a moment when the remaining charge on the capacitor is Q, the potential difference between the top and bottom plates of the capacitor is

$$\Delta V_C = \frac{1}{C}Q. \tag{8.47}$$

Assuming the wires connecting the resistor and the capacitor are ideal (i.e., their resistance is negligible, so that, even with a finite amount of current flowing through them, the potential difference between their two ends is negligible), ΔV_R and ΔV_C will be equal. Hence

$$IR = \frac{1}{C}Q. \tag{8.48}$$

But the charge Q on the capacitor, and the current I running through the resistor, are not independent quantities: the current measures the rate at which the capacitor is discharging. That is, if current I runs

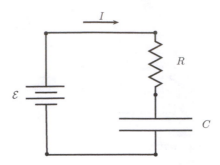

Figure 8.6: A charging capacitor.

for a time dt, the charge Q on the capacitor will change by $dQ = -I\,dt$. (The minus sign is because the diagram described positive current as flowing counterclockwise in the circuit, which implies a *decrease* in the charge Q on the capacitor.) So $I = -dQ/dt$ and we can rewrite Equation (8.48) as:

$$\frac{dQ}{dt} = -\frac{1}{RC}Q. \tag{8.49}$$

It is easy to check that this first-order differential equation for $Q(t)$ has the following solution:

$$Q(t) = Q_0 e^{-t/\tau} \tag{8.50}$$

where the "RC time constant" is just the product of the resistor's resistance and the capacitor's capacitance: $\tau = RC$.

The charge on the capacitor (and therefore also the Voltage across the capacitor, since these are proportional to each other) thus decays exponentially toward zero. In class we will have the chance to observe this exponential decay and determine the time constant τ as a way of accurately measuring the capacitance C of a capacitor.

Notice that, if R is very small (e.g., if one discharges the capacitor through a wire rather than a resistor), then τ will be very small, i.e., it will take hardly any time at all for the exponential decay to approach zero. The discharge, in short, will happen very rapidly.

What about the case of charging an initially-uncharged capacitor? This is depicted in Figure 8.6. If the battery maintains a constant EMF $\mathcal{E}$, this means (again assuming ideal wires) that the electric potential drops by $\mathcal{E}$ from the top of the resistor to the bottom of the capacitor. That is:

$$\mathcal{E} = \Delta V_R + \Delta V_C. \tag{8.51}$$

Using Ohm's law ($\Delta V_R = IR$), the Capacitor law ($\Delta V_C = Q/C$), and the relationship between the current I and the charge Q that is appropriate here (namey, $I = dQ/dt$, without the minus sign, because now a positive current corresponds to an increasing Q), this becomes

$$\mathcal{E} = R\frac{dQ}{dt} + \frac{1}{C}Q. \tag{8.52}$$

The solution of this differential equation can be written

$$Q(t) = \mathcal{E}C + (Q_0 - \mathcal{E}C)e^{-t/\tau} \tag{8.53}$$

where Q_0 is the charge on the capacitor at $t = 0$ and $\tau = RC$ as before. For the case $Q_0 = 0$ in which the capacitor is initially uncharged, we have

$$Q(t) = \mathcal{E}C\left(1 - e^{-t/\tau}\right) \tag{8.54}$$

which describes an "upside-down exponential decay" toward the value $\mathcal{E}C$, which is the "fully-charged" state of the capacitor when the potential difference across it is $\mathcal{E}$.

Note that, in the charging case, if R is very small (e.g., we just hook the capacitor up directly to the battery, using wires with negligible resistance), then τ is again very small, and the capacitor becomes "fully charged" in practically no time at all.

8.3.3   The Capacitor as an Energy Storage Device

A charged capacitor, in which lots of positive charge is collected together on one plate and lots of negative charge is collected together on the other plate, has a large electrical potential energy. How much, exactly? There are several ways to calculate an answer to this question, and it is worth thinking through several of them.

First, suppose a capacitor with capacitance C has been given an initial charge Q_0 and is then discharged through a resistor of resistance R. In terms of energy, what's happening during the discharge process is that the electrical potential energy that was originally stored in the capacitor is dissipated as heat (thermal energy) as the discharging current flows through the resistor. So we can compute the total amount of energy that was initially in the capacitor, by computing the total amount of energy that is dissipated by the resistor during the discharging process. With

$$Q(t) = Q_0 e^{-t/\tau} \tag{8.55}$$

and $I = -dQ/dt$, the current passing through the resistor at time t is

$$I(t) = \frac{Q_0}{\tau} e^{-t/\tau}. \tag{8.56}$$

But this means that the rate at which the resistor is producing thermal energy, at time t, is

$$P(t) = I^2 R = \left(\frac{Q_0^2}{RC^2} \right) e^{-2t/\tau}. \tag{8.57}$$

The thermal energy produced in time dt is just $P\,dt$, so the total thermal energy produced by the resistor during the discharge – equal to the potential energy initially stored in the charged capacitor – is

$$U = \int P(t)\,dt = \left(\frac{Q_0^2}{RC^2} \right) \int_{t=0}^{t=\infty} e^{-2t/\tau}\,dt = \frac{Q_0^2}{2C}. \tag{8.58}$$

Here is another way to derive that same correct result. Consider this time the charging of the capacitor. Suppose, at some intermediate moment when the capacitor is not yet fully charged, its charge is Q. This implies, from Equation (8.44), that the potential difference between the plates is $V = Q/C$. Now the next "baby step" in the charging process will involve the battery pushing an additional amount of charge dQ onto the capacitor plates. But moving a charge dQ from (say) a region where the electric potential is zero, to a region where the electric potential is V, requires work $dW = V\,dQ = Q\,dQ/C$ to be done by the battery. (Note that this is just the work done by the battery to charge the capacitor. Of course, really, to push an extra bit of charge onto the capacitor, the battery has to push that charge through the resistor, which dissipates some energy as heat. So the total work done by the battery, in the process of charging a capacitor in an RC circuit, is larger than this. But since we are here only interested in the potential energy stored in the capacitor, we set aside the additional work done that just gets radiated as heat by the resistor.) The total energy added to the capacitor by the battery during the entire charging process – i.e., the total potential energy stored in the capacitor when it reaches its full charge Q_0 – is therefore

$$U = \int dW = \frac{1}{C} \int_{Q=0}^{Q=Q_0} Q\,dQ = \frac{Q_0^2}{2C}. \tag{8.59}$$

This is, happily, the same result we found before.

Here, though, is another way we might think about the potential energy stored in the capacitor. We could just say: "Look, when the capacitor is fully charged, one plate has a charge $+Q_0$ and the other has a charge $-Q_0$. The electric potential at the positive plate is higher than the electric potential at the negative plate by $\Delta V = Q_0/C$. The zero of electric potential is arbitrary, so we might as well call the potential at the negative plate zero, so that the electric potential at the positive plate is Q_0/C. But then, remember that electric potential is just defined as the potential energy that a charge at that location would have, divided by the charge. That is, the potential energy of all of the charge sitting on the positive plate is just the total amount of charge there, Q_0, times the electric potential there, Q_0/C. And so the total potential energy of the charged capacitor is just

$$U = \frac{Q_0^2}{C} \tag{8.60}$$

since the negative charge on the negative plate has no potential energy since we defined the potential to be zero there."

That argument sounds good, but gives us an answer that is wrong – too big by a factor of two. You might think that the mistake has something to do with picking the zero of electric potential, but that is not the problem. We'd get the same (wrong) answer if we instead picked $V = 0$ at the positive plate, or halfway in between the plates (so that $V = +Q_0/2C$ at the positive plate and $V = -Q_0/2C$ at the negative plate), or any other place.

The problem instead has to do with the fact that the potential energy associated with a pair of charges is a property of the *pair*, not of each charge individually. For example, two particles with charge Q separated by distance d do not *each* have potential energy kQ^2/d, such that the total potential energy of the pair is $2kQ^2/d$, even though we could say that each particle is sitting at a location where the electric potential, set up by the other particle, is kQ/d. If we think of it this way, we would have double-counted the energy. Really, the potential energy kQ^2/d is possessed, jointly, by the pair.

Similarly, three charges (Q) arranged at the vertices of an equilateral triangle (side d) will have a total electric potential energy of $3kQ^2/d$ – not the $6kQ^2/d$ you might expect if you say that each charge is sitting at a location where the electric potential is $V = 2kQ/d$ (the sum of the contributions from each of the other two charges) and hence has its very own potential energy of $2kQ^2/d$.

We can think of the charged capacitor similarly. For example, pretend that the total charge Q_0 on the positive plate is carried by 10 particles with charge $q = Q_0/10$. We might then be tempted to say that the first particle, sitting in a location with electric potential V due to the other 9 charges, has potential energy qV. (This, of course, would be the sum of the potential energies associated with the following pairs of particles: 1 and 2, 1 and 3, 1 and 4, ..., 1 and 10.) Then we could continue like this for all the other particles. We would, for example, assign potential energy qV to particle 2 as well, this being the sum of the potential energies associated with the following pairs: 2 and 1, 2 and 3, 2 and 4, ..., 2 and 10. And so on. Adding up the total for all 10 particles, we would have arrived at a total potential energy equal to $10\,qV = Q_0V$. But we can see that this answer would be too big by precisely a factor of 2, because we would really have been counting the potential energy, associated with each pair, *twice*.

So that is why our third attempt to calculate the total potential energy of a charged capacitor, gave an answer that was too big by a factor of two. We inadvertently double-counted!

This is one of those somewhat subtle and interesting things that comes up every now and again, so it is worth being attuned to the possibility of this kind of mistake. But the importance of associating electric potential energy ultimately with *pairs* of particles, also raises some further difficult questions. For example, what does it mean, exactly, to say that a pair of particles (with charge Q and separation *d*) has potential energy $U = kQ^2/d$? Where, precisely, is this energy located? Does each particle carry half of this energy with it, at its location? But then what about a less symmetrical case in which, say, the charges are not the same? Or is the energy in some mysterious way possessed by the two separated particles, but without being located at any one particular place (or divided coherently between several particular places), since the pair is not located at any one particular place? But what would that even mean?

Since our introduction of (electric, as well as magnetic and gravitational) *fields* was first motivated by concerns about action-at-a-distance (i.e., the incomprehensibility of interactions between spatially-separated objects), it is very satisfying that these new questions about the localization of interaction *energy* can also be addressed in terms of fields.

To see the idea here, consider the formula we developed for the potential energy stored in a charged capacitor:

$$U = \frac{Q^2}{2C}. \tag{8.61}$$

Re-writing this using $Q = C\Delta V$ and $\Delta V = Ed$ and $C = \epsilon_0 A/d$, we arrive at the equivalent expression

$$U = \frac{1}{2}\epsilon_0 E^2 Ad. \tag{8.62}$$

But A times d is just the volume of the gap between the plates, i.e., the volume of the region where the electric field has the nonzero magnitude E. So the above formula suggests that we could associate an energy density

$$u_E = \frac{1}{2}\epsilon_0 E^2 \tag{8.63}$$

with the electric field, and then think of the potential energy of the charged capacitor as possessed by – as stored in – the electric field between the charged plates (rather than at the locations of the charges themselves).

This idea provides an alternative perspective on the potential energy of interaction between two charged particles. Of course, for a truly point-like charged particle, the electric field – and hence the electric field energy density – diverges as one approaches the charge. So the total energy stored in the field around an isolated point charge is not well-defined. One can elude this problem by attributing some small-but-finite size to the particle. But if our interest is in the interaction energy of a pair, we can set this complication aside in the following way: the total electric field $\vec{E}$ that exists in the vicinity of, say, a pair of charged particles, can be written, using the principle of superposition, as

$$\vec{E} = \vec{E}_1 + \vec{E}_2 \tag{8.64}$$

where $\vec{E}_1$ and $\vec{E}_2$ are, respectively, the separate contributions to the field from particles 1 and 2. But then the potential energy, stored in the electric field surrounding the charges, can be written (in terms of the energy density u_E) as follows:

$$\begin{aligned} U &= \int u_E \, dV \\ &= \int \frac{1}{2}\epsilon_0 |\vec{E}|^2 \, dV \\ &= \int \frac{1}{2}\epsilon_0 |\vec{E}_1|^2 \, dV + \int \frac{1}{2}\epsilon_0 |\vec{E}_2|^2 \, dV + \int \frac{1}{2}\epsilon_0 2\vec{E}_1 \cdot \vec{E}_2 \, dV. \end{aligned} \tag{8.65}$$

The first two terms may be ill-defined for true point charges, but they are otherwise not very interesting – for example, they do not change as the separation between the particles is varied. The term that will

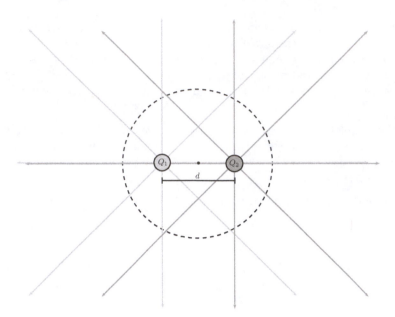

Figure 8.7: Two particles with charges Q_1 and Q_2 are separated by distance d. Some field lines associated with $\vec{E}_1$, the field contributed by Q_1, are shown in light gray, and some field lines associated with $\vec{E}_2$, the field contributed by Q_2, are shown in darker gray. In the immediate vicinity of the two charges (i.e., roughly inside the radius-d dashed circle) $\vec{E}_1$ and $\vec{E}_2$ are in the same direction in some places, perpendicular in some places, and in opposite directions in some places. So the volume integral of $\vec{E}_1 \cdot \vec{E}_2$ over that region roughly vanishes. Farther away from the charges, however, (i.e., roughly outside the dashed circle) the individual field contributions $\vec{E}_1$ and $\vec{E}_2$ are both approximately radial (relative to the black dot at the center) and hence approximately parallel to each other. So the major contribution to the volume integral of $\vec{E}_1 \cdot \vec{E}_2$ comes from this region.

vary with separation – and hence contains all the information about how the field energy associated with the particle pair *changes* as they are moved closer together or further apart – is the third term.

If the two particles are infinitely far apart, $\vec{E}_1$ will be negligibly small wherever $\vec{E}_2$ is big, and vice versa, so this crucial third term will simply vanish. By how much does the energy stored in the electric field change if the particles are moved closer together? Consider the situation depicted in Figure 8.7. Using a coordinate system with the origin at the midpoint between the charges (the black dot in the Figure) and with a standard horizontal/vertical orientation for the x/y axes (and the z axis coming out of the page toward us), the field $\vec{E}_1$, depicted in the Figure with the light gray field lines, can be written as

$$\vec{E}_1 = kQ_1 \frac{(x + d/2)\hat{i} + y\hat{j} + z\hat{k}}{((x + d/2)^2 + y^2 + z^2)^{3/2}}. \tag{8.66}$$

Similarly, the field $\vec{E}_2$, depicted in the Figure with the darker gray field lines, can be written as

$$\vec{E}_2 = kQ_2 \frac{(x - d/2)\hat{i} + y\hat{j} + z\hat{k}}{((x - d/2)^2 + y^2 + z^2)^{3/2}}. \tag{8.67}$$

It is possible, though not exactly trivial, to integrate the dot product of $\vec{E}_1$ and $\vec{E}_2$ over all space, and show that the third term of Equation (8.65) reduces to exactly the familiar potential energy formula:

kQ_1Q_2/d. But we can understand how and why it comes out in the claimed way with the following rough approximation.

As can be appreciated by contemplation of the Figure, the relationship between $\vec{E}_1$ and $\vec{E}_2$ is rather complicated in the immediate vicinity of the charges. In some places the two fields are in the same direction (so their dot product is positive), but in other places the fields are in opposite directions (so their dot product is negative), and in still others they are perpendicular (so their dot product vanishes). It seems reasonable, as a rough approximation, to assume that the integral of the dot product simply vanishes in this region.

Whereas, if we are far away from the particles (compared to their separation, d), then the expressions for the fields simplify to

$$\vec{E}_1 \approx \frac{kQ_1}{r^2}\hat{r} \tag{8.68}$$

and

$$\vec{E}_2 \approx \frac{kQ_2}{r^2}\hat{r}. \tag{8.69}$$

That is, far away from the charges, the individual contributions to the total field are both approximately *radial* (relative to the midpoint of the charges); hence they are parallel to each other and their dot product is easy to evaluate. So it is reasonable to approximate the crucial third term of Equation (8.65) in the following way:

$$
\begin{aligned}
\Delta U &= \int \epsilon_0\, \vec{E}_1 \cdot \vec{E}_2 \, dV \\
&= \epsilon_0 \int_{r=d}^{r=\infty} \frac{kQ_1}{r^2}\frac{kQ_2}{r^2}\, 4\pi r^2\, dr \\
&= 4\pi\epsilon_0 k^2 Q_1 Q_2 \int_d^\infty \frac{1}{r^2}\, dr \\
&= \frac{kQ_1Q_2}{d}
\end{aligned}
\tag{8.70}
$$

(where we have used that $4\pi\epsilon_0 = 1/k$). This is, of course, the same expression we arrived at earlier for the potential energy associated with a pair of charges, but now the spatial distribution of the energy is less mysterious: the energy is not really carried by either of the charged particles themselves, but is instead distributed around them, in the electric field to which they jointly contribute.

If you had any doubt about the electric field being a physically real thing (as opposed to merely a convenient calculation device), the fact that the field can be understood as the physical "seat" of the energy associated with charge distributions, might change your mind. Of course, we haven't really proved that the field can possess energy. We've just shown that the formula for the potential energy stored in the electric field gives us an alternative way of calculating the potential energy we had previously assigned to various situations (like a charged capacitor or a pair of charged particles). But the fact that it is possible to do this is a clue in support of the physical reality of the fields – one which will receive further evidentiary support next week when we explore electromagnetic radiation, i.e., the phenomenon whereby electromagnetic fields transport energy from one place to another at finite speed.

Before turning to that, however, we should explore the (largely parallel) case of *magnetic* energy.

§ 8.4 Inductors and Magnetic Energy

The closest magnetic analog to the parallel plate capacitor (which produces an electric field that is nonzero and relatively uniform in a certain region, namely, the gap between the charged plates) is, as we have

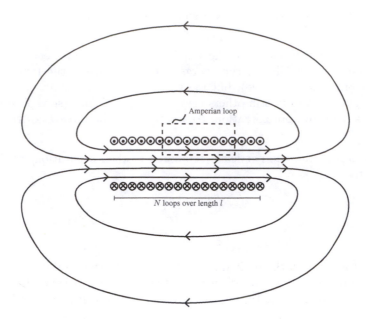

Figure 8.8: An inductor consisting of N loops of wire wrapped around a cylinder of length l. The magnetic field is quite uniform and has a relatively large magnitude on the interior of the cylinder. As the field lines spread out and loop around at the ends of the cylinder, however, the magnitude of the field is much smaller outside the cylinder. Approximating the field as having zero magnitude on the outside allows a simple calculation of the magnitude of the field on the interior using Ampère's law with the Ampèrian loop shown.

seen previously, a coil of wire – a so-called "solenoid" – through which electrical current flows.

8.4.1   The Inductor Law

An "inductor" is simply a solenoid – a coil of wire – used in an electrical circuit. Consider an inductor consisting of N loops of wire wrapped around (for example) a hollow cylinder of cross-sectional area A and length l. As sketched in Figure 8.8, the magnetic field has a large magnitude on the interior of the solenoid but, due to the spreading of the field lines, a much smaller magnitude outside. (It is, in this respect, just like the electric field produced by a charged parallel plate capacitor.)

Applying Ampère's law ($\oint \vec{B} \cdot d\vec{s} = \mu_0 I_{encl}$) to the width-$w$ Ampèrian loop shown in the Figure gives (taking the magnitude of the field outside the interior of the inductor to be zero) $Bw = \mu_0 w N I / l$, so the magnitude of the field along the axis of the inductor is

$$B = \frac{\mu_0 N I}{l} \tag{8.71}$$

where I is the current flowing through the wire loops.

That much we have seen before. But now we want to explore the question of what happens to an inductor in an electrical circuit – and, in particular, in a situation where the electric current I passing through it is *changing*. The important point here is that each of the N loops has a *magnetic flux*

$$\Phi_1 = BA = \frac{\mu_0 N A}{l} I \tag{8.72}$$

passing through it, where A is the area of a loop, i.e., the cross-sectional area of the inductor. The total magnetic flux for the inductor as a whole is therefore N times the flux associated with each loop, i.e.,

$$\Phi = N\Phi_1 = \frac{\mu_0 N^2 A}{l} I. \tag{8.73}$$

Now, according to Faraday's law (that we encountered briefly last week), a changing magnetic flux through some object causes an EMF which (like an electric potential difference across it) can drive current. The induced EMF is given by

$$\mathcal{E} = -\frac{d\Phi}{dt} \tag{8.74}$$

so that, for our inductor, the EMF associated with a changing electrical current will be

$$\mathcal{E} = -\frac{\mu_0 N^2 A}{l} \frac{dI}{dt}. \tag{8.75}$$

Qualitatively, this says that if current is flowing through the inductor in a certain direction and then the current *increases*, the increasing current causes a negative EMF – a so-called "back EMF" – which attemps to push current *backwards* through the inductor. It is as if the inductor doesn't want the current through it to change, so it resists the change by responding to the externally-imposed increase with its own induced current, flowing in the opposite direction and hence attempting to cancel or undo the change. Similarly, if the current is decreasing, so that dI/dt is negative, then the induced EMF pushes current in the forward (positive) direction, again "trying" to cancel the change. One can summarize by saying that an inductor resists changes in electrical current.

The proportionality constant between the induced EMF $\mathcal{E}$ and (minus) the rate of change of the current, dI/dt, is called the "inductance". So for our standard cylindrical solenoid, the inductance is given by

$$L = \frac{\mu_0 N^2 A}{l}. \tag{8.76}$$

We can then write the fundamental circuit law for inductors (the thing that is analogous to Ohm's law for resistors and the capacitor law for capacitors) as follows:

$$\mathcal{E} = -L\frac{dI}{dt}. \tag{8.77}$$

Let's develop our understanding by considering a couple of simple circuits, with inductors, in a little more detail.

8.4.2   LR Circuit

Suppose, at some time $t = 0$, we close a switch and thus connect a circuit containing a battery, an inductor (of inductance L), and a resistor (of resistance R), as shown in Figure 8.9.

The battery produces a constant EMF $\mathcal{E}_0$ which pushes current to flow clockwise in the circuit. But as the clockwise current begins to increase, the inductor develops its own counter-clockwise "back-EMF" that resists this increasing current flow. The net clockwise EMF driving current through the resistor is therefore $\mathcal{E}_0 - L\frac{dI}{dt}$. This net EMF should, according to Ohm's law, equal the electric potential difference across the resistor, i.e.,

$$\mathcal{E}_0 - L\frac{dI}{dt} = IR. \tag{8.78}$$

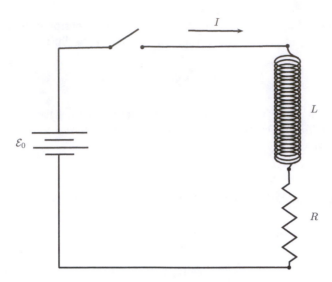

Figure 8.9: An inductor and resistor, in series, are connected to a battery by closing a switch at $t = 0$.

The solution of this first-order differential equation is

$$I(t) = \frac{\mathcal{E}_0}{R} + Ae^{-t/\tau} \tag{8.79}$$

where A is an arbitrary constant and $\tau = L/R$. Assuming that no current was flowing at $t = 0$, i.e., assuming $I(0) = 0$, we evidently need $A = -\mathcal{E}_0/R$ so that the solution is given by

$$I(t) = \frac{\mathcal{E}_0}{R} \left(1 - e^{-t/\tau} \right). \tag{8.80}$$

This describes an "upside-down exponential decay" in which the current begins at zero and then increases asymptotically toward the final value $I_f = \mathcal{E}_0/R$.

Qualitatively, this makes sense. The inductor is, after all, just a long coil of wire (which we idealize as having no resistance). So it makes sense that, at the end of the day, when some steady current is flowing and nothing is changing any longer, the current that flows is just the current we would expect in a circuit in which just the resistor was connected across the battery.

The inductor, though, "fights" the increase in the current from zero to this final steady-state value. So instead of just jumping immediately from zero to I_f, the current ramps up gradually, over a time-scale of order $\tau = L/R$, only really reaching I_f asymptotically. This all makes sense given our initial qualitative understanding that the inductor's role is to resist changes in current.

8.4.3   Magnetic Field Energy

Since the inductor "fights against" increases in the electrical current through it, the battery must do some work to get a current flowing through an inductor. To calculate the potential energy stored in an inductor (through which some current I_0 is flowing), therefore, we can simply calculate the amount of work done by a battery to achieve that current flow. (As with the similar case of the capacitor that we discussed above, in an RC circuit, we will ignore the part of the work done by the battery that goes into pushing charge through the resistor, i.e., the part that does not appear as potential energy in the inductor but instead gets radiated as heat by the resistor.)

Consider, then, a process in which the current through an inductor ramps up from zero to some final value I_0. At a moment when the current is some intermediate value I and is increasing at rate dI/dt, there will be a "back-EMF" across the inductor equal to

$$\mathcal{E} = -L\frac{dI}{dt}. \tag{8.81}$$

This back-EMF is morally equivalent to an electrical potential difference across which the next bit of charge dQ must be pushed by the battery. This requires that the battery do work

$$dW = L\frac{dI}{dt}\,dQ. \tag{8.82}$$

But the current I is the rate at which charge is being pushed through, i.e., $dQ = I\,dt$. So we can rewrite the work done by the battery (to push the next bit of charge through and simultaneously increase the current by dI) as

$$dW = L\,I\,dI. \tag{8.83}$$

It is then easy to add up the work done for each incremental increase to get the total work done by the battery to ramp the current up from zero to I_0, and re-name this the potential energy stored in the inductor:

$$W = U_L = \int_0^{I_0} L\,I\,dI = \frac{1}{2}LI_0^2. \tag{8.84}$$

That is a useful result, but as with the case of the potential energy of the charged capacitor, there is another perspective that is even more illuminating. Using Equation (8.71) to rewrite the current I_0 in terms of the magnetic field B inside the inductor, and then using Equation (8.76) to rewrite the inductance L in terms of the geometrical properties of the inductor, the potential energy expression can be rewritten as follows:

$$U_L = \frac{1}{2}\frac{\mu_0 N^2 A}{l}\left(\frac{Bl}{\mu_0 N}\right)^2 = \frac{B^2}{2\mu_0}\,A\,l. \tag{8.85}$$

But $A\,l$ is just the *volume* of the interior region of the inductor, i.e., the region where the magnetic field has the non-negligible magnitude B.

So there is again a hint, here, that we can consider the potential energy of the inductor to be stored, not in the current-carrying loops of wire, exactly, but rather in the magnetic field those loops surround. In particular, Equation (8.85) implies that we should associate an energy density

$$u_B = \frac{B^2}{2\mu_0} \tag{8.86}$$

with the magnetic field.

8.4.4   Aside on "Magnetic Potential"

For both gravitational and electric forces, we introduced the idea of an associated "potential" – a scalar-field that can be understood as the potential energy (of the appropriate type) that a particle of mass or charge m or q would have, if it were present at a given location, divided by m or q. You might be wondering if there is an analogous quantity – the "magnetic potential" – associated with magnetic forces.

It is possible to define such a quantity in the delimited domain in which it is an acceptable approximation to think about magnetic forces in terms of North and South magnetic poles. But as we have discussed, such isolated magnetic poles – "magnetic monopoles" – do not in fact exist. So it should be clear that the magnetic analog to the gravitational or electric potential, V, cannot really be defined.

There is another helpful perspective on this that is worth appreciating. Mathematically, the relationship between the (gravitational or electric) potential V and the (gravitational or electric) field $\vec{f}$ is that $\vec{f}$ is minus the gradient of V:

$$\vec{f} = -\vec{\nabla}V. \tag{8.87}$$

This means, recall, that the field $\vec{f}$ points, at a given location, in the direction along which the potential V *decreases most rapidly* (and the magnitude of $\vec{f}$ is just the rate of change of V along this direction). Qualitatively, if we walk along a path through space that follows the direction of the field $\vec{f}$ at each point, we will be walking continuously "downhill" – from regions of higher V to regions of lower V.

This is all perfectly happy and consistent for gravitational and electric fields, which terminate on masses and charges, respectively. So there is a well-defined direction that is "downhill" at each location. But – as sketched for example for the case of the inductor/solenoid in Figure 8.8 – magnetic field lines make continuous loops. And it is mathematically impossible that one could start at some point on such a loop, follow the local field direction all the way around the loop and end up back where one started, going "downhill" in magnetic potential all the while. The magnetic potential at the starting location would have to be lower than itself, which is impossible. So one can see in this way as well that the concept of "magnetic potential" is mathematically incoherent.

All of this is closely related to an important difference between electric and gravitational fields, on the one hand, and magnetic fields, on the other. Electric and gravitational fields (at least as we have encountered them so far) are "conservative" meaning that the path integral of the field, from some point A to some other point B, is the same, no matter what path one takes from A to B. Or, equivalently, for electric and gravitational fields,

$$\oint \vec{f} \cdot d\vec{s} = 0 \tag{8.88}$$

(where, as above, $\vec{f}$ means either $\vec{E}$ or $\vec{g}$).

Magnetic fields, by contrast, are not conservative. The path integral of $\vec{B}$, from one point to another, can be completely different depending on the path one takes. For example, the path integral from a point on one side of a current-carrying wire, to a point on the other side, could be either positive or negative depending on whether the path goes around one way, or the other way. Indeed, as we have seen, the path integral of the magnetic field around a closed loop – the thing that is always zero for electric and gravitational fields – is definitely not zero. It is instead given by Ampère's law:

$$\oint \vec{B} \cdot d\vec{s} = \mu_0 I_{encl}. \tag{8.89}$$

So this is a fundamental difference between electric and gravitational fields, on the one hand, and magnetic fields, on the other, which allows us to define an associated (electric or gravitational) potential, for those cases, but prevents the definition of a scalar magnetic potential whose gradient gives the magnetic field.

But there are a couple of further complications that we will mention. First, although the magnetic field cannot be derived by taking (minus) the gradient of a scalar magnetic potential, it is actually possible to derive the magnetic field from a different kind of (vector-valued) magnetic potential field. The relevant formula here is

$$\vec{B} = \vec{\nabla} \times \vec{A} \tag{8.90}$$

where $\vec{A}$ is the so-called "magnetic vector potential" and the right hand side means "the curl of $\vec{A}$". The "curl" is another type of spatial derivative (which acts on a vector field and spits out a new vector field) that you will learn more about in future math and physics classes. We will not really use it in this course, but I think it is fun to see things like this to help you understand why we want you to learn certain bits of more advanced math!

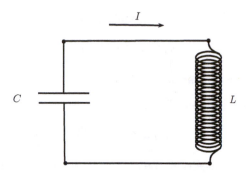

Figure 8.10: *A capacitor (with some initial charge, say) is connected to an inductor.*

And second, everything we've been saying about electric and gravitational fields here – namely that they are "conservative" and can therefore be derived from an associated scalar potential field – turns out to be true only in the context we have encountered so far, in which the fields are *static*, i.e., unchanging. But an important aspect of Faraday's law of induction (which we have carefully avoided raising so far!) is that changing magnetic fields produce an entirely new kind of electric field whose field lines swirl around in closed loops like the magnetic field lines we've encountered from the very beginning.

We will discuss all of this in more detail in Chapter 9.

8.4.5   LC Circuit

Before ending our discussion of inductors and field energy, though, let's consider one last type of circuit: the LC circuit, shown in Figure 8.10, which is just an inductor connected to a capacitor. Of course, if the capacitor is initially uncharged, and no current is initially flowing through the inductor, nothing will happen. So let's assume that, when the circuit is hooked up at $t = 0$, there is a nonzero charge $Q(0) = Q_0$ on the capacitor. The inductor – which remember is just a wire loop – provides a path through which the positive charge from the upper capacitor plate can flow to reach, and cancel, the negative charge on the lower plate, so current will begin to flow. But, on the other hand, the inductor fights against this (changing!) current. So it is not entirely clear what might happen.

To predict what will happen, we better use math! The potential drop across the capacitor is, using the capacitor law, $Q(t)/C$ at some arbitrary time t when the charge on the capacitor is $Q(t)$. Since the capacitor is just connected directly to the inductor, this should match the magnitude of the EMF produced by the inductor, which is $L\dfrac{dI}{dt}$. And finally, as we saw earlier in the case of the RC circuit, the current I is related to the charge Q on the capacitor by $I = -\dfrac{dQ}{dt}$. The rate of change of the current, therefore, is $\dfrac{dI}{dt} = -\dfrac{d^2Q}{dt^2}$. And so the charge $Q(t)$ on the capacitor should obey the following differential equation:

$$\frac{d^2Q(t)}{dt^2} = -\frac{1}{LC}Q(t).$$
(8.91)

This differential equation is structurally identical to the equation that one would get by applying Newton's second law to a mass m attached to a spring of stiffness k, namely

$$\frac{d^2x(t)}{dt^2} = -\frac{k}{m}x(t),$$
(8.92)

whose solution is sinusoidal oscillations with some amplitude A:

$$x(t) = A\cos(\omega t) \tag{8.93}$$

where the angular frequency of the oscillations is given by $\omega = \sqrt{k/m}$.

So, in exactly the same way, the solution of Equation (8.91) is

$$Q(t) = Q_0 \cos(\omega t) \tag{8.94}$$

with $\omega = \sqrt{1/LC}$. This solution implies that the current $I = -dQ/dt$ is given by

$$I(t) = \omega Q_0 \sin(\omega t) \tag{8.95}$$

which makes sense: no current is flowing at $t = 0$ (when the circuit is first hooked up); the capacitor "wants" to push current through the inductor, but the inductor's resistance to changes in currents prevents the current from jumping up to some large value right away; they compromise, in effect, and the current starts increasing, from zero, at a finite rate.

After a while, current has been flowing (at an increasing rate) for enough time that the capacitor is completely discharged ($Q = 0$). At this moment, the capacitor is perfectly happy and is no longer attempting to push current through the inductor. But some finite amount of current is flowing at this moment, and the inductor resists changes in current flow. So the current just keeps flowing, which means that the capacitor "overshoots" its equilibrium, uncharged state, and starts getting charged up now in the opposite sense (a negative charge on the top plate and a positive charge on the bottom plate). The capacitor is now trying to push current back around the other way (counter-clockwise in the Figure) to discharge, but this would require changing the direction of current flow; the inductor fights this change, but they compromise and the rate of clockwise current flow slows, eventually stopping and reversing.

Anyway, you can see how it makes qualitative sense that we get never-ending cyclic oscillations of the charge $Q(t)$. The process is really analogous to what happens in a mass-on-a-spring mechanical system: the mass, initially displaced from its equilibrium position where the spring force is zero, is pushed toward that equilibrium position by the (say) initially compressed spring; but, with no friction to slow it down, the mass is moving quite fast when it reaches the equilibrium position, so it overshoots the equilibrium and relies on the now-stretched spring to slow it down on the other side, and the cycle repeats endlessly.

In the mechanical case, we are perhaps accustomed to thinking of this process in terms of energy in the following way: the initially-compressed spring has potential energy stored in it; as the spring relaxes back toward its equilibrium length, pushing the mass as it does so, this elastic potential energy is converted into kinetic energy for the mass; the kinetic energy is then converted back into elastic potential energy (of the now-stretching spring); and so on. We could say that the energy "sloshes back and forth" between these two forms, spring potential energy and kinetic energy of the mass.

And the same way of thinking about it applies to our LC circuit. Initially, all of the energy of the circuit is in the form of electric field energy, concentrated in the gap between the charged capacitor plates. But a quarter of a cycle later, when the capacitor's charge is zero but there is a large current running through the inductor, the energy is now stored in the magnetic field inside the inductor. And, as the cyclical oscillations proceed, we can think of the energy as again "sloshing back and forth", between two forms: electric field energy and magnetic field energy.

If you are wondering "How, exactly, does the energy get from the region between the capacitor plates, over to the region inside the inductor?" that is an excellent question that we will address in the next Chapter!

Questions:

Q1. Two particles with masses m_1 and m_2 are separated by a distance R, with nothing else nearby. What is the total gravitational potential energy of the pair?

Q2. Three particles of mass m are placed at the vertices of an equilateral triangle with sides of length R. What is the total gravitational potential energy of the system?

Q3. Which has more resistance, a "100 Watt" light bulb, or a "60 Watt" light bulb? Hint: the nominal power of a light bulb corresponds to the actual rate at which it produces energy – mostly in the form of light – when connected across a standard US electrical socket whose terminals maintain a potential difference of 120 Volts.

Q4. We mentioned that one can burn one's fingers by connecting a wire across a battery, due to the large rate at which thermal energy is produced (according to $P = (\Delta V)^2/R$) when R is small. But then, if lightning hits your house, why is it better for the electrical current to go through an iron lightning rod instead of, say, the wood frame of the house (whose resistance R is significantly greater than that of the iron rod)?

Q5. We discussed the analogy between an LC circuit, and an idealized mass-on-a-spring system in which there is no friction or air drag. What do you think happens if a resistor with resistance R is added to an LC circuit, to make an LRC circuit? If such a circuit is started (at $t = 0$) with $I(0) = 0$ and $Q(0) = Q_0$, what do you think a graph of $Q(t)$ would look like?

Q6. We have explored the idea that gravity and electrostatics are in some sense "the same" – we just replace $k = 1/4\pi\epsilon_0$ with $-G$, and all the charges Q_i with masses M_i. So perhaps we can also think of gravitational potential energy as being stored in the gravitational field $\vec{g}$, just like in the electric case? What do you think would be the formula for the gravitational field energy density, u_{grav}?

Q7. In the schematic circuit digrams we've been drawing, e.g., Figures 8.5 and 8.9, the schematic representations of capacitors and inductors are pretty self-explanatory: the inductor looks like a coil of wire, and the capacitor looks like two parallel plates (seen edge-on). But the zig-zag line used to indicate a resistor doesn't look much like a real resistor. Why do you think this particular schematic depiction is used?

Exercises:

E1. A pendulum is made from a bob of mass m and a (very light) string of length L. If the pendulum is released (from rest) with the string horizontal (i.e., 90° away from its downward equilibrium position), what is the speed of the bob when it passes through the equilibrium point at the bottom of its swing? What is the tension in the string at that moment?

E2. A 60 kg skier skis, starting from rest, along a 200-meter-long slope that makes a 20° angle with the horizontal. The coefficient of kinetic friction between her skis and the snow is $\mu_k = .05$. Calculate the works done by all three of the forces that act on her during the process, and use the Work - KE theorem to determine her speed at the end of the slope.

E3. A particle of mass m moving with speed v_0 collides with another particle of mass m which is initially at rest. If the collision is perfectly elastic (so both translational momentum and kinetic energy are conserved) show that the angle between the post-collision directions of motion of the two particles is exactly 90°. (This principle is very useful in pool to anticipate and avoid scratching.)

E4. Two particles, each with charge Q, are located along the x-axis at $x = +d$ and $x = -d$. What is the electric potential V at a point a distance y away from the origin along the y-axis? To assess your answer, compute the electric field at this same point using $\vec{E} = -\vec{\nabla}V = -\frac{\partial V}{\partial y}\hat{j}$, and show that the result is the same as what you would get by calculating $\vec{E}$ directly, using $\vec{E} = \frac{kQ}{r^2}\hat{r}$ and superposition.

E5. Two particles, each with mass M, are located along the x-axis at $x = +d$ and $x = -d$. What is the gravitational potential V at a point a distance y away from the origin along the y-axis? To assess your answer, compute the gravitational field at this same point using $\vec{g} = -\vec{\nabla}V = \frac{\partial V}{\partial y}\hat{j}$, and show that the result is the same as what you would get by calculating $\vec{g}$ directly, using $\vec{g} = -\frac{GM}{r^2}\hat{r}$ and superposition.

E6. The electric potential V at a distance $r < R$ from the center of a solid sphere of radius R and total charge Q (uniformly distributed throughout the volume) can be calculated as follows. Think of the sphere as a set of concentric spherical shells of radius s and thickness ds, and hence $dQ = \frac{3Q}{R^3}s^2\,ds$. The shells with $s < r$ each contribute a potential $dV = k\,dQ/r$, while the shells with $s > r$ each contribute a potential $dV = k\,dQ/s$. Add up the contributions to find $V(r)$. To assess your result, calculate $\vec{E} = -\vec{\nabla}V = -\frac{\partial V}{\partial r}\hat{r}$ and show that you get the same $\vec{E}$ that you can also find directly using Gauss' Law.

E7. The gravitational potential V at a distance $r < R$ from the center of a solid sphere of radius R and total mass M (uniformly distributed throughout the volume) can be calculated as follows. Think of the sphere as a set of concentric spherical shells of radius s and thickness ds, and hence $dM = \frac{3M}{R^3}s^2\,ds$. The shells with $s < r$ each contribute a potential $dV = -G\,dM/r$, while the shells with $s > r$ each contribute a potential $dV = -G\,dM/s$. Add up the contributions to find $V(r)$. To assess your result, calculate $\vec{g} = -\vec{\nabla}V = \frac{\partial V}{\partial r}\hat{r}$ and show that you get the same $\vec{g}$ that you can also find directly using Gauss' Law for Gravity.

E8. The "effective resistance" R_{eff} of a group of resistors, connected to one another in some way, can be defined, following Ohm's Law, as the ratio of the potential difference ΔV across the whole group, to the total current I flowing through the group. Prove that, for two resistors in series, $R_{eff} = R_1 + R_2$ (where R_1 and R_2 are the individual resistances of the two resistors). And prove also that, for two resistors connected in parallel, $R_{eff} = 1/(1/R_1 + 1/R_2)$. Explain how this relates to Equation (8.37). (For example, two identical resistors connected in series is rather similar to a single resistor that is twice as long.)

E9. The "effective capacitance" C_{eff} of a group of capacitors, connected to one another in some way, can be defined, following Equation (8.44), as the ratio of the total charge Q that flowed into the group, to the potential difference ΔV across the whole group. Prove that, for two capacitors in parallel, $C_{eff} = C_1 + C_2$ (where C_1 and C_2 are the individual capacitances of the two capacitors). And prove also that, for two capacitors connected in series, $C_{eff} = 1/(1/C_1 + 1/C_2)$. Explain why the rules for adding in series and parallel are opposite, for capacitors, compared to resistors. Which way does it work for inductors? For example, is $L_{eff} = L_1 + L_2$ the effective inductance of two inductors in series, or in parallel?

E10. A hollow sphere of radius R has a total charge Q uniformly distributed over its surface. The electric potential at the surface of the sphere will be $V = kQ/R$. What is the electric potential energy of the charged sphere? (You should answer this in a way that takes practically no calculation. Just be careful about factors of $1/2$...) Now write down an expression for the magnitude $E(r)$ of the electric field at a distance $r > R$ from the center of the sphere, and plug this into Equation (8.63) to find the electric field energy density at distance r from the center. What is the electric field energy density for $r < R$? What do you get if you integrate this energy density over all space?

E11. Derive a formula for the "escape velocity", from the surface of a body of mass M and radius R, using energy conservation: the escape velocity is the speed that an object would need at the surface of the body so that (ignoring air drag or other losses) it has just enough energy to get infinitely far away. Calculate the escape velocities for the Earth and the Moon.

E12. Energy conservation makes it (relatively) easy to see, analytically, that Newton's theory of gravity predicts Kepler's first law (i.e., elliptical orbits with the Sun at one focus). Let's walk through this. Describe the planet's orbit (about the fixed Sun) using polar coordinates r and θ. The statement that the total energy is constant throughout the orbit can be written

$$E = -\frac{GMm}{r} + \frac{1}{2}m\left(\frac{dr}{dt}\right)^2 + \frac{1}{2}mr^2\left(\frac{d\theta}{dt}\right)^2$$

where we have used the fact that the radial and tangential components of the velocity are respectively $\frac{dr}{dt}$ and $r\frac{d\theta}{dt}$. But the orbital angular momentum $L = mr^2\frac{d\theta}{dt}$ is also a constant of the motion, and by the chain rule the radial component of the velocity can be rewritten $\frac{dr}{dt} = \frac{dr}{d\theta}\frac{d\theta}{dt}$. Use these to re-write the energy conservation equation in terms of r and $\frac{dr}{d\theta}$. You should then be able to show that the (polar coordinates) equation for an ellipse (about one of the focus points),

$$r(\theta) = \frac{c}{1 - \epsilon\cos(\theta)}, \tag{8.96}$$

is a solution, as long as the constants c and ϵ are defined in a certain way in terms of the constants (E, L, G, M, and m) that appear in the differential equation.

Projects:

P1. Make an "I-V curve" (i.e., a plot of the electric current I through the object as the electric potential ΔV across it is varied) for several different objects: a manufactured resistor, a light bulb, and a diode. (It can also be interesting to produce I-V curves for batteries and solar cells.)

P2. Acquire data for the potential difference ΔV over time, for an initially-charged capacitor connected to a resistor of known resistance R. Find an empirical value for the decay time constant τ and use it to compute the capacitance C of the capacitor.

P3. Use a function generator and an oscilloscope to observe the current vs. time in an LR circuit and infer a value for τ, from which the inductance L of the inductor can be computed if R is known. Do this for (at least) two different R values and see if the results (for L) are consistent. They might not be, if the function generator has an appreciable internal resistance, but we may be able to find a value for the internal resistance which renders the L values compatible and hence believable.

P4. Play around with an electric circuit containing both a capacitor and an inductor. Unlike the idealized LC circuit discussed in the text, your circuit will have some resistance: even if you don't add a "resistor", the inductor, which remember is just a long coil of wire, will have some resistance. So your circuit will be an RLC circuit rather than a pure LC circuit. Still, we can see oscillations in the circuit and confirm that they have the frequency we would expect based on the values of L and C. The RLC circuit also provides a nice electrical example of the concept of "resonance", which we can explore.

CHAPTER 9

Radiation

O<small>UR</small> topic this week is the mathematical and physical synthesis of electromagnetic theory that was provided largely by the Scottish physicist James Clerk Maxwell in the mid- to late-1800s. As we will see, Maxwell brought considerable mathematical organization and clarity to the theoretical concepts that had been enunciated by his predecessors. But he also made profound new discoveries of his own, including the recognition that electric and magnetic fields could support the propagation of electromagnetic waves which propagate, as it turns out, at the speed of light. Maxwell's most important insight, then, was that light *is* electromagnetic radiation.

Figure 9.1: James Clerk Maxwell (1831-1879)

§ 9.1 Induced Electric Fields

Let us begin by clarifying something that has been lurking in the shadows for a while now. In Chapter 7 we discussed Faraday's law of induction, according to which an electro-motive force – an EMF – is produced whenever the magnetic flux through a circuit changes. An EMF, recall, is the net strength (with dimensions of energy per unit charge) of an electromagnetic "push" that can make electric charges flow in the form of an electric current.

The first EMF we encountered was simply the electric potential difference produced by the chemical reactions in a battery – this potential difference, of course, being understandable in terms of the integral of the electric field along a path connecting the battery terminals. In a circuit powered by a battery, then, the current flow is maintained (in the face of the internal friction-like processes that would otherwise

231

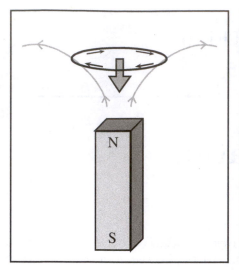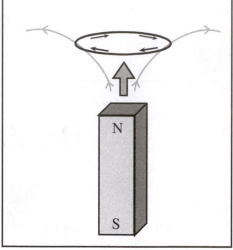

Figure 9.2: In the left panel, a wire loop is moved toward a permanent magnet; magnetic forces on the (moving) mobile charges within the wire loop generate an EMF and cause an induced current to flow (clockwise, as shown by the black arrows). In the right panel, the relative motion of the magnet and the loop is the same. But with the magnet moving toward the stationary wire loop, the mobile charges in the loop remain at rest and hence cannot experience a magnetic force. So although a current-inducing EMF does still occur in this case, it is somewhat mysterious what kind of force, exactly, is responsible.

make it cease) by *electric* forces ($\vec{F}_{elec} = q\vec{E}$) exerted by *electric fields* on the mobile charged particles within the wires and other circuit elements.

An important aspect of Faraday's law of induction is that an EMF can also be provided by *magnetic* forces ($\vec{F}_{mag} = q\vec{v} \times \vec{B}$). This typically happens when a circuit (perhaps as simple as just a loop of wire) moves into or out of – or rotates in the presence of – a magnetic field. The motion of the wire loop – which of course implies a nonzero velocity for the mobile charges contained within it – *through* the magnetic field, gives rise to magnetic forces on those mobile charges. In the right circumstances, these magnetic forces can point along the wire and hence produce/sustain a flowing electric current.

We analyzed some simple examples of this phenomenon in previous weeks. The left panel of Figure 9.2 shows another very simple case, in which a circular wire loop is moved toward the North pole end of a bar magnet. In this situation, the downward velocity of (hypothetical, mobile) positive charges in the wire, and the radially-outward component of the magnetic field (associated with the "spreading out" of the magnetic field lines emerging from the North pole end of the magnet), implies (by the right hand rule) that there are magnetic forces on those (hypothetical mobile) positive charges in the directions indicated by the black arrows in the Figure. The magnetic forces, in particular, push positive charges to flow clockwise (as seen from above, as in the Figure) and so a clockwise electric current is induced in the wire loop.

(As an interesting aside, note that the wire loop, with an induced current flowing clockwise as seen in the Figure, acts like a little bar magnet with its South pole end above the loop and its North pole end below the loop. Since their North pole ends are facing each other, this means that the permanent magnet and the wire loop *repel* one another as the wire loop approaches the magnet. So, for example, to maintain a constant downward velocity for the wire loop, you would have to exert a downward force. This is yet another perspective on the anthropomorphic idea, introduced before, that the wire loop doesn't "want" the magnetic flux through it to change. You can make it change, by moving it downward, closer to the magnet – but you have to push it. It won't move that way spontaneously.)

Anyway, the EMF which drives the induced current flow in this kind of case, can be understood as the integral, all the way around the loop, of the magnetic force that would be felt by a mobile charged particle, divided by that particle's charge:

$$\mathcal{E} = \frac{1}{q} \oint \vec{F}_{mag} \cdot d\vec{s} = \oint \vec{v} \times \vec{B} \cdot d\vec{s}. \tag{9.1}$$

That's what the EMF *is* for this kind of case, where a circuit *moves* in a fixed background magnetic field. And the *size* of the EMF is given, according to Faraday's law and as we discussed before, by the rate of change of the magnetic flux through the circuit:

$$\mathcal{E} = -\frac{d\Phi_{mag}}{dt} = -\frac{d}{dt} \iint \vec{B} \cdot d\vec{A} \tag{9.2}$$

where the integral is over an open two-dimensional surface that has the circuit as its boundary. Recall that the minus sign in Faraday's law tells us the direction or "sense" of the EMF and hence the direction or "sense" of induced (conventional) current flow. In this case, if we talk about the upward flux through the loop (i.e., if we define the area vector $d\vec{A}$ in the surface integral to point upward) then the magnetic flux is *positive* and *increasing* as the loop approaches the North pole end of the magnet. The right hand side of Faraday's law is therefore negative, which tells us that the EMF is so as to push (conventional) current to flow opposite the direction that is associated with "up" via the right hand rule.

Note also that although Faraday discovered and explored the phenomenon of induction, it was Maxwell who first put Faraday's law into something like this modern, mathematical form.

But there is something very puzzling. If a wire loop moves through a fixed background magnetic field in such a way that the magnetic flux through the loop changes, an EMF (constituted by magnetic forces) will be produced and an induced current will flow. But this is not the only way that the magnetic flux through a loop can change. The flux can also change if, for example, the loop remains at rest and the magnetic field at its location changes, e.g., by moving the permanent magnet that is the source of the field. Experiment shows that in this kind of case, there is also an induced EMF – given still by Equation (9.2) – that gives rise to an induced electric current in the loop.

This is probably what one should expect, on the grounds of applying some of Galileo's and Newton's insights about motion to this kind of situation. Everything would look and feel the same, Galileo reminded us, if we were locked inside the hold of a ship that was at rest, or instead locked inside the hold of a ship that was cruising at constant velocity. The reason is that the *relative motion* of the various contents of the ship (fish and butterflies and water drops and so forth) are the same in both cases, and, evidently, it is only this relative motion which dictates how things will behave. Or, to put the same point in a somewhat more Newtonian way, the two situations can be thought of as the same one situation, viewed from two different inertial reference frames (one in which the ship is at rest, and one in which the ship is moving). Anyway, from this point of view, it would be extremely surprising if the observable effect – a clockwise induced current of a certain magnitude in the wire loop – were different depending on whether the loop approached the magnet, or instead the magnet approached the loop. So in this way it makes perfect sense that the EMF is the same, whether the magnetic flux through the loop changes at a certain rate because the loop approaches the magnet, or instead because the magnet approaches the loop.

But a little contemplation reveals that, when the magnet approaches the loop, the EMF cannot be associated with magnetic forces. After all,

$$\vec{F}_{mag} = q\vec{v} \times \vec{B} \tag{9.3}$$

immediately implies that if $\vec{v} = 0$, then $\vec{F}_{mag} = 0$. So regardless of what exactly is happening with $\vec{B}$ at the location of the mobile charges in the wire loop, as the magnet approaches, a magnetic force simply cannot cause them to accelerate, from rest, and begin flowing as an electric current.

And yet, when the magnet is moved toward the loop, a current does indeed begin to flow!

It was Maxwell who sorted out the resolution of this puzzle: when magnetic fields change in time (as happens, for example, at the location of the charges in the wire loop as the permanent magnet is brought closer) the changing magnetic fields cause – call into existence – *electric* fields with a novel character. In particular, the electric field lines associated with these electric fields that are induced by changing magnetic fields, do not begin and end on positive and negative charges, respectively. (Indeed, the novel induced type of electric field can exist even in a region where no electric charges are present.) Instead, the induced electric field lines make complete closed loops (like magnetic fields lines do).

The EMF associated with such spiraling electric fields can be defined just like the EMF associated with the electric potential difference, namely, in terms of the integral of the electric field: $\mathcal{E} = \oint \vec{E} \cdot d\vec{s}$. The only difference is that the integral is now taken over a complete closed loop rather than just from (say) one battery terminal to the other. And so, for this case (namely, a case in which the circuit, around which we are calculating the EMF, is at rest) Faraday's law can be written:

$$\mathcal{E} = -\frac{d}{dt} \iint \vec{B} \cdot d\vec{A} = \oint \vec{E} \cdot d\vec{s} \tag{9.4}$$

where the surface integral on the left is over a two-dimensional (open) surface whose boundary is the closed path we are integrating around on the right.[1]

But we should not let the crucial qualitative physics principle get lost in the (admittedly slightly overwhelming!) mathematical expression. The crucial point here is that *changing magnetic fields give rise to electric fields (with a new, swirly, character)*. We will explore one extremely important practical application of this idea in the following section.

§ 9.2 Transformers

Consider a solenoid that consists of N_1 radius-R_1 circular loops of wire, wound around a cylinder of length l. As we have seen previously, the magnitude of the axial magnetic field inside the cylinder is given (approximately) by

$$B = \frac{\mu_0 N_1}{l} I \tag{9.5}$$

---

[1]If your head is spinning, due to the several different-looking equations for, supposedly, the same thing, in Equations (9.1), (9.2), and (9.4), I wouldn't blame you. Here is an attempt to clarify the situation. The crucial point is that the magnetic flux is the integral of $\vec{B}$ over some surface S: $\Phi_{mag} = \iint_S \vec{B} \cdot d\vec{A}$. Faraday's Law involves the rate of change of the flux, $d\Phi_{mag}/dt$. But there are two distinct ways the flux can change: it can change because $\vec{B}$ itself is changing in time, *or* because the surface S over which we are integrating changes! So $d\Phi_{mag}/dt$ has two terms: one involving the rate of change of $\vec{B}$ itself, and the other involving the motion of the perimeter, P, of the surface S. (In general, there is actually a third term as well, having to do not with changes in the perimeter of S, but with changes in the shape of the surface inside of the perimeter. But since magnetic field lines make complete loops, a distortion of the shape of the surface which doesn't change P won't change the flux. So this third term vanishes in the present case.) The first term, involving $\partial \vec{B}/\partial t$ can be re-written in terms of the path integral, around P, of the electric field: this is Maxwell's idea that changing magnetic fields call swirling electric fields into existence. And the second term can be written as a path integral, around P, involving the magnetic field and the velocity with which different parts of the perimeter are moving. In equation form, at the end of the day, we have that $-\frac{d}{dt} \iint_S \vec{B} \cdot d\vec{A} = \oint_P \vec{E} \cdot d\vec{s} + \oint_P (\vec{v} \times \vec{B}) \cdot d\vec{s}$. So the two different expressions for the EMF are really just what (minus) the rate of change of the magnetic flux reduces to, in the two special cases we considered: a fixed loop in the presence of a changing magnetic field, and a moving loop in the presence of a fixed magnetic field. The detailed vector calculus involved in all of this, though, is a little beyond the level of this course!

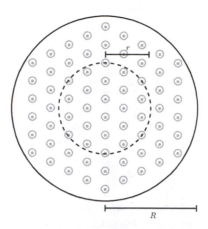

Figure 9.3:   Cross-sectional view through a cylindrical solenoid of radius R. The gray encircled dots indicate a magnetic field that is pointing out of the page toward us. If, for example, the magnitude of that magnetic field is increasing, there will be an increasing outward magnetic flux through a loop like the radius-r dashed circle indicated. Faraday's law then implies that the path integral (taken counter-clockwise, since that is the direction that correlates with "outward" via the right hand rule) of $\vec{E}$ around the radius-r loop is negative *– i.e., the induced electric field will swirl around* clockwise, *as shown explicitly in Figure 9.4.*

where I is the current running through the solenoid. This means that the outward magnetic flux, through for example the dashed circular loop of radius r shown in Figure 9.3, will be given by

$$\Phi_{mag} = \iint \vec{B} \cdot d\vec{A} = B\pi r^2 = \frac{\mu_0 N_1 \pi r^2}{l} I. \tag{9.6}$$

Now suppose the current I is changing. Faraday's law tells us that the path integral of $\vec{E}$, around the radius-r circular path (and going counter-clockwise, since that is the direction that corresponds to "outward" by the right hand rule) is given by $-\partial \Phi_{mag}/\partial t$. So

$$2\pi r E_t(r) = -\frac{\mu_0 N_1 \pi r^2}{l} \frac{dI}{dt} \tag{9.7}$$

where E_t is the (counter-clockwise) tangential (i.e., $\hat{\phi}$) component of the electric field at a distance r from the center. Rearranging, we find that

$$E_t(r) = -\frac{\mu_0 N_1 r}{2l} \frac{dI}{dt}. \tag{9.8}$$

The minus sign tells us that the counter-clockwise component of $\vec{E}$ is negative, i.e., the induced $\vec{E}$ swirls around in a *clockwise* direction.

The structure of the induced electric field – spiraling around clockwise and with a magnitude that increases with increasing r – is indicated in Figure 9.4.

So far we've just been talking about the induced electric field within the solenoid. But of course, if we were to place a real wire loop (say, of area A_2) in that region, the induced electric field implies an EMF that could drive current to flow through the loop. To amplify the effect, instead of placing a single loop inside the solenoid, we could insert a whole coil consisting of some large number N_2 of such loops. Then

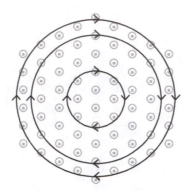

Figure 9.4: If the magnetic field (indicated in gray) inside a cylindrical solenoid is pointing toward us and increasing in magnitude, there will be an induced electric field which spirals around in a clockwise direction and with a magnitude that increases with increasing distance from the center.

the total EMF generated across the secondary coil is

$$\mathcal{E}_2 = -N_2 \frac{\partial}{\partial t}(BA_2) = -N_2 A_2 \frac{\mu_0 N_1}{l}\frac{dI}{dt}. \tag{9.9}$$

A device such as this – basically just two solenoids whose wires do not actually connect, but which are nevertheless coupled by the electric and magnetic fields inside them – is called a "transformer". The previous equation says that a changing electrical current through the primary coil will induce an EMF in the secondary coil.

The relationship between what we put into the primary coil, and what we get out of the secondary coil, is made a little more transparent by remembering that the input coil is just an inductor, the EMF $\mathcal{E}_1$ across which is given by

$$\mathcal{E}_1 = -L_1 \frac{dI}{dt} \tag{9.10}$$

where the inductance L_1 is given by $\mu_0 N_1^2 A_1/l$. Solving for dI/dt and substituting into Equation (9.9) gives

$$\mathcal{E}_2 = \frac{N_2 A_2}{N_1 A_1}\mathcal{E}_1 \tag{9.11}$$

which is called the "transformer equation".

Suppose, for example, that the input to the primary coil is a sinusoidally-oscillating signal (such as the output we would get from an electric generator) with some amplitude V_1 and some angular frequency ω_1:

$$\mathcal{E}_1 = V_1 \sin(\omega_1 t). \tag{9.12}$$

Then the transformer equation says that the EMF output of the secondary coil will also be a sinusoidal signal (with the same angular frequency)

$$\mathcal{E}_2 = V_2 \sin(\omega_1 t) \tag{9.13}$$

but with amplitude

$$V_2 = \frac{N_2 A_2}{N_1 A_1}V_1. \tag{9.14}$$

This is how transformers are most often used in practice. They are used to convert a sinusoidal (i.e., "alternating current", "AC") electrical signal from one voltage to another. Very often the areas A_1 and

A_2 of the individual loops in the primary and secondary coils are about the same, in which case the transformer just amplifies the voltage by a factor equal to the ratio of the number of loops in each coil, N_2/N_1. If this ratio is greater than unity, the transformer is referred to as a "step-up transformer" because it steps the signal voltage up from a lower to a higher value. The opposite case, in which N_2/N_1 is smaller than unity, is a "step-down transformer", which reduces the signal voltage.

It may seem like there is some free lunch involved in stepping the amplitude of a sinusoidal voltage up or down, but this is not the case. The input power – energy per unit time – will always match (or, in practice, slightly exceed, if there are some small inefficiencies) the output power. These powers can be expressed in terms of the product of the voltages and the currents. So if we use a step-up transformer to convert a low-voltage signal to a higher-voltage signal, we also necessarily reduce the amount of electrical current that flows.

This principle can help one appreciate one of the most important practical applications of transformers, which is the transmission of electrical energy across great distances. Suppose, for example, one wants to build a power station several miles outside of a big city, and transmit the electrical energy to the city using some power lines. Well, those power lines will have some electrical resistance, so electrical current running through them will generate heat – which is, from the point of view of trying to get electrical energy to the city, just wasted energy. As we have seen previously, the rate at which heat is produced in the transmission wires – i.e., the rate at which energy is being wasted – is given by

$$P = I^2 R \tag{9.15}$$

where R is the resistance in the wires and I is the electrical current. One can, of course, control R in various ways – thicker wires, for example, will have lower resistance, and will thereby reduce waste. But thick wires are heavy and extremely expensive.

The solution that is thus widely adopted today – and which was originally suggested by the enigmatic Serbian-American genius Nicola Tesla – is to instead control the current I using transformers. In particular, the electricity produced at the power station is stepped up to an extremely high voltage using a step-up transformer near the power station. It is then transmitted to the city along not-unreasonably-thick transmission wires with extremely low current I – and hence very small thermal losses – made possible by the high voltage. Then, inside the city, the voltage is reduced, at local sub-stations, using step-down transformers, to voltages that are safer for residential neighborhoods and more in line with what household applicances need.

You should have the chance to play around with transformers in class. Although they are fairly simple devices (literally nothing more than two coils of wire) and their function is relatively straightforward to understand, they represent a beautiful engineering application of an incredibly fundamental physical idea: changing magnetic fields give rise to electric fields.

§ 9.3 Ampère's Law and Displacement Current

One of Maxwell's most important contributions to electromagnetic theory was recognizing that Ampère's law (as we have formulated it so far), namely,

$$\oint \vec{B} \cdot d\vec{s} = \mu_0 I_{through}, \tag{9.16}$$

is mathematically inconsistent and hence needs to be fixed up.

The left hand side of Ampère's law is the path integral of the magnetic field around some closed loop. The "$I_{through}$" on the right hand side means the amount of electrical current that goes through the

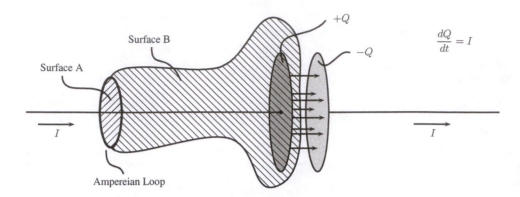

Figure 9.5:   *Electric current I flowing in a long straight wire with a capacitor inserted at some point. A standard Ampèreian Loop around the wire is shown, with two different surfaces, A and B, which have that Loop as their perimeter. Surface A, which is just the flat disk whose edge is the circular loop, has the current I in the wire piercing through it. But Surface B, which balloons out to the right and passes between the plates of the capacitor, has no electric current piercing through it. This is the mathematical inconsistency in Ampère's law as it had been formulated prior to Maxwell, who recognized that consistency could be achieved by treating the changing electric flux through the surface – the so-called "displacement current" – as equivalent to an electric current through it.*

loop (around which we integrate on the left). More precisely, "$I_{through}$" denotes the amount of electrical current that pierces through an open two-dimensional surface whose perimeter is the loop we integrate along on the left. The problem that Maxwell recognized arises because, for a given loop, there are many different surfaces which have that loop as their perimeter, and the amount of electrical current piercing through the different surfaces need not be the same.

This is easiest to see in the following kind of case. Consider a long straight wire with electrical current I running through it. But suppose that at some point there is a break in the wire where a parallel plate capacitor is located. See Figure 9.5. We can learn about the magnetic field in the vicinity of the wire by applying Ampère's law in the standard way. Consider, for example, a radius-r Ampèreian loop, centered on the wire. The left hand side of Ampère's law will give, in the familiar way, $2\pi r B(r)$, where $B(r)$ is the magnitude of the magnetic field at distance r from the wire.

To evaluate the right hand side of Ampère's law, we must pick a surface whose edge is the circular loop we integrated around on the left. The obvious choice – Surface A in the Figure – is just the flat circular disk of radius r. The current in the wire obviously pierces right through this surface, so the right hand side of Ampère's law is $\mu_0 I$, and, solving for the magnitude of the magnetic field we find

$$B(r) = \frac{\mu_0 I}{2\pi r} \tag{9.17}$$

as we have seen before.

But we could also have chosen a different surface on the right hand side of Ampère's Law. Consider, for example, the surface labelled "Surface B" in the Figure, which balloons out to the right from the Ampèreian loop and passes between the capacitor plates so that no electrical current pierces through it. This implies that the right hand side of Ampère's law is zero, which implies

$$B(r) = 0. \tag{9.18}$$

But Equations (9.17) and (9.18) obviously cannot both be true. So there is an inconsistency in Ampère's law as it had been formulated previously.

Maxwell's clever idea was to add a term to the right hand side of Ampère's law to restore its mathematical consistency. In particular, Maxwell recognized that although there is no electrical current piercing through Surface B, there will be, in a case like this, a changing electric flux through that surface, because the electric charge flowing through the wire will pile up on the capacitor plates, and a steadily increasing charge Q on the plates implies an electric field with a steadily increasing amplitude between the plates. In particular, we have seen before that the magnitude E of the electric field between the plates of a capacitor with charge Q is

$$E = \frac{Q}{\epsilon_0 A} \tag{9.19}$$

where A is the area of the plates.

The electric field flux $\iint \vec{E} \cdot d\vec{A}$ through Surface B will just be EA. So we can relate the charge Q on the capacitor to the electric flux through Surface B like this:

$$Q = \epsilon_0 \iint_B \vec{E} \cdot d\vec{A}. \tag{9.20}$$

But the rate dQ/dt at which the charge on the capacitor increases will match I, the current flowing in the wire. So evidently the current I flowing in the wire is proportional to the rate of change of the electric flux through Surface B:

$$I = \epsilon_0 \frac{d}{dt} \iint_B \vec{E} \cdot d\vec{A}. \tag{9.21}$$

And so, if we replace the "$I_{through}$" on the right hand side of Ampère's law with $I_{through} + \epsilon_0 \frac{d}{dt} \iint \vec{E} \cdot d\vec{A}$, then we will get the same thing when we evaluate that right hand side, whether we use Surface A or Surface B from the Figure.

Maxwell, that is, proposed replacing Ampère's law with the following modified law, which we could call the Ampère-Maxwell law:

$$\oint \vec{B} \cdot d\vec{s} = \mu_0 I_{through} + \mu_0 \epsilon_0 \frac{d}{dt} \iint \vec{E} \cdot d\vec{A}. \tag{9.22}$$

Maxwell called $\epsilon_0 \frac{d}{dt} \iint \vec{E} \cdot d\vec{A}$ the "displacement current". It is easy to understand why he would think of it as some new kind of current, since the idea is that the changing electric flux acts just like a regular electric current in so far as its generation of magnetic field is concerned. "Displacement" is a little harder to understand. This funny name has to do with the fact that Maxwell was primarily thinking of capacitors in which the gap between the plates was filled with some insulating material. In the presence of an electric field, such a material will polarize electrically (that is, the positive and negative microscopic charges will "stretch" slightly, in opposite directions) and Maxwell initially thought of the "displacement current" not as associated with the changing electric field itself, but instead with the increasing polarization that such a changing field would produce.

Maxwell's own thinking about this issue was also affected by his belief that even what we call "empty space" is filled with some highly structured material (but invisible) medium. Indeed, Maxwell famously invented detailed mechanical models for this medium – the "aether" – in which, for example, electric and magnetic fields represented the rotation of various sorts of gears and bearings.

These mechanical models genuinely helped Maxwell compile the various principles discovered by his predecessors into a coherent and mathematically consistent set, by giving him a concrete visualizable way of understanding their physical significance and inter-relations. But in the end, even Maxwell realized

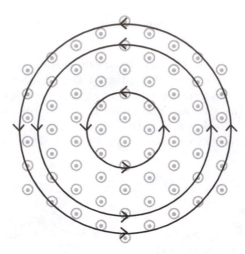

Figure 9.6:   The gray encircled dots here represent an electric field pointing out of the page toward us.   If the magnitude of that electric field in increasing, there will be an increasing outward electric flux through any circular loop.   The Ampère-Maxwell equation then implies that the path integral (taken counter-clockwise, since that is the direction that correlates with "outward" via the right hand rule) of $\vec{B}$ is positive – i.e., the induced magnetic field lines will swirl around counter-clockwise, as indicated with the solid black curves.

the heuristic character of these models – in a letter to a colleague he wrote that "[t]he nature of this mechanism is to the true mechanism what an orrery is to the solar system" – and let the mathematical laws, relating the various integrals and derivatives of the electric and magnetic fields, stand on their own.

In particular, with the so-called "displacement current", we now understand that a material medium (whose charged microscopic parts increasingly "displace" from their equilibrium positions in the presence of an increasing electric field) is unnecessary. The new term is not just a way of describing some perhaps subtle but still completely standard electrical current associated with the motion of (perhaps microscopic) electric charges.

Instead, the new term is describing a genuinely new sort of process in which *a changing electric field produces a magnetic field*.

Note that this is the opposite – or complement – of the process described previously (in our discussion of Faraday's law) in which a changing magnetic field produces an electric field. The similarities and differences between the two processes can perhaps be captured by re-using an earlier Figure – with one small change – to depict the new process. See Figure 9.6, which depicts (in black) the magnetic field that would be induced, in for example the region between the capacitor plates in the earlier Figure, by a changing electric field.

Figure 9.7: Maxwell's equations printed on a T-shirt.

§ 9.4 Maxwell's Equations and Electromagnetic Waves

Let's collect together all of the basic laws describing electric and magnetic fields that we have encountered:[2]

$$\text{Gauss' Law:} \qquad \oiint \vec{E} \cdot d\vec{A} = \frac{Q_{encl}}{\epsilon_0}$$

$$\text{Gauss' Law for Magnetism:} \qquad \oiint \vec{B} \cdot d\vec{A} = 0$$

$$\text{Faraday's Law:} \qquad \oint \vec{E} \cdot d\vec{s} = -\frac{d}{dt} \iint \vec{B} \cdot d\vec{A} \qquad (9.23)$$

$$\text{Ampère-Maxwell Law:} \qquad \oint \vec{B} \cdot d\vec{s} = \mu_0 I_{through} + \mu_0 \epsilon_0 \frac{d}{dt} \iint \vec{E} \cdot d\vec{A}$$

These four equations together are called "Maxwell's Equations" and they – along with the law describing how electric and magnetic fields affect matter: $\vec{F} = q\vec{E} + q\vec{v} \times \vec{B}$ – constitute the fundamental description of how electric and magnetic fields interact with electric charge (and with one another!).

It is incredible that a fundamental description of such a wide variety of phenomena – from charged glass rods picking up scraps of paper and permanent magnets picking up scraps of iron, to the functioning of electric circuits involving batteries, resistors, capacitors, and inductors, to electrical generators and transformers – can be provided in just a few lines of mathematics. Indeed, Maxwell's theory of electricity and magnetism is so elegantly essentialized that it can literally fit on a T-shirt: see Figure 9.7.

But that is only the beginning. As soon as Maxwell had fixed up Ampère's law, he realized that the laws of electricity and magnetism had a profound and previously unrecognized implication. If changing electric fields can call into existence magnetic fields, and changing magnetic fields can call into existence electric fields, the laws would seem to allow a self-sustaining alternation in which a changing electric field at one location induces a magnetic field nearby, the coming-into-existence of which in turn induces a

---

[2]Note that the version of "Faraday's Law" written here assumes a fixed (non-moving) surface S, so the term involving $\oint (\vec{v} \times \vec{B}) \cdot d\vec{s}$ does not appear.

further electric field at a new location a little further down the line, and so on. In short, with the new "displacement current" term added, Maxwell's equations seem to allow for the existence of propagating electromagnetic waves.

With a little vector calculus, it is possible to re-write Maxwell's equations in a new form – so-called "differential form" (as opposed to the "integral form" in which we've been working with them) – which makes it relatively easy to show that (in empty space, where there are no electric charges or currents) the electric and magnetic fields obey a differential equation called the "wave equation" which is known to have propagating wave-like solutions. The mathematics involved in that, however, is a little beyond the level of this course. (But ask me if you're interested and we can talk through it!)

But we can still see that Maxwell's equations – in integral form – are consistent with propagating wave solutions. Let's start by writing down the simplified form of Maxwell's equations that apply when there is no electric charge or electric current in the vicinity:

$$\text{Maxwell's Equations in Empty Space:} \begin{cases} \text{(i)} & \oiint \vec{E} \cdot d\vec{A} = 0 \\[2ex] \text{(ii)} & \oiint \vec{B} \cdot d\vec{A} = 0 \\[2ex] \text{(iii)} & \oint \vec{E} \cdot d\vec{s} = -\dfrac{d}{dt} \iint \vec{B} \cdot d\vec{A} \\[2ex] \text{(iv)} & \oint \vec{B} \cdot d\vec{s} = \mu_0 \epsilon_0 \dfrac{d}{dt} \iint \vec{E} \cdot d\vec{A} \end{cases} \tag{9.24}$$

This is the same set of equations as before, obviously, just with the terms involving Q_{encl} and $I_{through}$ omitted because these will vanish in empty space where there is no charge or current.

Now, our conjecture is that the following expressions for $\vec{E}$ and $\vec{B}$ – which describe sinusoidal waves, with a wavelength λ, propagating in the +z-direction with speed c – are compatible with Maxwell's equations in empty space:

$$\vec{E}(z,t) = \hat{i}\, E_0 \sin\left[\frac{2\pi(z - ct)}{\lambda}\right] \tag{9.25}$$

and

$$\vec{B}(z,t) = \hat{j}\, B_0 \sin\left[\frac{2\pi(z - ct)}{\lambda}\right]. \tag{9.26}$$

See Figure 9.8 for a depiction of the structure of these waves.

Note that the electric field points in the x-direction and the magnetic field points in the y-direction. So the electric and magnetic fields are perpendicular to each other, and also to the direction of propagation of the wave itself: the waves are "transverse". Note also that the expressions for $\vec{E}$ and $\vec{B}$ depend on z and t, but not on x or y. That is, the electric and magnetic fields are identical everywhere on planes of constant z (at a given time t): the waves are so-called "plane waves". Such plane waves are idealizations, obviously; no real wave can have a nonzero amplitude across an infinitely wide area perpendicular to the direction of propagation. Still, such waves are excellent approximations to the structure of real waves as long as we are not too near the edges of the waveform where the field amplitudes drop off to zero.

OK, so let's try to prove that Equations (9.25) and (9.26) are indeed valid solutions to Maxwell's equations in empty space.

To begin with, note that the first two Maxwell equations – (i) and (ii) in Equation (9.24) are automatically respected. Consider first (i) – Gauss' Law – for some closed rectangular region with faces perpendicular to the x-, y-, and z-axes. Since the $\vec{E}$ field points everywhere in the x-direction, the surface integral over –

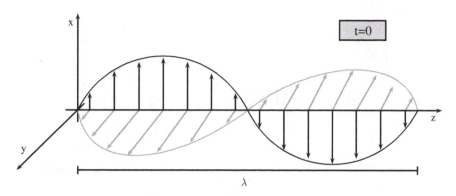

Figure 9.8: Depiction of the sinusoidal electromagnetic wave described in the text. The wave – shown here at $t = 0$ – propagates in the $+z$ direction, meaning that the structure depicted here simply slides to the right, at speed c, as time progresses. The structure involves an electric field, pointing in the x-direction, and also a magnetic field, pointing in the y-direction, both of which vary sinusoidally with position z along the direction of propagation. Note also that the electric and magnetic fields are the same everywhere on planes of constant z: there is, in fact, nothing special about the z-axis. We have just drawn some field vectors (electric in black, magnetic in gray) along the z-axis to illustrate how the fields vary with z at a given t. Finally, note that the sinusoidal-shaped curves, along the tips of the field vectors, do not correspond to anything like a sinusoidal-shaped trajectory through space. There is no such sinusoidal-shaped trajectory involved in these waves.

that is, the flux through – the faces perpendicular to the y- and z-axes will vanish. There will, in general, be a nonzero electric flux through the faces perpendicular to the x-axis, but since the electric field (for a given z) is the same for all x, the inward (negative) electric flux through one of these faces will perfectly cancel the outward (positive) electric flux through the opposite face. So the total flux through the closed surface will vanish.

The same thing is true for the magnetic field, just exchanging the roles of the x- and y-directions. So (ii) –Gauss' Law for magnetism – is also satisfied by our conjectured formulas for $\vec{E}$ and $\vec{B}$.

So far so good. Now let's consider (iii) – Faraday's law. The left hand side is a path integral of $\vec{E}$ around a closed loop. This will simply vanish for any loop lying in a plane perpendicular to the x- or z-axes, as will the right hand side. So the non-trivial question is whether Faraday's law is respected for a loop lying in a plane perpendicular to the y-axis. We can consider, actually without any loss of generality, the rectangular loop shown in Figure 9.9, at $t = 0$, two of whose opposite corners are at the origin and at the point $z = a$, $x = b$. For this loop, we have

$$\oint \vec{E} \cdot d\vec{s} = b E_0 \sin\left[\frac{2\pi a}{\lambda}\right]. \tag{9.27}$$

The right hand side of Faraday's law is $-\frac{d}{dt} \iint \vec{B} \cdot d\vec{A} = -\iint \frac{\partial \vec{B}}{\partial t} \cdot d\vec{A}$. Differentiating Equation (9.26) with respect to time and evaluating the result at $t = 0$ gives

$$\frac{\partial \vec{B}}{\partial t} = \hat{j} B_0 \left(-\frac{2\pi c}{\lambda}\right) \cos\left[\frac{2\pi z}{\lambda}\right]. \tag{9.28}$$

Integrating this over the rectangular surface of width a and height b (and tacking on the minus sign)

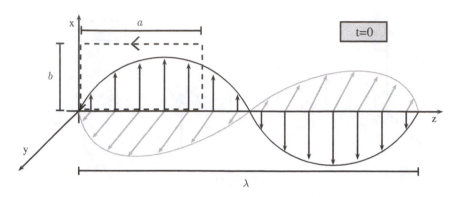

Figure 9.9: *Same as the previous Figure, but now including the rectangular loop (of width a and height b, in the y = 0 plane) that we use to check for consistency between Equations (9.25) and (9.26) and our third Maxwell equation, Faraday's law.*

gives

$$-\iint \frac{\partial \vec{B}}{\partial t} \cdot d\vec{A} = B_0 \left(\frac{2\pi c}{\lambda} \right) b \int_{z=0}^{z=a} \cos\left[\frac{2\pi z}{\lambda} \right] dz$$

$$= b B_0 c \sin\left[\frac{2\pi a}{\lambda} \right]. \tag{9.29}$$

Good. Finally, equating the left and right hand sides of Faraday's law, i.e., Equations (9.27) and (9.29), and cancelling the common factors, we arrive at

$$E_0 = c B_0. \tag{9.30}$$

So evidently Faraday's law is respected so long as this relationship between the maximum amplitudes of the electric and magnetic fields, and the wave propagation speed c, is respected.

That just leaves (iv), the Ampère-Maxwell law. We can proceed in a very similar way, using the rectangular loop shown now in Figure 9.10. On the left hand side, we have

$$\oint \vec{B} \cdot d\vec{s} = b B_0 \sin\left[\frac{2\pi a}{\lambda} \right]. \tag{9.31}$$

The right hand side involves $\frac{d}{dt} \iint \vec{E} \cdot d\vec{A} = \iint \frac{\partial \vec{E}}{\partial t} \cdot d\vec{A}$. Differentiating Equation (9.25) with respect to time and evaluating at $t = 0$ we have

$$\frac{\partial \vec{E}}{\partial t} = \hat{i} E_0 \left(-\frac{2\pi c}{\lambda} \right) \cos\left[\frac{2\pi z}{\lambda} \right]. \tag{9.32}$$

Note that, since we integrated clockwise around the rectangular loop (as shown in Figure 9.10), the $d\vec{A}$ vector in the surface integral is downward, i.e., in the negative x-direction. We then have that

$$\mu_0 \epsilon_0 \iint \frac{\partial \vec{E}}{\partial t} \cdot d\vec{A} = \mu_0 \epsilon_0 E_0 b \left(\frac{2\pi c}{\lambda} \right) \int_{z=0}^{z=a} \cos\left[\frac{2\pi z}{\lambda} \right] dz$$

$$= \mu_0 \epsilon_0 E_0 b c \sin\left[\frac{2\pi a}{\lambda} \right]. \tag{9.33}$$

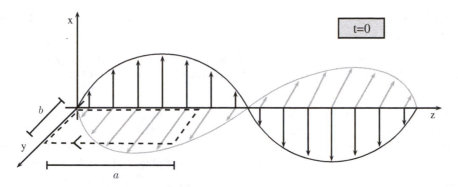

Figure 9.10: Same as the previous Figure, but now showing the rectangular loop (of width a and height b, in the x = 0 plane) that we use to check for consistency between Equations (9.25) and (9.26) and our fourth and final Maxwell equation, the Ampère-Maxwell law.

Equating the left and right hand sides, using the relationship $E_0 = cB_0$ that we arrived at before, and cancelling the common factors gives

$$1 = \mu_0 \epsilon_0 c^2 \tag{9.34}$$

which tells us that the wave propagation speed must be given, in terms of the fundamental electric and magnetic constants, by

$$c = \frac{1}{\sqrt{\mu_0 \epsilon_0}}. \tag{9.35}$$

Plugging in the known values for the constants ($\epsilon_0 = 8.85 \times 10^{-12}$ C^2/Nm^2 and $\mu_0 = 4\pi \times 10^{-7}$ Ns^2/C^2) gives

$$c \approx 3 \times 10^8 \text{ m/s} \tag{9.36}$$

which is an extremely interesting number because it matches the independently measured speed of light.

You probably saw this coming because somebody spoiled the drama by telling you, in some earlier science class, that light *is* an electromagnetic wave. (Heck, I mentioned it in the opening paragraph of this Chapter!) So of course electromagnetic waves propagate at the speed of light.

But imagine the shock of profound discovery Maxwell must have felt when he first recognized that the corrected equations describing electric and magnetic fields implied the existence of propagating waves, with a propagation speed given by the strange-looking formula, Equation (9.35), and plugged in the values, for the first time, and got out a familiar but unanticipated speed. As he wrote, "We can scarcely avoid the inference that *light consists in the transverse undulations of the same medium which is the cause of electric and magnetic phenomena.*"

§ 9.5 Energy Flow

Maxwell had introduced not only the full electric and magnetic field equations, but also the idea, discussed in Chapter 8, that it is possible to locate electric and magnetic energy not in the charges and currents themselves, but rather in the surrounding fields. And he seemed to exhibit a kind of philosophical preference for the latter view. Whereas the idea that energy was concentrated at the charges and currents was, Maxwell wrote, "the natural expression of the theory which supposes the [charges and] currents to act upon each other directly at a distance", the alternative field energy view was the form "appropriate to the theory which endeavours to explain the action between [charges and] currents by means of some intermediate action in the space between them".

But in practice Maxwell didn't seem to take the idea of continuous, local energy conservation all that seriously. This task thus fell to some of Maxwell's followers, in particular John Henry Poynting and Oliver Heaviside.

Poynting stressed explicitly that, if the electromagnetic fields can store energy, "we are naturally led to consider the problem: How does the energy about an electric current pass from point to point – that is, by what paths and according to what law does it travel from the part of the circuit where it is first recognizable as electric and magnetic to the parts where it is changed into heat and other forms?"

Poynting addressed the question by showing (using some vector calculus that is slightly beyond the level of this course) that the rate of change of the total electromagnetic field energy stored within some region can be expressed as the integral of a certain vector $\vec{S}$ over the surface of that region:

$$\frac{d}{dt} \iiint \left[\frac{1}{2}\epsilon_0 E^2 + \frac{1}{2\mu_0} B^2 \right] dV = - \oiint \vec{S} \cdot d\vec{A}. \tag{9.37}$$

What we now call the "Poynting vector", $\vec{S}$, can thus be understood as giving the direction and rate of electromagnetic field energy flow. For example, if the total energy stored in the fields in the region in question is decreasing (making the left hand side negative), that is because there is a net outflow of energy across the boundary, as captured by a positive $\oiint \vec{S} \cdot d\vec{A}$.

Poynting showed that the energy flow vector $\vec{S}$ could be expressed in a beautifully simple way in terms of the electric and magnetic fields:

$$\vec{S} = \frac{1}{\mu_0} \vec{E} \times \vec{B}. \tag{9.38}$$

The cross-product makes perfect sense, at least for the electromagnetic plane wave we considered in the previous section: at places and times where $\vec{E}$ is in the positive x-direction and $\vec{B}$ is in the positive y-direction, the Poynting vector $\vec{S}$ will point, by the right hand rule, in the positive z-direction. And at places/times where $\vec{E}$ is in the negative x-direction and $\vec{B}$ is in the negative y-direction, $\vec{S}$ will still be in the positive z-direction! So for an electromagnetic wave propagating in the positive z-direction, the Poynting vector implies that the energy is flowing, everywhere, in the positive z-direction. This seems reasonable.

To understand the *magnitude* of the Poynting vector, it is helpful to again consider the case of the electromagnetic plane wave. Plugging Equations (9.25) and (9.26) into

$$u = \frac{1}{2}\epsilon_0 E^2 + \frac{1}{2\mu_0} B^2 \tag{9.39}$$

and using the relationships between E_0 and B_0 that we found in the previous section gives

$$u = \epsilon_0 E_0^2 \sin^2 \left[\frac{2\pi(z - ct)}{\lambda} \right]. \tag{9.40}$$

(Interestingly, $E_0 = cB_0$ and $1 = \epsilon_0\mu_0 c^2$ together imply that the electric and magnetic contributions to the total energy density of the electromagnetic wave are exactly the same!)

On the other hand, plugging Equations (9.25) and (9.26) into Equation (9.38) (and again using $E_0 = cB_0$ and $1 = \epsilon_0\mu_0 c^2$) we find

$$\vec{S} = \epsilon_0 c E_0^2 \sin^2 \left[\frac{2\pi(z - ct)}{\lambda} \right] \hat{k}. \tag{9.41}$$

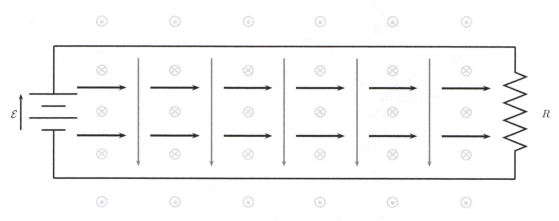

Figure 9.11: An electric circuit in which a battery pushes current through a resistor. The electrical current flowing to the right through the top wire, and flowing to the left through the bottom wire, produce a magnetic field with the structure indicated in light gray. In particular, in the region between the two wires, the magnetic field is into the page, away from us. The electric potential difference maintained between the battery terminals (and hence between the top and bottom wires) implies that there is an electric field from the higher potential wire (on top) toward the lower potential wire (below). This electric field is indicated by the darker gray arrows. The Poynting vector, $\vec{S} = \frac{1}{\mu_0} \vec{E} \times \vec{B}$, therefore (by the right hand rule) points to the right, in the region between the two wires, as indicated by the solid black arrows. Thus, the energy is transported, from the battery to the resistor (where it is dissipated as heat), not along the wires themselves (as one might have naively expected), but rather by the fields in the otherwise-empty space surrounding the wires.

The magnitude of $\vec{S}$ can thus be understood as the energy density, u, multiplied by the speed – here c – with which the energy is flowing (in the direction of $\vec{S}$). The Poynting vector, that is, has units of energy per unit volume times speed – or, equivalently, energy per unit area per unit time. That is, of course, perfectly consistent with our starting point: the integral of the Poynting vector over a two-dimensional surface tells us the rate (in Joules per second or some equivalent unit) at which electromagnetic field energy is passing through the surface.

Poynting's formula thus helps us understand how an electromagnetic wave conveys field energy from place to place. But the formula can shed light on other situations as well. For example, consider a simple electric circuit in which a battery pushes electrical current through a resistor. We know, in principle, that the chemical reactions in the battery produce some electrical energy which is eventually dissipated, as heat, at the resistor. But how does the energy get from the battery to the resistor?

To answer using the Poynting vector, we must first determine what the electric and magnetic fields will look like in the vicinity of the circuit. This is sketched in Figure 9.11. The top wire is maintained at a relatively high electrical potential due to its connection to the positive battery terminal, whereas the electric potential along the bottom wire is lower. So there is an electric field in the region between the wires. In the Figure this electric field points downward, as indicated by the dark gray arrows. The electric current flowing to the right in the top wire, and the return current flowing to the left in the bottom wire, combine to produce a magnetic field pointing into the page in the region between the wires.

And so, in that region between the wires, the Poynting vector, Equation (9.38), is to the right, indicating that electromagnetic field energy is flowing to the right, from the general vicinity of the battery to the general vicinity of the resistor. The interesting implication is that the energy is not transported from the battery to the resistor along the wires, as one might naively expect on the basis of a bias toward material objects which can actually be seen and touched. In fact, though, it seems that the energy is transported from the battery to the resistor by the (invisible) electric and magnetic fields, in the otherwise empty space between the wires.

In class you'll have a chance to work through a similar example problem, in which the geometry allows us to be a little bit more quantitative and precise and to show not just that the Poynting vector indicates energy flow in the right direction, but, moreover, that the integral of the Poynting vector, across a surface perpendicular to the flow, captures the entirety of the energy which is being dissipated by the resistor. We can also apply the idea to the case of the oscillating LC circuit that was raised at the end of Chapter 8.

For now, though, let us close this section with a short poem, written by Oliver Heaviside, who also worked on electromagnetic energy conservation and indeed independently discovered Poynting's energy flow vector less than a year after Poynting. The couplet captures the essential lesson of this section about as well as anything possibly could:

> "When energy goes from place to place,
> it traverses the intermediate space."

§ 9.6 Producing and Detecting Radiation

We have seen that electromagnetic waves are compatible with Maxwell's equations and we have just been discussing how to describe the transport of electromagnetic field energy (by electromagnetic waves, and also in other situations) using the Poynting vector. But how does one *produce* electromagnetic radiation in the first place?

The basic idea was inherent in our initial qualitative description of the waves: a changing electric field can produce a magnetic field which can produce an electric field, and so on. So, really, to create a propagating electromagnetic wave with a certain frequency, one just needs to produce an oscillating electric field. The simplest such device – that is, the simplest type of electromagnetic broadcast *antenna* – is simply a source of alternating current (AC) connected to a capacitor or, even more simply, a broken wire, as shown in Figure 9.12.

The top end of the antenna (i.e., the upper black dot in the Figure) will have a sinusoidally oscillating electric charge

$$Q_{upper}(t) = Q_0 \sin(\omega t) \tag{9.42}$$

while the bottom end will have, at each moment, the opposite charge:

$$Q_{lower}(t) = -Q_0 \sin(\omega t). \tag{9.43}$$

The antenna thus constitutes an electric dipole, whose dipole moment $p = Qd$ (where d is the length of the antenna, i.e., the distance between the two dots in the Figure) oscillates sinusoidally in time: $p(t) = p_0 \sin(\omega t)$ where the amplitude $p_0 = Q_0 d$.

A detailed analysis of the radiation emerging from such an oscillating electric dipole is a little beyond what we have space and time for here. But there is a simple formula, due to Larmor, for the total rate at which energy is converted into electromagnetic radiation, which we can derive using the technique of dimensional (i.e., unit) analysis.

Here's how that works. Let's assume that the total power P radiated by the electric dipole antenna depends on the amplitude p_0 of the dipole oscillations, the frequency ω, and the fundamental electromagnetic constants ϵ_0 and c. And let us furthermore assume that the formula for P involves nothing

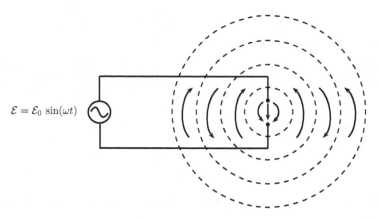

Figure 9.12: The simplest type of electromagnetic broadcast antenna is really just a broken wire connected to a source of alternating current. As the power source pushes electric current alternately clockwise and counterclockwise, the top wire end (represented by the upper black dot) becomes charged alternately positive and negative. The bottom wire end (represented by the lower black dot) has, at each moment, a charge that is opposite that of the upper wire end. So the two black dots constitute an electric dipole with a sinusoidally-oscillating electric dipole moment. This produces an electric field with a sinusoidally oscillating character, which field propagates outward, with greatest intensity around the "equator" (i.e., in the plane perpendicular to the electric dipole antenna).

more complicated than multiplying each of these quantities raised to some appropriate power. Thus, the formula should have the structure

$$P = k\, p_0^\alpha\, \omega^\beta\, \epsilon_0^\gamma\, c^\delta \tag{9.44}$$

where k is a pure-number (i.e., dimensionless) proportionality constant, and α, β, γ, and δ are as-yet unknown powers.

The crucial point is that we can determine α, β, γ, and δ by requiring that the formula has sensible units, i.e., is dimensionally coherent. The left hand side is supposed to represent a power, i.e., an energy per unit time. Thus

$$[P] = \frac{\text{Joules}}{\text{second}} = \frac{\text{kg}\,\text{m}^2}{\text{s}^3}. \tag{9.45}$$

(The square brackets here mean: "the units of".) Similarly, we have that

$$[p_0] = \text{C}\,\text{m}, \tag{9.46}$$

$$[\omega] = \frac{1}{\text{s}}, \tag{9.47}$$

$$[\epsilon_0] = \frac{\text{C}^2\,\text{s}^2}{\text{kg}\,\text{m}^3}, \tag{9.48}$$

and, of course,

$$[c] = \frac{\text{m}}{\text{s}}. \tag{9.49}$$

And so the dimensional coherence of Equation (9.44) requires that

$$\frac{\text{kg}\,\text{m}^2}{\text{s}} = (\text{C}\,\text{m})^\alpha \left(\frac{1}{\text{s}}\right)^\beta \left(\frac{\text{C}^2\,\text{s}^2}{\text{kg}\,\text{m}^3}\right)^\gamma \left(\frac{\text{m}}{\text{s}}\right)^\delta. \tag{9.50}$$

Requiring that the powers of each of our four basic units (C, m, s, and kg) match on both sides gives four equations in the four unknowns (α, β, γ, and δ).

But if we approach things in the right order, we can avoid a messy algebra problem. For example, since "kg" only appears on the right in one place, we can see immediately that we need $\gamma = -1$. But then the units of ϵ_0 to the -1 power give two factors of "C" in the denominator on the right. These must cancel with two "C"s in the numerator, since there are no "C"s on the left. And so evidently we need $\alpha = 2$. Plugging those in and simplifying what remains, we are left with

$$\frac{\text{kg}\,\text{m}^2}{\text{s}^3} = \frac{\text{kg}\,\text{m}^5}{\text{s}^2} \left(\frac{1}{\text{s}}\right)^\beta \left(\frac{\text{m}}{\text{s}}\right)^\delta. \tag{9.51}$$

From here it is clear that, in order to get the "m"s right, we need $\delta = -3$. And then, finally, to get the "s"s right, we need $\beta = 4$.

Putting that all together, we are left with

$$P = k\frac{p_0^2\,\omega^4}{\epsilon_0\,c^3}. \tag{9.52}$$

Unfortunately, we can never determine the dimensionless constant k using units.

A more detailed analysis gives the precise Larmor formula to be

$$P = \frac{1}{6\pi}\frac{p_0^2\,\omega^4}{\epsilon_0\,c^3} \tag{9.53}$$

so, evidently, $k = 1/6\pi$. Still, it is extremely cool that we can determine the overall structure of the formula – including, for example, the fact that the radiated power grows as the 4th power of the frequency – just from common sense and units.

(In class you can use this same technique to determine the rate of *gravitational* radiation from binary systems such as inspiraling black hole pairs and we will use the results to understand the recent detection of gravitational radiation by LIGO.)

Maxwell's consolidation and completion of electromagnetic theory – including his demonstration that the theory supported the existence of electromagnetic waves – occurred in the 1860s, while he was still a young man. But as is typical of important theoretical advances, Maxwell's ideas were not universally adopted right away. The obstacles included the complexity and obscurity of Maxwell's mathematical expression (basically, modern vector notation had not yet been invented, so Maxwell's own versions of "Maxwell's equations" were far more complicated than the four only-somewhat-intimidating vector equations we have been writing) and also lingering controversy over the status of the "displacement current" (and in particular whether it required a polarizable material substrate or was the genuinely novel field phenomenon we have described).

In addition, Maxwell's prediction of electromagnetic waves (and his identification of them with light) was regarded as somewhat speculative and difficult to accept. Unfortunately, Maxwell died of cancer in 1879, at the age of only 48, before the existence of electromagnetic waves had been experimentally demonstrated. But that experimental proof did come, about a decade after Maxwell's death.

The proof was provided by the German physicist Heinrich Hertz, a student (and then post-doc) under Hermann von Helmholtz. Hertz used a dipole antenna like the one discussed just above, powered by a variant of the LC circuit we discussed earlier, oscillating with a frequency of about one hundred million cycles per second, i.e., 100 MHz. Maxwell's theory of course predicted that this rapidly oscillating electric dipole would generate outgoing electromagnetic waves, and Hertz found that he could detect those waves using a simple coil of wire with a very narrow gap between the two ends; the passing electromagnetic wave would induce a large EMF across the narrow gap and cause a visible and audible spark.

To establish that the electromagnetic waves emitted by the dipole antenna had the detailed properties predicted by Maxwell's theory, Hertz placed a large metal wall on the other side of the room. This wall would reflect the electromagnetic waves that were incident on it so that, in the region between the antenna and the wall, electromagnetic waves of the same 100 MHz frequency were propagating in two opposite directions, producing the phenomenon of "standing waves" in which the wave intensity varies sinusoidally with position. By measuring the distance between adjacent "nodes" in this standing wave pattern – that is, the distance between points of zero wave intensity where the detector failed to generate sparks – Hertz was able to determine the wavelength of the waves being emitted by the antenna.

But the frequency f of a wave multiplied by its wavelength λ gives the propagation speed c. So Hertz was able to show that the electromagnetic waves emitted by his dipole antenna propagated at precisely the speed predicted by Maxwell's theory – i.e., the speed of light. In a series of follow-up experiments, Hertz would go on to demonstrate that the electromagnetic waves also exhibited a number of other behaviors characteristic of light, for example refraction and polarization.

The waves that Hertz generated and studied lie in what we now describe as the "radio wave" part of the electromagnetic spectrum. He was therefore, in some sense, the first to intentionally broadcast and receive radio signals. But Hertz was surprisingly oblivious to the revolutionary possibilities afforded by his work. Asked by a student about possible practical applications, he replied:

> "It's of no use whatsoever. This is just an experiment that proves Maestro Maxwell was right – we just have these mysterious electromagnetic waves that we cannot see with the naked eye. But they are there."

Still, although he was obviously quite wrong about electromagnetic radiation being "of no use", Hertz was correct that his work showed clearly that "Maxwell was right". In the wake of Hertz's experimental demonstrations, Maxwell's theory took its place, alongside Newtonian mechanics and gravitational theory, as one of the great pillars of pre-20th-century physics.

Questions:

Q1. If a wire loop is moved toward a stationary magnet, as depicted on the left of Figure 9.2, the EMF which drives the induced current is associated with magnetic forces. On the other hand, if a magnet is moved toward a stationary wire loop, as depicted on the right of Figure 9.2, the EMF which drives the induced current is associated with electric forces. What if the wire loop and the magnet are both moving (say, with equal speeds) toward one another? Is the EMF in that case electric or magnetic? Both? Neither?

Q2. Figure 9.4 depicts the clockwise-swirling electric field that would be produced by a magnetic field that points toward us and is increasing in magnitude. What direction would the electric field swirl in if the magnetic field were pointing toward us but instead *decreasing* in magnitude? What if the magnetic field were pointing *away* from us and increasing in magnitude?

Q3. Our derivation of Equation (9.14) assumed that the secondary coil, whose loops have area A_2, was *inside* the primary coil (so $A_2 < A_1$). How would the transformer equation need to change if the secondary coil is outside of – and hence bigger than – the primary coil?

Q4. Explain why it would have been impossible to recognize the possibility of electromagnetic waves without first identifying the previously-missing "displacement current" term in Ampère's law.

Q5. An electric charge Q is located next to (or if you prefer, for symmetry reasons, is embedded in the center of) a permanent bar magnet. Sketch the electric and magnetic fields in the vicinity, and describe the flow (if any) of electromagnetic field energy.

Q6. Consider again the situation depicted in Figure 9.4: a magnetic field that points toward us and has an increasing magnitude will induce a clockwise-swirling electric field. What direction is the Poynting vector in? Explain why a flow of field energy in that direction makes sense.

Q7. In Chapter 7, Exercise E8 you considered a situation in which the magnetic forces that two current elements (which could also just be two moving charged particles) exert on one another, do not obey Newton's third law. As you probably realized then, this implies a failure of momentum conservation for the (say) two-particle system! Explain qualitatively, using ideas from this new chapter, how the idea of momentum conservation can be saved.

Exercises:

E1. A parallel plate capacitor consisting of two radius-R circular plates separated by a distance h has a steadily-increasing charge $Q = I t$. Write down an expression for the magnitude $E(t)$ of the electric field between the plates at time t, then use the Ampère-Maxwell law to determine the magnitude $B(t)$ of the magnetic field around the perimeter of the gap between the plates. What are the direction and magnitude of the Poynting vector at the perimeter of the gap? Show that the total rate at which energy is flowing inward through this perimeter matches the rate at which the electric field energy stored in the capacitor is increasing.

E2. The primary coil of a transformer is a 15 cm long, 3 cm diameter cylindrical solenoid consisting of 500 circular loops of wire. The secondary coil has the same length, but contains only 100 circular loops with a 2 cm diameter. If the primary coil is connected to a standard electrical outlet (which produces a sinusoidally varying potential difference of amplitude 170 Volts and frequency 60 Hz), what will be the amplitude and frequency of the sinusoidally varying voltage at the output of the secondary coil?

E3. Use the technique of dimensional analysis to derive a formula for the period T of a simple pendulum, in terms of the mass m of the bob, the length L of the string, and the strength g of the gravitational field.

E4. Equation (9.53) gives the rate at which energy, in the form of electromagnetic radiation, is emitted by a simple dipole antenna. A closely-related expression (more commonly called "the Larmor formula") gives the rate of energy emission from a single particle of charge Q in terms of its acceleration a. Use dimensional analysis to derive this more standard version of the Larmor formula (up to a dimensionless constant).

E5. You have previously seen (perhaps in Chapter 8, Exercise E11) the formula

$$v_{esc} = \sqrt{\frac{2GM}{R}} \tag{9.54}$$

for the escape velocity from the surface of a body of mass M and radius R. Note that if M is big and R is small, v_{esc} can equal or exceed the speed of light – i.e., even light is not moving fast enough to escape the surface. Such an object is called a "black hole" and the surface (which must be fully outside of the object) at which the escape velocity is equal to c is called the "event horizon". Basically, since nothing can go faster than the speed of light, nothing can emerge, to the outside, from inside the event horizon. Calculate the radius of the event horizon of a one-solar-mass black hole. It is illuminating to compare to the radius of a one-solar-mass neutron star, which is about 12 km.

E6. In Chapter 17 of Volume II of his famous Lectures, Richard Feynman poses the following paradox:

"Imagine that we construct a device like that shown in [the figure below]. There is a thin, circular plastic disc supported on a concentric shaft with excellent bearings, so that it is quite free to rotate. On the disc is a coil of wire in the form of a short solenoid concentric with the axis of rotation. This solenoid carries a steady current I provided by a small battery, also mounted on the disc. Near the edge of the disc and spaced uniformly around its circumference are a number of small metal spheres insulated from each other and from the solenoid by the plastic material of the disc. Each of these small conducting spheres is charged with the same electrostatic charge Q. Everything is quite stationary, and the disc is at rest. Suppose now that by some accident – or by prearrangement – the current in the solenoid is interrupted, without, however, any intervention from the outside. So long as the current continued, there was a magnetic flux through the solenoid more or less parallel to the axis of the disc. When the current is interrupted, this flux must go to zero. There will, therefore, be an electric field induced which will circulate around in circles centered at the axis. The charged spheres on the perimeter of the disc will all experience an electric field tangential to the perimeter of the disc. This electric force is in the same sense for all the charges and so will result in a net torque on the disc. From these arguments we would expect that as the current in the solenoid disappears, the disc would begin to rotate. If we knew the moment of inertia of the disc, the current in the solenoid, and the charges on the small spheres, we could compute the resulting angular velocity.

"But we could also make a different argument. Using the principle of the conservation of angular momentum, we could say that the angular momentum of the disc with all its equipment is initially zero, and so the angular momentum of the assembly should remain zero. There should be no rotation when the current is stopped. Which argument is correct? Will the disc rotate or will it not?"

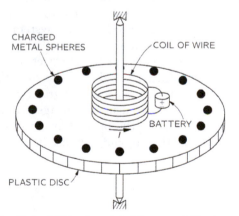

Fig. 17–5. Will the disc rotate if the current I is stopped?

What do you think? Will the disk rotate when the current turns off? And if so, doesn't this violate the principle of angular momentum conservation?

Projects:

P1. Use a function generator and oscilloscope to explore the ratio of input/output voltage amplitudes for a transformer consisting of two cylindrical solenoids, as various factors are varied: the frequency of the (sinusoidal) input voltage, the presence or absense of an "iron core", and whether the bigger/outer solenoid is used as the primary or the secondary.

P2. A battery that produces EMF $\mathcal{E}$ is connected to a resistor of resistance R by a long straight perfectly-conducting co-axial cable. The positive terminal of the battery is connected to the inner conductor of the cable, which is a solid cylinder of radius a. The negative battery terminal is connected to the outer conductor, which is a thin hollow cylinder of radius b. (We can assume for simplicity, even though it is not very realistic, that there is no material – just empty space – between the inner and outer conductors of the co-axial cable. (a) Use Ampère's law to find the magnetic field $B(r)$ in the gap between the conductors. (b) Find the electric field $E(r)$ in the gap. (c) Find the magnitude and direction of the Poynting vector in the gap. (d) Integrate the Poynting vector over the annular region $a < r < b$ to find the total rate at which energy is flowing down the cable.

P3. Reproduce Hertz's demonstration of (microwave frequency) standing waves, and determine the propagation speed of the waves by observing the spacing between nodes and anti-nodes.

P4. Develop the theory called "gravito-electromagnetism", which is a full gravitational analog to Maxwell's electromagnetic theory. We have already developed the first fundamental equation of this theory, namely "Gauss' Law for Gravity". See if you can write down a full set of four "Maxwell equations" for this gravitational theory based on the assumption that gravity basically works like electromagnetism (including supporting gravitational waves that propagate at the speed of light), and try to come up with at least one concrete (if qualitative) prediction for a surprising effect involving the gravitational analog of magnetism. (As it turns out, this theory, although perfectly relativistic, is not the correct theory of gravity. But it is extremely similar to the correct theory, namely Einstein's general relativity, in a certain limit, and its predictions for "gravito-magnetic" effects are qualitatively correct.)

P5. In an astrophysical binary system, the two objects (stars, black holes, etc.) orbit about their mutual center of mass and the (gravitational) dipole moment remains zero by virtue of the definition of the center of mass. Such a system, however, does have a sinusoidally-varying *quadrupole* moment. The quadrupole moment is a quantity, closely related to the moment of inertia, with units $\mathrm{kg\,m^2}$. And the varying quadrupole moment allows the binary system to emit gravitational radiation. Use dimensional analysis to develop a formula for the rate at which such a system radiates energy, and use this (as well as some orbital mechanics) to construct a differential equation describing how the frequency of the emitted gravitational radiation varies with time as the two bodies spiral in toward one another. Use data from the recent LIGO gravitational radiation detection to estimate the masses of the two black holes involved in that merger, and some other interesting properties of the event. (This Project is based on the paper "An Analysis of the LIGO discovery based on Introductory Physics" by Mathur, Brown, and Lowenstein.)

Part III

The Atomic Theory

CHAPTER 10

Atoms and Gases

T HE broad topic of our third and final unit is *the atomic theory of matter*. In this Chapter, we'll focus on two threads that provided early evidence for the idea that matter is composed of atoms – one thread has to do with the physical properties of *gases*, and the other has to do with certain quantitative regularities that were observed to hold in *chemical reactions*. As we will see, the two threads become entwined by observations pertaining to the role of specifically gaseous reactants and products in chemical reactions.

§ 10.1 Properties of Gases

You might recall that the Ancient Greeks posited four basic material elements: Earth, Water, Air, and Fire. We of course no longer recognize these four substances as, in any meaningful sense, elemental. But we might interpret the Greek taxonomy as corresponding roughly to what we now describe as the different *phases* of matter: solid, liquid, and gas (and plasma?).

Solids and liquids are tangible and hence relatively easy to study. So it is not surprising that science had moved rather quickly beyond the Ancient Greek taxonomy on those fronts: by the time of Newton, for example, it was clearly recognized that there were many chemically distinct solid and liquid substances.

But air – being invisible and considerably less tangible – was harder to study and was indeed still regarded as a single monolithic substance around the time of Newton. Let us briefly review some of the important developments that provided insight into the physical properties of air (and, eventually, its chemically distinct components).

10.1.1   The concept of Pressure

The pump – a mechanical device used to move water or another fluid – was one of the earliest and most important practical inventions. Pumps were used already in ancient times to, for example, bring water from wells to the surface.

Let's consider the simplest type of pump, which is essentially just a modern syringe: see Figure 10.1. If the open end is placed in some water, say, and the moveable piston is pulled back, water will rush into the cylinder. (Note that this is basically what happens when you drink through a straw, with your flexing diaphragm muscle playing the role of the moving piston.)

Why does the fluid rush in to fill the expanding cavity? The traditional answer – which is still a commonly-heard saying – was that "nature abhors a vacuum". That is, if the piston receded and the fluid didn't flow in, a void – a little pocket of empty nothingness would be created. And, for reasons that were never

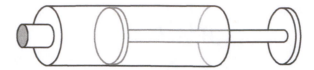

Figure 10.1: A simple pump, consisting of a hollow cylinder with an opening on one end and an airtight piston (which can be slid back and forth through the cylinder) on the other.

articulated too carefully (perhaps something along the lines of "There's no such thing as nothing!"?) that was regarded as physically impossible. In effect, the fluid was *pulled* into the cavity by the threat of something forbidden (namely the formation of a vacuum) at the location of the receding piston.

But this way of thinking about a pump's ability to lift water was difficult to reconcile with an empirical fact which had probably been known about much earlier, but which was first recorded in the scientific literature by Galileo:

> "I once saw a cistern which had been provided with a pump under the mistaken impression that the water might thus be drawn with less effort or in greater quantity than by means of the ordinary bucket... This pump worked perfectly so long as the water in the cistern stood above a certain level; but below this level the pump failed to work. When I first noticed this phenomenon I thought the machine was out of order; but the workman whom I called in to repair it told me the defect was not in the pump but in the water, which had fallen too low to be raised through such a height; and he added that it was not possible, either by a pump or by any other machine working on the principle of attraction, to lift water a hair's breadth above eighteen cubits [i.e., about 34 feet]."

This observation is illustrated schematically, in terms of our simple syringe-type pump, in Figure 10.2. The idea is just that, if one were to use a (very long!) syringe to try to pull a column of water up out of a reservoir, it would work perfectly until the height of the column reached 34 feet above the surface of the reservoir. But then any further upward motion of the piston would fail to lift the water column even "a hair's breadth" further.

What is left in the space between the top of the 34-foot-high column of water and the receding piston? Evidently nothing – i.e., a vacuum. But then it can't be true that "nature abhors a vacuum", and can't be the avoidance of a vacuum that pulled the water into the cylinder before its height reached 34 feet. Indeed, the phenomenon suggests that it is not accurate to think of the water as being pulled in at all – for nothing about the water-piston interface (where such a pulling force would have to originate) changes as the column height passes through 34 feet. Instead, the phenomenon suggests that the water is *pushed* into the cylinder, from the outside, with a certain finite force (which is sufficient to raise its level only by a certain finite amount). The idea, in particular, that this suggests is that the air outside has a finite *pressure*, and is hence pushing on everything it touches – including the water at the base of the column – with a certain associated force.

The first clean laboratory demonstration of this idea was produced by the 17th century Italian scientist Evangelista Torricelli. Torricelli's experiment consisted of filling a glass tube, sealed at one end, with Mercury, and then inverting the filled tube into a dish of Mercury. The result was that the Mercury level in the tube dropped to a height of 30 inches above the level in the dish. That is, without any air bubbling up into the top of the tube, a configuration like that pictured in the third frame of Figure 10.2 was reached, the only difference being that the height of the Mercury column that could be supported by the outside atmospheric pressure was only 30 inches – about 14 times shorter than the corresponding 34 foot high column of water.

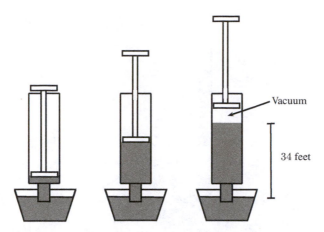

Figure 10.2: One can raise the water in the cylinder by pulling back on the piston – but only until the water level reaches 34 feet above the level in the dish.

This ratio is significant because Mercury is 14 times more dense than water. The 30 inch Mercury column is thus clearly demonstrating the same expansive power of the atmosphere that was demonstrated by the 34 foot column of water. Note also that Torricelli's simple apparatus is also the first *barometer*, i.e., the first device for *measuring pressure*. Indeed, still today, scientists measure pressure in Torricellean units – referring, for example, to atmospheric pressure P_{atm} in terms of the height of a column of Mercury that the pressure in the atmosphere can support:

$$P_{\text{atm}} = 30 \text{ inches of Mercury} = 76 \text{ cm of Mercury}. \tag{10.1}$$

Not long after Torricelli's laboratory demonstration of (and, really, measurement of) atmospheric pressure, the French scientist Blaise Pascal reasoned that atmospheric pressure should decrease with altitude, just as the water pressure increases dramatically as one dives to deeper depths, and just as the people on the bottom level of a human pyramid are more squeezed, i.e., simultaneously suffering and exerting more force than those above. In all these cases, each "layer" of material (be it air or water or people) supports the weight of all the layers above it, and so must possess the corresponding pressure.

Pascal thus sent, in 1648, an assistant to climb a nearby mountain and test (using a Torricellean barometer) whether the pressure did indeed decrease somewhat with increasing altitude. Here is the assistant's self-explanatory report:

"The weather was chancy last Saturday...[but] around five o'clock that morning ... the Puy-de-Dome was visible ... so I decided to give it a try. Several important people of the city of Clermont had asked me to let them know when I would make the ascent... I was delighted to have them with me in this great work.... [A]t eight o'clock we met in the gardens of the Minim Fathers, which has the lowest elevation in town.... First I poured sixteen pounds of quicksilver...into a vessel... then took several glass tubes...each four feet long and hermetically sealed at one end and opened at the other...then placed them in the vessel [of quicksilver]...I found the quick silver stood at 26 [inches] and 3 $\frac{1}{2}$ lines above the quicksilver in the vessel... I repeated the experiment two more times while standing in the same spot...[and] produced the same result each time... I attached one of the tubes to the vessel and marked the height of the quicksilver and...asked Father Chastin, one of the Minim Brothers...to watch if any changes should occur through the day... Taking the other tube and a portion of the quick silver...I walked to the top of Puy-de-Dome, about 500 fathoms higher than the monastery, where upon experiment...found that the quicksilver reached a height of only 23 [inches] and 2 lines... I

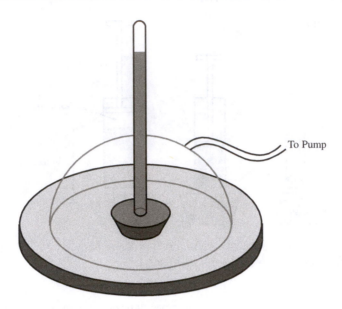

To Pump

Figure 10.3: Boyle's apparatus to demonstrate explicitly that the pressure in the surrounding atmosphere is what pushes the column of Mercury up into the inverted Torricellean tube.

repeated the experiment five times with care...each at different points on the summit...found the same height of quicksilver...in each case..."

An even more dramatic experiment in a similar vein was carried out by the English scientist Robert Boyle – a contemporary of Newton's – in about 1660. Boyle created a large chamber (basically an upside-down glass bowl, sealed around the edges) from which the air could be pumped out, and put a Torricellian barometer inside it. The apparatus is sketched in Figure 10.3. Here is Boyle's description of the experiment and its results:

"Proceed we now to the mention of that experiment, whereof the satisfactory trial was the principal fruit I promised myself from our engine [i.e., the air pump he had instructed his assistant, Robert Hooke, to build after hearing about a similar device invented by von Guericke], it being then sufficiently known, that in the experiment *de vacuo*, the quicksilver [i.e., Mercury] in the tube is wont to remain elevated, above the surface of that whereon it leans, about [30 inches]. I considered, that, if the true and only reason why the quicksilver falls no lower, be, that at that altitude the mercurial cylinder in the tube is in an aequilibrium with the cylinder of air supposed to reach from the adjacent mercury to the top of the atmosphere; then if this experiment could be tried out of the atmosphere, the quicksilver in the tube would fall down to a level with that in the vessel, since then there would be no pressure upon the subjacent, to resist the weight of the incumbent mercury. Whence I inferred (as easily I might) that if the experiment could be tried in our engine, the quicksilver would subside below [34 inches], in proportion to the exsuction of air, that should be made out of the receiver. For, as when the air is shut into the receiver, it doth (according to what hath above been taught) continue there as strongly compressed, as it did whilst all the incumbent cylinder of the atmosphere leaned immediately upon it; because the glass, wherein it is penned up, hinders it to deliver itself, by an expansion of its parts, from the pressure wherewith it was shut up. So if we could perfectly draw the air out of the receiver, it would conduce as well to our purpose, as if we were allowed to try the experiment beyond the atmosphere. ...."

"All things being thus in a readiness, the sucker was drawn down; and, immediately upon the egress of a cylinder of air out of the receiver, the quicksilver in the tube did, according to expectation, subside: and notice being carefully taken (by a mark fastened to the outside) of the place where it stopt, we caused him that managed the pump to pump again, and marked how low the quicksilver fell at the second exsuction; but continuing this work, we were quickly hindered from accurately marking the stages made by the mercury, in its descent, because it soon sunk below the top of the receiver, so that we could henceforward mark it no other ways than by the eye. And thus, continuing the labour of pumping for about a quarter of an hour, we found ourselves unable to bring the quicksilver in the tube totally to subside; because, when the receiver was considerably emptied of its air, and consequently that little that remained grown unable to resist the irruption of the external, that air would (in spight of whatever we could do) press in at some little avenue or other; and though much could not get in, yet a little was sufficient to counterbalance the pressure of so small a cylinder of quicksilver, as then remained in the tube."

Boyle then allowed air from the atmosphere to flow back into the sealed vessel:

"Now (to satisfy ourselves farther, that the falling of the quicksilver in the tube to a determinate height, proceedeth from the equilibrium, wherein it is at that height with the external air, the one gravitating, the other pressing with equal force upon the subjacent mercury) we returned the key [i.e., opened the valve] and let in some new air; upon which the mercury immediately began to ascend (or rather to be impelled upwards) in the tube, and continued ascending, till, having returned the key, it immediately rested at the height which it had then attained: and so, by turning and returning the key, we did several times at pleasure impel it upwards, and check its ascent. And lastly, having given a free egress at the stop-cock to as much of the external air as would come in, the quicksilver was impelled up almost to its first height: I say almost, because it stopt near a quarter of an inch beneath the paper-mark formerly mentioned; which we ascribed to this, that there was (as is usual in this experiment) some little particles of air engaged among those of the quicksilver; which particles, upon the descent of the quicksilver, did manifestly rise up in bubbles towards the top of the tube, and by their pressure, as well as by lessening the cylinder by as much room as they formerly took up in it, hindered the quicksilver from regaining its first height."

Boyle then reversed the pump and pumped air *into* the vessel, so as to produce a pressure inside even greater than the outside atmospheric pressure:

"Lastly, we also observed, that if (when the mercury in the tube had been drawn down, and by an ingress permitted to the external air, impelled up again to its former height) there were some more air thrust up by the help of the pump into the receiver, the quicksilver in the tube would ascend much above the wonted height of [30 inches], and immediately upon the letting out of that air would fall again to the height it rested at before."

10.1.2   The spring of air

In addition to the above experiments which helped to establish and clarify the pressure concept, Boyle designed an apparatus which allowed him to measure quantitatively how the pressure of a given sample of air varied with density (or equivalently, volume). The apparatus is sketched in Figure 10.4. It is a J-shaped tube of glass, sealed on the short end and open to the atmosphere on the long end. A little bit of air is initially trapped in the closed end by some Mercury which is poured in through the open end. The clever thing is that we can "read off" the pressure of the trapped air by comparing the Mercury levels on the two sides.

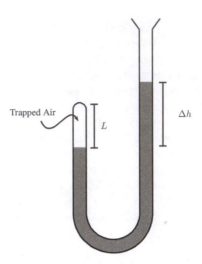

Figure 10.4: Boyle's apparatus for measuring the "spring" of air – i.e., determining the quantitative relation between pressure and volume now known as Boyle's Law.

Here is how that works. We define pressure, formally, as the force-per-unit-area that a fluid (gas or liquid) exerts on whatever surfaces it is in contact with (and perpendicular to those surfaces):

$$P = \frac{F}{A}. \tag{10.2}$$

Think, for example, about a glass of water. Suppose the column of water in the glass has a height h and a cross-sectional area A and is open to the atmosphere at the top. Thus, the top surface (of area A) of the water, which is in contact with the atmosphere, is subject to a downward force (exerted by the air it is touching) with magnitude $F = P_{atm}A$. The water itself is also subject to a downward gravitational force of magnitude $mg = \rho Ahg$, where ρ is the density of water. What prevents the water in the glass from accelerating downward in response to these two forces? There is also an upward force, exerted, let's say, by the bottom layer of water on the rest of the column. Equilibrium requires that this upward force have a magnitude equal to $P_{atm}A + \rho Ahg$, i.e., equilibrium requires that the water pressure at the bottom of the glass – i.e., at a depth h below the surface – be

$$P_h = P_{atm} + \rho g h. \tag{10.3}$$

That is a convenient formula for how the pressure in a fluid varies with depth below the surface, and also a nice illustration of the method of using Equation (10.2) and the Newtonian concept of equilibrium (i.e., zero net force) to learn about the pressure at different points in a fluid.

Boyle used this same technique to determine the pressure of the sample of trapped air in his apparatus, Figure 10.4. First, the pressure P_{air} of the trapped air must match the pressure of the Mercury that it is directly in contact with, i.e., the top layer of the Mercury on the left in the Figure. If these two pressures weren't equal, the top layer of Mercury would experience a net force up or down and would hence start moving! Second, the pressure of the Mercury at that same level, but on the right, must be the same, i.e., also equal to P_{air}. If this weren't the case, the U-shaped portion of Mercury (i.e., all of the Mercury except the portion whose length is Δh in the Figure) would be pushed downward harder on one side than the other, and would hence move. But, third and finally, the Mercury on the right that is at the same level as the top of the Mercury on the left is also a depth Δh below the top level on the right, which is in contact with the atmosphere. So its pressure should be $P_{atm} + \rho g \Delta h$, where now ρ is the density of

Mercury. Thus, the pressure of the trapped air sample can be read off from the difference, Δh, in the heights of the Mercury columns on the two sides:

$$P_{air} = P_{atm} + \rho g \Delta h \tag{10.4}$$

The height difference Δh, and hence P_{air}, could be systematically varied by simply pouring more Mercury into the open tube on the right. And of course the volume V of the trapped air was simply proportional to the length L of the visible cylinder of trapped air.

Boyle found that the pressure of the trapped air was *inversely proportional* to the length of the column, i.e., to the air's volume:

$$P \sim \frac{1}{V} \tag{10.5}$$

or equivalently

$$PV = \text{constant.} \tag{10.6}$$

This relationship between pressure and volume is now known as "Boyle's Law".

Boyle conceived of this as a measurement of the air's "spring" (i.e., springiness) and summarized his results as follows:

> "It is evident, that as common air, when reduced to half its wonted extent, obtained near about twice as forcible a spring as it had before; so this thus comprest air being further thrust into half this narrow room, obtained thereby a spring about as strong again as that it last had, and consequently four times as strong as that of the common air. And there is no cause to doubt, that if we had been here furnished with a greater quantity of quicksilver and a very strong tube, we might, by a further compression of the included air, have made it counterbalance the pressure of a far taller and heavier cylinder of mercury. For no man perhaps yet knows, how near to an infinite compression the air may be capable of, if the compresisng force be competently increased."

10.1.3   Thermal Expansion

The concept of "temperature" denotes the quantitative axis along which the hotness and coldness of objects can vary continuously. The earliest thermometers – devices for measuring temperature – relied on the phenomenon of thermal expansion: most substances, when heated, increase somewhat in volume.

The simplest thermometer is simply an inverted tube, with some trapped air, not unlike Torricelli's barometer. See Figure 10.5. When the temperature increases, the trapped air expands, and so the length of the visible cylinder of air increases.

In practice, though, this type of thermometer is rather inconvenient. A more practical and probably more familiar type of thermometer, first created in the 1600s, relies on the thermal expansion of a liquid such as Mercury. Although liquids generally expand by a much smaller amount, for a given temperature change, compared to gases, the expansion can be "amplified" with the design sketched in Figure 10.6: the liquid expands, from a large-volume reservoir or "bulb", into a narrow tube, so that even a tiny fractional increase in the volume of the liquid can register a large change in the visible height of the column in the tube.

As indicated in Figure 10.6, the particular degree of hotness – the temperature – can be read off of a given thermometer by making a sequence of equally-spaced marks. A standard calibration scheme (defining the so-called Celcius or Centigrade scale) involves placing a mark at the level corresponding to the temperature of ice-water and another mark at the level corresponding to boiling water. The space between is then divided into 100 equally spaced "degrees", with the temperature of ice-water called "zero degrees" and the temperature of boiling water called "100 degrees".

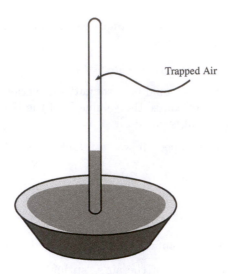

Figure 10.5: A simple thermometer: some trapped air in a tube expands or contracts as it is made respectively hotter or colder, thus changing the height of the column of liquid in the tube. Technically, such a device registers not only the temperature but also the ambient pressure, and so is properly termed a "barothermoscope."

10.1.4   Charles' Law

Just as Boyle's U-tube apparatus allowed him to determine how the volume of a sample of trapped air varies with changes in its pressure, so technical advances in thermometry allowed scientists to study how the volume of air varied with temperature. Actually, by the time such studies were undertaken in the late 1700s and early 1800s, "air" had been recognized to be a mixture of several chemically-distinct species of gas (what we now call nitrogen, oxygen, etc.). So it is more precise to say that, around this time, scientists studied the temperature-dependence of the volume of several different gases.

The first published study of this sort was undertaken by the French chemist Joseph Louis Gay-Lussac. He found that the relationship between volume and temperature was, when the pressure was held constant, *linear*, i.e.,

$$V = aT + b \tag{10.7}$$

where the constants, a and b, depend on the amount of gas present, the pressure, and (of course) the particular scheme that has been adopted for defining temperatures.

This linear dependence of volume on temperature, however, is usually called "Charles' Law" because, as it turned out, Gay-Lussac's fellow Frenchman, Jacques Charles, had performed (but failed to publish) similar work a few years earlier.

One interesting thing about Charles' Law is that it implies the existence of a (very low) temperture at which, if the linear dependence of volume on temperature persisted, the volume of the gas would go to zero:

$$T_0 = -b/a. \tag{10.8}$$

We now call this temperature, which turns out to be about $T_0 \approx -273$ degrees on the Celcius scale, "absolute zero" . It is then rather convenient to define a new so-called "absolute" temperature scale in which T_0 (rather than the temperature of ice water, or some other arbitrary fixed point) is assigned the value zero. That is, we may simply shift the temperature scale so that what used to be called $T = -273$ is now called $T = 0$, what used to be called $T = 0$ is now called $T = +273$, and so on. As you know, this

Figure 10.6: A simple liquid thermometer consists of a bulb of liquid (which might be water or alcohol or Mercury) with a thin tube into which the liquid can expand. The space above the liquid in the tube is empty (as opposed to containing, say, some trapped air). In principle, a simple sealed tube of liquid (without the bulb) would function as a thermometer, but the fractional increase in most liquids' volume with temperature is quite small (compared, say, to air). The bulb-and-tube design allows such a small fractional increase in the volume of the liquid to register a noticable change in the height of the liquid in the tube. Note that one can mark off "degrees" on the tube, as indicated on the diagram. Note also that, because the device is completely sealed off from the outside, the height of the column of liquid is independent of the outside air pressure, and registers exclusively the temperature.

new scale is called the Kelvin scale, and it makes the mathematical expression of Charles' Law a little simpler. With $T = T_K + T_0$ and $T_0 = -b/a$, Equation (10.7) becomes

$$V = aT_K. \tag{10.9}$$

That is, the volume of a sample of gas is just *proportional* to its (absolute) temperature (i.e., the temperature on the "absolute", Kelvin scale).

10.1.5   The Ideal Gas Law

As we saw, Boyle himself speculated that the inverse relationship between P and V would persist at higher pressures than he was able to achieve with the equipment available to him. And, to a good approximation, this is true. But Boyle's Law does begin to break down eventually, when the pressure gets very high. Similarly, although we defined the Kelvin temperature scale by extrapolating the linear volume-temperature relationship to a very low temperature at which the volume of the gas would go to zero, in actual fact gases turn into liquids and some low temperature (but, for most species, not nearly as low as absolute zero), and their volume basically stops decreasing as the temperature is further decreased.

Both Boyle's law and Charles' law, that is, are in some sense idealizations that reflect the actual behavior of real gases only imperfectly. Still, the idealization involved is important and empirically meaningful: the flip side of saying that these relationships break down at high pressures and low temperatures, is that the relationships are obeyed perfectly for low pressures and high temperatures.

This brings us to the ideal gas law, which can be understood as a combined statement of Boyle's and Charles' laws, regarded as a summary of the observed behavior of real gases whose pressures are not too high and whose temperatures are not too low. With the volume V being inversely proportional to the pressure P (that's Boyle) and also proportional to the absolute temperature T (that's Charles), we have

$$PV \sim T. \tag{10.10}$$

The proportionality constant depends on the quantity of gas that we are working with, and also on the type of gas involved (nitrogen, oxygen, hydrogen, etc.). It is a further (rather obvious) empirical fact that, for a given type of gas, the volume (at some fixed temperature and pressure) is proportional to the mass M of the gas, which is a convenient way of measuring the quantity of the gas involved. So we can write the ideal gas law a little more explicitly as

$$PV = gMT \tag{10.11}$$

where the proportionality constant g will be different for each different chemical species.

Note, finally that these proportionality constants can be determined easily from observation. For example, it is an empirical fact that one gram of Hydrogen gas, at a pressure of one atmosphere and at "room temperature" ($T = 300\,\text{K}$), occupies a volume of about 12.3 liters. Hence

$$g_H = \frac{12.3}{300}\frac{\text{L} \cdot \text{atm}}{\text{gram} \cdot \text{K}}. \tag{10.12}$$

By contrast, one gram of Nitrogen gas (also at atmospheric pressure and room temperature) occupies a much smaller volume of only about 1.8 liters. So

$$g_N = \frac{1.8}{300}\frac{\text{L} \cdot \text{atm}}{\text{gram} \cdot \text{K}}. \tag{10.13}$$

We will return to these strange proportionality constants (and the probably-unfamiliar way I've written the ideal gas law here!) shortly.

§ 10.2 The Chemical Atom

The idea that matter was composed of small not-further-reducible particles was first proposed by the Ancient Greek philosophers Leucippus and Democritus. (The word "atomos" literally means "uncuttable".) The Greek philosophers, however, postulated the existence of atoms not for any reasons we would today regard as empirical-scientific. Instead, they proposed atoms more as a possible solution to philosophical puzzles having to do with the nature of change: if a thing changes, it's different, and yet still the same... But isn't that contradictory? How can it be both the same and different?

Despite being not-particularly-well-grounded, scientifically speaking, the atomic view of matter was attractive to many thinkers, including scientists. Isaac Newton, for example, held that matter was fundamentally corpuscular:

> "It seems probable to me that God in the beginning formed matter in *solid, massy, hard, impenetrable, movable* particles, of such *sizes* and *figures*, and with such other *properties*, and in such proportion to space as most conduced to the end for which he formed them; and that these primitive particles being solids, are incomparably harder than any porous bodies compounded of them; even so very hard as never to wear or break in pieces; no ordinary power being able to divide what God Himself made *One*, in the first creation."

And many other scientists adopted a similar "corpuscularian" or "atomist" view, perhaps largely on the authority of Newton.

But, despite being rather popular among scientists throughout the 1600s and 1700s, the idea that matter is made of invisibly-tiny particles remained largely a kind of philosophical speculation that neither led to, nor found support in, scientific advances.

10.2.1   The law of definite proportions

That changed, however, in the early 1800s. A controversy had developed between two French chemists – Joseph-Louis Proust (1754-1826) and Claude Berthollet (1748-1822) – about the possible compounds that could be formed from two or more chemically pure substances. Berthollet held that two substances could combine in almost any of the continuum of possible ratios, to form a continuum of distinct compounds. Proust, on the other hand, held that two substances should combine in only a single, fixed ratio by weight – i.e., a *definite proportion*.

For example, suppose there are two pure substances, A and B, which can combine chemically to form a third substance, C. Suppose in particular that one observes the following chemical reaction:

$$8 \text{ grams A} + 5 \text{ grams B} \rightarrow 13 \text{ grams C}. \tag{10.14}$$

The question is now: what will happen if one attempts to react the same 8 grams of A with, say, *6 grams* of B?

According to Berthollet, the result should be 14 grams of a pure substance whose properties (such as color, density, hardness, chemical affinity, etc.) are relatively close to those of C – but perhaps a bit more B-like. For example, if B is a yellow liquid and C is a blue solid, perhaps the result of this proposed reaction will be a slightly malleable solid that is a slightly greenish shade of blue.

Proust, by contrast, suggested that the extra gram of substance B should simply go *unused*, so that the final products of the reaction will be *the same 13 grams of C as before* plus the leftover one gram of B – the idea being that A and B can only combine in the fixed mass ratio of $8 : 5$, so that the reactants will react *in that definite proportion* until one of them is used up – at which point the reaction will simply cease.

This may seem like the kind of dispute that can be straightforwardly settled by just performing the relevant experiments. But actually the controversy is more subtle and difficult to decide because of background conceptual issues that make it unclear precisely which experiments one should look to. Berthollet cited considerable experimental evidence to support his side of the argument. For example, he discussed cases of a solid dissolving in a liquid, or an alloy being produced by combining several metals. He also discussed the fact that iron can combine with oxygen in an apparent continuum of ratios to produce an apparent continuum of rust-like substances.

Slowly, however, it was recognized that solutions and alloys were not chemically pure substances, but were instead better classified as *mixtures*. And it was similarly recognized that "rust" is a mixture of two chemically distinct pure substances – each, incidentally, with a different fixed ratio of iron to oxygen by weight – which allowed those observations to be reconciled with the ideas of Proust.

It would be a little misleading to say that the concensus, in favor of Proust's view, developed because chemists found independent, clearer, and less ambiguous ways of distinguishing pure substances from mere mixtures, thus converting the otherwise-ambiguous experimental data into a clear experimental proof that Proust was right and Berthollet was wrong. That *is* part of the story. But another part of the story is that chemists began to recognize that "Proust's law" did definitely apply to a wide range of clear-cut cases of two (or more) substances combining to form a new pure substance – and so chemists subsequently began to treat this as part of the *definition* of a chemically pure substance. That is, Proust's law became part of the criteria by which scientists would distinguish genuinely *chemical* reactions from cases of mere mixing of two or more substances.

10.2.2   Dalton's Atoms

In 1803, the English chemist John Dalton ushered in the modern chemical atomic theory by proposing an atom-based *explanation* of the law of definite proportions. As he wrote:

> "We endeavoured to show that matter, though divisible in an *extreme degree*, is nevertheless not *infinitely* divisible. That there must be some point beyond which we cannot go in the division of matter. The existence of these ultimate particles of matter can scarcely be doubted, though they are probably much too small ever to be exhibited by microscopic improvements."

The basic assumptions of Dalton's theory were that:

1. Each chemical element is composed of a certain distinct type of atom.

2. All atoms of a given type are identical (and in particular had the same *mass*).

3. The smallest particles of compound substances are combinations (called "molecules") of two or more atoms of different types.

On these assumptions, it is clear how the law of definite proportions comes about. Suppose, to return to our earlier example, that the 8 grams of substance A consists of some large number (call it N) of A-atoms, and that the 5 grams of B consists of that same number, N, of B-atoms, and that the smallest particle (or "molecule") of the compound substance C consists of one A-atom attached to one B-atom. Then the reaction above can be written symbolically as follows:

$$N\,Ⓐ + N\,Ⓑ \to N\,Ⓐ\text{--}Ⓑ \tag{10.15}$$

where the "Ⓐ–Ⓑ" on the right stands for a single *molecule* of the compound substance C. Just to make sure it's clear, one should read the equation here as follows: N atoms of A and N atoms of B combine to make N molecules of C (with a molecule of C looking like Ⓐ–Ⓑ, i.e., with C having "molecular formula" AB).

The idea, of course, is that if the chemical reaction between the macroscopic sample of A and the macroscopic sample of B can be analyzed as just so many microscopic joinings of A atoms with B atoms, then the law of definite proportions will be naturally explained. If the mass ratio of the two reactants is just $8 : 5$ this will mean that the same number of atoms of each sort are present, which will mean that all of the atoms can combine to make molecules of C. On the other hand, if there is some other ratio – like $8 : 6$ – then there will be some atoms left over with nothing to pair with. They won't be able to distribute themselves evenly over the C molecules already produced (to make some alternative substance D) because there aren't enough of them and they can't divide! That is, you can't have each of the leftover atoms shatter into 5 equal pieces, with each C molecule getting an extra fifth of a B atom. The atoms, after all, are supposed to be indivisible (at least in so far as chemical reactions are concerned). So that one gram of leftover A atoms will simply have to remain left over – just as is in fact observed experimentally in this sort of case.

Note that if the previous chemical equation is correct (in particular, if the molecules of the substance C contain one Ⓐ and one Ⓑ each) then we have also learned something about the *relative atomic weights* of A and B – namely:

$$m_Ⓐ : m_Ⓑ = 8 : 5. \tag{10.16}$$

That is, an Ⓐ weighs 1.6 times as much as a Ⓑ. This follows because, by hypothesis, 8 grams of A contains the same number of A-atoms as 5 grams of B contains B-atoms.

This is a very important feature of Dalton's theory. The fact that it was put forward as an explanation for the empirical law of definite proportions allowed a real physical property of the proposed atoms (namely their relative masses) to be argued for on solidly empirical grounds. Unfortunately, as we will discuss in due course, the relative atomic weights assigned by Dalton turned out to be *completely wrong*. Nevertheless, the path forward did crucially involve determining the relative atomic weights for the different elements, and Dalton's theory helped focus chemists' attention on this property of atoms (as opposed for example to their shapes or sizes).

Dalton himself clearly recognized the importance and novelty of determining the relative atomic weights of the elements based on evidence from the masses of combining elements in chemical reactions:

> "An enquiry into the relative weights of the ultimate particles of bodies is a subject, as far as I know, entirely new: I have lately been prosecuting this enquiry with remarkable success."

Or as he elaborated elsewhere:

> "In all chemical investigations, it has justly been considered an important object to ascertain the relative *weights* of the simples which constitute a compound. But unfortunately the enquiry has terminated here; whereas from the relative weights in the mass, the relative weights of the ultimate particles or atoms of the bodies might have been inferred, from which their number and weight in various other compounds would appear, in order to assist and to guide future investigations, and to correct their results. Now it is one great object of this work, to show the importance and advantage of ascertaining *the relative weights of the ultimate particles, both of simple and compound bodies, the number of simple elementary particles which constitute one compound particle, and the number of less compound particles which enter into the formation of one more compound particle.*"

Having now stressed the importance of the fact that Dalton's theory allowed for some empirically-grounded (if for the most part ultimately wrong!) assignment of relative atomic weights to the different types of atoms, let us clarify why it is the *relative* atomic weights that we are talking about here. By "relative" here we just mean: the mass of one type of atom *relative to* another type – for example, that an A atom weighs 1.6 times as much as a B atom. Why not instead just determine and report the "absolute" weight of the atoms of a given type – that is, for example, the weight of a single A atom *in grams*?

The answer is that the number we called N above – the number of A atoms in 8 grams of substance A – is completely unknown, and there is no obvious way even to try to determine it! All we can say on the basis of Dalton's theory is that the absolute atomic weight of A is:

$$m_{\textcircled{A}} = \frac{8\,\text{grams}}{N} \tag{10.17}$$

where N is the number (whatever it might be) of A atoms in 8 grams of substance A. Which is not really saying anything, since this is equivalent to the stipulative definition of N.

10.2.3 Atom models

In the passage quoted above, Dalton suggests that, in addition to learning about the "relative weights of the ultimate particles", we might also learn about "the number of simple elementary particles which constitute one compound particle". Let us consider that aspect more carefully.

So far, again returning to our toy example, we've just *assumed* that the numbers of A and B atoms that combine to form C are equal – i.e., we've assumed that a molecule of C looks like this: $\textcircled{A}\text{–}\textcircled{B}$. But where did this hypothesis come from? A little reflection shows that there was no real basis for this at all in the chemical reaction data. We could just as easily have guessed – as the atomic theory explanation of the fact that 8 grams of A will combine with 5 grams of B to make 13 grams of the pure substance C – the following reaction:

$$2N\textcircled{A} + N\textcircled{B} \rightarrow N\textcircled{A}\text{–}\textcircled{B}\text{–}\textcircled{A}. \tag{10.18}$$

That chemical formula would account for the empirically observed "definite proportions" relation between A and B *just as well* as the one hypothesized above. And if this alternative chemical formula turns out to be the correct one, then there would evidently be twice as many A-atoms in the 8 grams of A as there are B-atoms in the 5 grams of B. And so the relative atomic weights will not be in the ratio 8 : 5 as we inferred above, but will instead be related this way:

$$m_{\circledA} : m_{\circledB} = 4 : 5. \tag{10.19}$$

If you follow that, you will see immediately that we could actually get *any* arbitrarily desired ratio for the relative weights of A and B atoms, simply by picking just the right molecular formula for C. Or vice versa – that is, we could also arrange for the molecular formula for C to be anything we arbitrarily desired, simply by carefully selecting the relative atomic masses to make it work out!

And so there are two lessons here. First, the atomic theory provides a simple way of understanding the law of definite proportions, i.e., understanding why substances should combine chemically with a fixed mass ratio (as opposed to forming compounds with any of a continuum of mass ratios). But second, the atomic theory's explanation of this fact leaves certain seemingly crucial things undecided and indeed empirically undecideable. The fact that all the atoms of a given element have identical masses plays a crucial role in the theory's ability to explain the observed phenomena – hence Dalton's particular interest in being able to infer, from experiment, something about what those masses are. (And remember here, we're only talking about the *relative* masses of the atoms.) Yet the experiments only allow us to say anything about what those masses actually *are*, if we make a seemingly arbitrary assertion about the atomic composition of the molecules of the compounds involved. And likewise, if only we knew the relative atomic masses, then we could really infer something about the structures of the molecules from the experimental data.

Dalton's advertisement is thus a little misleading. It's not that his theory allows us to learn about relative atomic weights *and* the atomic composition of compounds. Rather, it allows us to learn about relative atomic weights *or* the atomic composition of compounds – once we make a seemingly arbitrary assumption about the *other* one! Which, if we are going to be honest about it, really means the theory (as presented so far) doesn't allow us to learn about *either* of these properties. Which rather severely undercuts what Dalton claimed as one of the main virtues of his theory.

Dalton, however, understood this perfectly well and had a ready answer: he insisted that we should always adopt the *simplest* possible scheme for the atomic composition of compound molecules. This meant, at least in the simplest sorts of cases, that all compounds formed from two elements were assumed to be *binary*, meaning that their molecules consisted of one atom each of the combining elements.

For Dalton, though, this was not an arbitrary assumption, and was not even a mere assumption of necessity – i.e., one made (reasonably, if tentatively) in the absense of some other, more empirically justifiable, basis for assigning molecular structures to compounds. Dalton, from the very beginning, was particularly interested in the properties of gases, and believed that certain empirical facts about gases supported his ideas. Most importantly, Dalton took the fact that gases *expand to fill the space available to them* as evidence that their basic particles *repel* one another, as sketched (very schematically!) in Figure 10.7

This, actually, was an idea that had been in the air (pun intended!) for some time. Newton, for example, had shown that Boyle's Law could be explained by repulsive forces, between the constituent corpuscles, with a certain mathematical character (not identical to, but comparable to, the attractive inverse-square-law gravitational forces he had argued explained the motions of heavenly bodies).

But if the individual atoms of, for example, hydrogen gas repel one another, it seems unlikely that two or more of them could be found *right next to each other* in (say) a molecule of water. And likewise for the

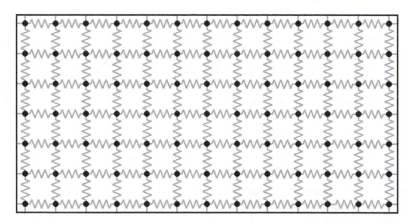

Figure 10.7: Schematic version of the Newton-Dalton model of the microscopic origins of macroscopic gas pressure. The gas is conceived as a lattice of particles which exert repulsive forces (indicated schematically here by the springs) on their neighbors. Thus, the gas as a whole possesses an expansive power because each individual constituent particle would prefer to get a little further away from its neighbors. Note that the springs are not to be taken literally, but only as a convenient symbol for some repulsive force of an as-yet undetermined character. For example, in the Newtonian model required to account for Boyle's Law, the inter-particle forces are not linear functions of the inter-particle separations (as would be the case for literal Hooke's Law springs), but rather vary inversely with the separation.

oxygen. We know from the fact that the hydrogen and oxygen can combine to form water that hydrogen and oxygen particles have some "affinity" or "attraction" to one another – but, argues Dalton, emprical observation of the properties of pure hydrogen gas and pure oxygen gas require that a water molecule must contain just one hydrogen atom and just one oxygen atom, i.e., water must be a "binary" compound of hydrogen and oxygen. And of course that was just the assumption Dalton's theory required in order to then, continuing with this example, infer the relative atomic weights of hydrogen and oxygen.

Dalton explained this point as follows:

> "When an element A has an affinity for another B, I see no mechanical reason why it should not take as many atoms of B as are presented to it, and can possibly come into contact with it (which may probably be 12 in general) *except so far as the repulsion of the atoms of B among themselves are more than a match for the attraction of an atom of A.* .... It is evident then ... that, as far as powers of attraction and repulsion are concerned (and we know of no other in chemistry) *binary* compounds must first be formed in the ordinary course of things..."

It is important that this reasoning implies (but in a way also assumes!) that elemental gases are *monatomic* – i.e., that the smallest physical particles of the gas (the gas "molecules") are just single, individual atoms. Readers will no doubt have learned in (say) high school chemistry that, in fact, many gases are *diatomic* (with the gas molecules consisting of two conjoined atoms of that element). Dalton, for the physical reasons explained here, would have none of that. According to Dalton, a diatomic gas molecule would necessarily explode!

Dalton also incorporated another then-prevalent idea into his model of atoms and their interactions. Based on the familiar observation that heat can and will spontaneously *flow* from a hotter object into a cooler one, most scientists prior to, and well into, the 19th century accepted the so-called caloric theory of heat according to which heat consisted of an invisible fluid called the "caloric" fluid. Temperature, in this view, corresponded to the density of caloric fluid contained within a body, and the spontaneous flow of heat from hotter to cooler objects was accounted for with the assumption that the caloric fluid, like a gas, had a self-repulsive character, i.e., preferred to expand to the greatest extent possible. Thus, the relatively high-density caloric fluid in a hot object will expand outward, decreasing the caloric density within the body and increasing the caloric density outside, until the densities are the same inside and out. This, of course, is the same process we would describe by saying that heat flows from the hotter object into the cooler surrounding objects until thermal equilibrium – i.e., equality of temperatures – is achieved.

Dalton cleverly adapted the caloric theory of heat to his atomic theory. He suggested, in particular, that the caloric fluid formed a kind of diffuse atmosphere around the individual atoms, so that adding caloric fluid (i.e., heat) to a substance had the effect of increasing the thickness of the atoms' caloric atmospheres, i.e., increasing the effective *size* of the atoms. In Dalton's picture, neighboring atoms of a particular chemical species repelled one another because and to the extent that their surrounding caloric atmospheres began to overlap. So this picture was in qualitative harmony with the observed temperature-dependence of the pressure and volumes of gases, summarized in Charles' Law and/or the ideal gas law: increasing the temperature of a gas (at constant pressure, say) causes it to expand, because adding caloric fluid increases the effective size of the individual mutually-repelling atoms constituting the gas.

So you can see how, for Dalton, the idea that the individual atoms of a given chemical species repel one another – thus requiring us to assume that the smallest chemically-operative particles of elements are single atoms, and that most chemical compounds are binary – was not baseless and arbitrary. Instead, it was harmoniously integrated with reasonable-seeming and widely-held ideas about how to physically understand and explain the previously-discovered properties of, especially, gases.

10.2.4   The law of multiple proportions

Dalton's ideas do raise some puzzling questions, though. For example, if two adjacent hydrogen atoms (in a sample of hydrogen gas) repel one another because of, and to the extent of, the overlapping of their caloric atmospheres, and if the same is true for two adjacent oxygen atoms (in a sample of oxygen gas), how can we understand the chemical *affinity* that allows a hydrogen atom and an oxygen atom to bond to one another to form a (binary!) water molecule?

There is no convincing answer available, although probably it is not too unreasonable to suspect that chemical reactions involve additional (not yet understood) forces, between atoms, beyond those that are responsible for the macroscopic physical behavior of gases as described in, for example, Charles' Law.

So, Dalton's theory does hang together nicely, but there is at least a little bit of tension between some of its parts. We mention that here because there is one final aspect to the theory which, on the one hand, constituted a brilliant empirical-scientific success – but which, on the other hand, simultaneously exacerbated the tension.

In particular, Dalton actually *predicted*, on the basis of his atomic account of the law of definite proportions, another very important law which was subsequently verified experimentally: *the law of multiple proportions*. This law concerns cases when two (or more) substances can chemically react to produce *two (or more) different compounds*. The law states that the masses of one substance which combine with a *fixed* mass of some second substance, will be in *small whole number ratios*. That is a little hard to process, so let us illustrate with an example: suppose, as above, that 8 grams of A can combine completely with 5 grams of B to form 13 grams of pure substance C. We follow Dalton's ideas as explained so far in assuming that the basic chemical process involved looks like this:

$$Ⓐ + Ⓑ \rightarrow Ⓐ–Ⓑ. \tag{10.20}$$

Dalton's new idea was that there could be a different compound, D, each of whose molecules contained, say, one Ⓐ and *two* Ⓑs. At the atomic level, the formation of D from A and B would proceed like this:

$$N\text{Ⓐ} + 2N\text{Ⓑ} \rightarrow N\text{Ⓓ} = N\text{Ⓑ–Ⓐ–Ⓑ} \tag{10.21}$$

where I have introduced the symbol "Ⓓ" to represent a single *molecule* of substance D.

Equivalently, re-written not in terms of individual atoms, but in terms of the macroscopically-observable masses of reactants and products, this reaction will read

$$8 \text{ grams A} + 10 \text{ grams B} \rightarrow 18 \text{ grams D}. \tag{10.22}$$

This (if such a reaction really exists) would represent an example of the law of multiple proportions: the two different masses of B (5 grams and 10 grams respectively) which react with the same 8 grams of A to make the two distinct chemical compounds C and D respectively, are in the small whole number ratio of $5{:}10 = 1{:}2$.

That, hopefully, clarifies both (a) how the law of multiple proportions is a *prediction* of Dalton's chemical atomic theory, and (b) what the law *says*.

Of course, you might feel slightly uncomfortable about the idea that an atom of A can combine with not one, but two atoms of B to form the compound D. Even leaving aside the question, mentioned above, about how the individual A and B atoms can attract each other, wouldn't the *two* B atoms in a single molecule of D repel one another and cause the D molecule to explode? It's a fair question, to be sure, but on the other hand, maybe the chemical affinity between Ⓐ's and Ⓑ's is sufficiently strong to outweigh the repulsion between the two Ⓑ's in a D molecule? (Note that when I wrote the atomic formula for D in the above reaction, I wrote it as "Ⓑ–Ⓐ–Ⓑ" rather than "Ⓐ–Ⓑ–Ⓑ" for just this reason: it seems plausible that if the two Ⓑ's repel one another, but are still sufficiently attracted to the Ⓐ for the D molecule to hang together, they should at least position themselves on opposite sides of the central Ⓐ.)

Anyway, despite whatever concerns we might have about how it could possibly work, Dalton made this prediction – and it was subsequently found to be correct in a wide variety of reactions. For example, it is observed that 3 grams of Carbon and 8 grams of Oxygen can combine to form one pure gas compound (which plants love but which kills birds and other animals); and, under different conditions, the same 8 grams of Oxygen can combine with 6 grams of Carbon to form a distinct gas compound (which even plants don't like). Thus, the masses of Carbon which react with a fixed mass (8 grams) of Oxygen to form these two different compound substances are in the small whole number ratio $1{:}2$.

Here's another example. We mentioned before that there are several distinct compounds of oxygen and iron, some of which are components of (the mixture usually referred to as) "rust". One iron oxide, called wüstite, is an explosive black solid. It contains iron and oxygen in the ratio of $3.48{:}1$ by mass. Another, sometimes called rouge, contains the same two elements in the ratio of $2.32{:}1$. And $3.48{:}2.32 = 3{:}2$. So here the masses of iron combining with the same fixed mass of oxygen to form these two distinct compounds are in the ratio $3{:}2$.

And there are many, many more examples.

The empirically supported law of multiple proportions thus constitutes some additional evidence for Dalton's atomic theory. Indeed, it could be regarded as a brilliantly-successful empirical prediction of the theory. In simple terms, the fact that the masses of combining elements have these curious small-whole-number numerical relations makes perfect sense if one is thinking about the chemical reactions in terms of atoms and molecules, and would be a weird coincidence otherwise.

But the extended theory also underscores the troubling questions about how the physical interactions among atoms work and hence about the believability of the relative atomic weights that Dalton proposed. For example, in a case (like our toy model involving A and B above, or the real interactions involving

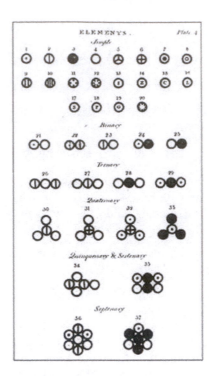

Figure 10.8: Figure reproduced from Dalton's 1808 A New System of Chemical Philosophy *showing his symbols for the known elements as well as a number of molecules grouped by binary, ternary, etc.*

Carbon and Oxygen, or Oxygen and Iron, that we just described) where two elements can combine in two distinct ratios, by mass, to form two distinct chemical compounds, Dalton would insist that one of the compounds be regarded as *binary*. And from that, the relative atomic weights of the two elements could be determined. Dalton advises as follows:

"1st. When only one combination of two bodies can be obtained, it must be assumed to be a *binary* one, unless some cause appear to the contrary.

"2d. When two combinations are observed, they must be presumed to be a *binary* and a *ternary*.

"3d. When three combinations are obtained, we may expect one to be a *binary*, and the other two *ternary*. [etc..]

But this is not terribly helpful. How do we decide which of (say) two combinations is the binary one, and which the ternary? The relative atomic weights will depend on the answer, but there is no clear empirical grounds for deciding.

To summarize, it is hard to disagree with what the Swedish chemist Berzelius wrote in a letter to Dalton: "[t]he theory of multiple proportions is a mystery but for the Atomic Hypothesis, and ... all the results so far obtained have contributed to justify this hypothesis". Dalton's theory held out the alluring possibility of identifying, based on genuine scientific data, the relative atomic weights of the different elements. Indeed, this possibility was the main feature distinguishing Dalton's chemical atomism from the speculative philosophical atomism of earlier centuries. But, in practice, the actual determination of these relative atomic weights involved questionable, even downright suspicious, assumptions and hence remained rather unconvincing.

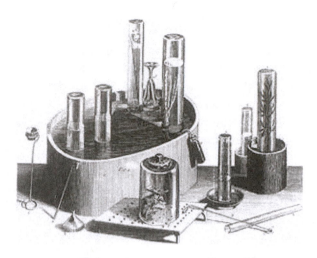

Figure 10.9: Priestley's pneumatic trough and other chemical apparatus. Note the various living organisms in the containers in the foreground used to test respirability of various gases. From Experiments and Observations on Different Kinds of Air (1774).

§ 10.3 Evidence for Atoms in the Chemical Reactions of Gases

In the first section of this chapter, we surveyed the discoveries about the physical behavior of air that were summarized, ultimately, in the ideal gas law. But advances were also made regarding the *chemical* properties of air – one of the most important such advances being, of course, the recognition that air is a mixture of several chemically distinct gases. The late 18th century British chemist, Joseph Priestley, wrote in 1775:

> "There are, I believe, very few maxims in philosophy that have laid firmer hold upon the mind, than that air, meaning atmospherical air (free from various foreign matters, which were always supposed to be dissolved, and intermixed with it) is a *simple elementary substance*, indestructible, and unalterable, at least as much so as water is supposed to be."

Despite the firm hold the idea of air's purity had held on his mind, however, Priestley and other chemists managed to isolate and study the several distinct gases (such as nitrogen and oxygen) that composed ordinary air, and also identify other gases (such as hydrogen) which were not part of ordinary air but which could be produced in certain chemical reactions.

A key technological advance which contributed to these discoveries was the introduction of the "pneumatic trough". See Figure 10.9. This is a device, a modern version of which you have probably used in your chemistry classes, consisting essentially of a water-filled upside-down bottle resting in a trough of water, which could be used to capture and contain the gaseous products of chemical reactions. Prior to its invention, chemical reactions were generally performed and studied with the reactants being open to the atmosphere, so it was impossible to recognize when one of the constituent pure substances from the atmosphere was being used up as a reactant in the reaction, and equally impossible to recognize when a gaseous product was given off. The use of the pneumatic trough allowed the existence of such products to be recognized – and perhaps more importantly, it allowed them to be subjected to further careful study. Thus, in the 1770s and subsequent decades, many different types of gas were differentiated based on their physical properties (such as density or reactivity in the presence of fire) or the effects they produced on plants or animals.

10.3.1   Gay-Lussac's Law of Combining Volumes

The French chemist Joseph Louis Gay-Lussac (1778-1850) was the first to carefully study the *volumes* of gases participating in chemical reactions. He noticed a curious empirical regularity which, when interpreted theoretically in terms of atoms, challenged Dalton's own schemes for determining the relative atomic weights.

Gay-Lussac was attracted to the study of gases because, as we have discussed earlier, the way they expand and contract on changes in pressure and temperature were not only very simple (compared to the corresponding behavior of liquids and solids) but also quite *regular*: different chemical species of gas seemed to expand and contract *in exactly the same way*. Gay-Lussac and others had taken this as evidence that, while in general the atoms in a substance exerted some complicated and unknown forces on one another, in the gaseous phase, the distance between neighboring atoms was quite large (and the forces correspondingly weak) such that the overall behavior of the gas was somehow independent of the unknown details of the short-range forces. As Gay-Lussac explained it,

> "The same pressure applied to all solid or liquid substances would produce a diminution of volume differing in each case, while it would be equal for all elastic fluids [i.e., gases]. Similarly, heat expands all substances; but the dilations of liquids and solids have hitherto presented no regularity, and it is only those of elastic fluids which are equal and independent of the nature of each gas. The attraction of the molecules in solids and liquids is, therefore, the cause which modifies their special properties; and it appears that it is only when the attraction is entirely destroyed, as in gases, that bodies under similar conditions obey simple and regular laws."

The empirical regularity of interest to us here, which Gay-Lussac articulated in about 1808, was that there was a simple relationship between the volumes of the gaseous participants in a given chemical reaction – namely: when all the reactants and products were held at the same temperature and pressure, the volumes of the several different gases involved in the reaction were in *small integer ratios*. For example, if oxygen gas and hydrogen gas are chemically combined to form water (and assuming a temperature such that the water, too, is a gas) the volumes of the three substances will be in the ratio 1:2:2. That is: one unit volume of oxygen combines with two unit volumes of hydrogen to make 2 unit volumes of water vapor.

Here is an excerpt from Gay-Lussac discussing this example and some others:

> "Suspecting, from the exact ratio of 100 of oxygen to 200 of hydrogen [by volume], which M. Humboldt and I had determined for the proportions of water, that other gases might also combine in simple ratios, I have made the following experiments. I prepared fluoboric, muriatic, and carbonic gases, and made them combine successively with ammonia gas. 100 parts of muriatic gas saturate precisely 100 parts of ammonia gas, and the salt which is formed from them is perfectly neutral, whether one or other of the gases is in excess.

> "Thus we may conclude that muriatic, fluoboric, and carbonic acids take exactly their own volume of ammonia gas to form neutral salts.... We might even now conclude that gases combine with each other in very simple ratios; but I shall still give some fresh proofs.

> "According to the experiments of M. Berthollet, ammonia is composed of 100 of nitrogen, 300 of hydrogen, by volume....

Note that the "law of combining volumes" applies *only* to those reactants and products which are gaseous. Nothing is implied, for example, about the volume of the "salt" produced when muriatic gas reacts with ammonia. Or here is another example: carbon can be burned in oxygen to produce a purely gaseous product – the gas that makes plants thrive which was mentioned earlier. Assuming the oxygen and this product gas are measured at the same temperature and pressure, it turns out that their volumes are *equal*, i.e., in the simplest-possible "small integer ratio" – 1:1. But the volume of the (solid) carbon that was burned need (and will) have no simple relation to the volumes of oxygen or this product gas.

Like the law of definite (and multiple) proportions, Gay-Lussac's law of combining volumes could be easily explained on the basis of the atomic theory: we need only assume that *the numbers of atoms contained in equal volumes of different gases, are themselves in small integer ratios*. For then, if the volumes are in small whole number ratios, the total numbers of the different types of atoms/molecules involved in a given reaction will themselves be in small whole number ratios, too – which is just what is required by the atomic theory to account for the ability of the reaction to proceed to completion.

For example, we have seen that, according to Dalton, the chemical reaction involved in the combination of oxygen and hydrogen to form water is as follows:

$$\text{Ⓞ} + \text{Ⓗ} \rightarrow \text{Ⓞ}–\text{Ⓗ}. \tag{10.23}$$

This is easily reconciled with the data about the volumes of the three gases (namely, that the three volumes are in the ratio of 1:2:2) if we assume that oxygen gas contains (say) $2N$ atoms per unit volume, hydrogen gas contains N atoms per unit volume, and water vapor contains N molecules per unit volume – where N is again some large and unknown number. Then the one unit volume of oxygen and the two unit volumes of hydrogen will contain the *same number* ($2N$) of atoms, which will allow the two substances to combine completely and form precisely $2N$ water molecules according to the above equation. And this $2N$ water molecules will then occupy two unit volumes, just as is observed experimentally.

However, just as with Dalton's accounts of chemical reactions, we could change some of the numbers and assumptions here, and still get a seemingly valid "explanation" for the experimentally observed facts. For example, one could assume a different molecular structure for the water molecule – say, $\text{Ⓞ}–\text{Ⓗ}–\text{Ⓞ}$ – and then find some way of assigning relative number-densities to the three gases to account for the observations.

The basic principle here is just that the total number of particles is equal to the number density times the volume. So we need the relative number densities, when multiplied by the (observed) relative volumes, to equal the relative numbers of particles involved. This last follows from the molecular structure we assume for the product. Thus, with the funny ternary structure for water mentioned in the previous paragraph, the basic microscopic reaction would be

$$2\text{Ⓞ} + \text{Ⓗ} \rightarrow \text{Ⓞ}–\text{Ⓗ}–\text{Ⓞ} \tag{10.24}$$

from which it is clear that the relative numbers of particles involved for the three substances are 2:1:1. Thus, the relative number densities of basic particles for the three substances would have to be 4:1:1 because then the relative number densities (4:1:1) times the relative volumes (1:2:2) equals the relative numbers of particles involved (2:1:1). That is:

$$4 : 1 : 1 \times 1 : 2 : 2 = 2 : 1 : 1. \tag{10.25}$$

The lesson is that we have here the same sort of ambiguity we had with Dalton. The hypothesis of atoms (and in particular the hypothesis that the number densities of gases are in small whole number ratios) provides a compelling explanation for the empirical law of combining volumes. But, you can make any assumption you want about the molecular structures, and find some nice (simple integer ratio) way of assigning number densities to the gases that will account for the empirical observations of combining volumes. Or vice versa: you can make some sort of assumption about the relative number densities and then work out molecular structures for the compounds to make everything consistent with the empirical observations. And so, if we're going to take advantage of the law of combining volumes to determine molecular structures (and hence then be able to assign relative atomic weights), we're again going to have to get the whole thing started with some sort of (dubious) "rule of simplicity."

Here, in the case of gases, the obvious such rule would be to assign to all gases *the same* number density – i.e., to posit that equal volumes of all different species of gas (at the same temperature and pressure) contain *the same* number of particles. This was precisely the hypothesis posited in 1811 by the Italian physicist Amedeo Avogadro.

10.3.2 Avogadro's Hypothesis

Here is Avogadro's own summary of his famous hypothesis:

> "M. Gay-Lussac has shown in an interesting Memoir that gases always unite in a very simple proportion by volume, and that when the result of the union is a gas, its volume also is very simply related to those of its components. But the proportions by weight of substances in compounds seem only to depend on the relative number of molecules which combine, and on the number of compound molecules which result. It must then be admitted that very simple relations also exist between the volumes of gaseous substances and the numbers of [atoms or] molecules which form them. The first hypothesis to present itself in this connection, and apparently even the only admissible one, is the supposition that the number of molecules in any gas is always the same for equal volumes, or always proportional to the volumes."

One could restate Avogadro's hypothesis in more concrete terms as follows. Consider a number of large – say, 24.5 liter – containers, each filled with a distinct chemical species of gas (e.g., one container is filled with hydrogen gas, one with nitrogen gas, one with oxygen gas, one with helium, one with methane, etc.). Suppose all the different gases are under the same conditions – say, atmospheric pressure and room temperature. Then Avogadro's hypothesis is that each of the containers contains *the same number of molecules*.

This of course immediately raises the question: what number is this? How many gas molecules are there in 24.5 liters? Nobody – Avogadro himself emphatically included – had a clue.

Anyway, now that we know what Avogadro's hypothesis was, it should be pointed out that Avogadro was not actually the first person to propose, on the basis of the law of combining volumes, that equal volumes of different gas might contain equal numbers of particles. This hypothesis is indeed "the first hypothesis to present itself" once one recognizes that the atomic theory account of the law of combining volumes requires the number densities of different gases to be in small whole number ratios. And so it is maybe not surprising that Gay-Lussac himself had already suggested this simplest possibility (that the number densities are simply all *the same*).

10.3.3 Controversy

Avogadro's real achievement, then, is not to have been the first to propose this simplest possible account of the law of combining volumes. It is, rather, to have been the first to understand how to respond to a seemingly devastating criticism against this idea which Dalton had leveled at Gay-Lussac. So let us first understand that controversy.

The important point to appreciate is that Dalton's "rule of simplicity" and Gay-Lussac's "rule of simplicity" led to conflicting assignments of molecular structures to molecules, and also to conflicting assignments of relative atomic weights to the different elements. They both couldn't be true. Let us illustrate with (to begin with) a "toy" example.

Suppose there is a chemical reaction in which two gaseous substances (A and B) combine to form a third gaseous substance (C). And suppose that both the masses and volumes of all three substances have been carefully measured, so that the reaction can be written either (by mass) as

$$8 \text{ grams A} + 5 \text{ grams B} \rightarrow 13 \text{ grams C} \tag{10.26}$$

or (by volume) as

$$100 \text{ liters A} + 200 \text{ liters B} \rightarrow 100 \text{ liters C}. \tag{10.27}$$

Now what will the two theorists (Dalton and Gay-Lussac) want to say?

Dalton will appeal to his "rule of simplicity" and begin by assuming that the compound C is *binary* – i.e., that a single molecule of C contains one Ⓐ and one Ⓑ. This immediately implies that there are the same number of A-atoms in the reactants as there are B-atoms – i.e., that there are the same number of A-atoms in 8 grams of A as there are B-atoms in 5 grams of B. Which of course implies that the atomic masses themselves are in this same ratio: $m_{Ⓐ} : m_{Ⓑ} = 8 : 5$. Incidentally, note that this Daltonian scheme – in particular the assumption that the same numbers of A and B atoms are involved in the reaction – implies (if we now consider the volume data for the reaction) that the number density of gas A is twice as large as that of gas B (since there the same number of atoms will be packed into half as much volume for A, as compared to B).

Gay-Lussac, on the other hand, wants to give priority to the volumes, and in particular to assume that all gases have the same number density. This (looking again at the volume data for the reaction) immediately implies that there will be *twice as many* B-atoms involved in the reaction, as A-atoms (since there is twice as much B by volume). And so the simplest possible molecular formula for C will be *ternary*, namely: Ⓒ = Ⓑ–Ⓐ–Ⓑ. Also, if there are twice as many B-atoms involved in the reaction as A-atoms, that means the 8 grams of A consists of half as many atoms as the 5 grams of B, which means that the ratio $m_{Ⓐ} : m_{Ⓑ}$ will not be 8:5, but 16:5.

In short, Dalton and Gay-Lussac will come to different (contradictory!) conclusions about the relative atomic weights of A and B, from the same empirical data. And they will also disagree about the molecular structure for C, and the number densities of the gases A and B.

Both of their schemes account equally well for the data. What they disagree about is, essentially, *where* to impose an arbitrary rule of simplicity. Dalton wants to insist that the molecular structures for compounds should be as simple as possible (namely, whenever possible, binary), and is happy to allow the number densities of gases to vary. By contrast, Gay-Lussac prefers to insist that the number densities of gases should be as simple as possible (namely, they should all be the same), and is happy to allow the molecular structures of compounds to come out less-than-perfectly simple.

We have already discussed one argument that Dalton gave for his side of this controversy – namely, that (as apparently evidenced by the gross properties of gases) the atoms of a given species of gas repel one another, thus making it relatively difficult to understand how the molecules of a compound could contain two or more atoms of the same element.

Gay-Lussac had an argument of a comparable caliber on his side, too. Citing again the fact that the gross properties of gases – such as the relationships they exhibited between volume, pressure, and temperature – were all the same, independent of which particular species of gas one considered, Gay-Lussac argued that this showed all gases to be somehow, structurally, identical. It is then quite reasonable to suppose in particular that the number densities of all gases might be (under identical conditions) the same. Or at least, if we are going to arbitrarily assume some notion of "simplicity" somewhere in the system, there is a much stronger motivation for Gay-Lussac's version than for Dalton's.

That, at any rate, is what Gay-Lussac would have argued. But Dalton had a ready – and indeed apparently fatal – reply.

To understand Dalton's reply, consider again the example of hydrogen and oxygen gas combining to make water. By weight, the reaction can be written

$$1 \text{ gram Hydrogen} + 8 \text{ grams Oxygen} \rightarrow 9 \text{ grams Water Vapor}. \tag{10.28}$$

Whereas, by volume, the same reaction is:

$$200 \text{ liters Hydrogen} + 100 \text{ liters Oxygen} \rightarrow 200 \text{ liters Water Vapor} \qquad (10.29)$$

From these two sets of data we can immediately infer (by simply dividing the corresponding coefficients from the two ways of writing the reaction) the relative *mass densities* of the three gases involved:

$$\rho_H : \rho_O : \rho_W = \frac{1}{2} : \frac{8}{1} : \frac{9}{2} = 1 : 16 : 9. \qquad (10.30)$$

Note in particular that Water Vapor has a smaller density than Oxygen gas.

But if we suppose that equal volumes of every type of gas contain the same number of ultimate particles (atoms or molecules), this ratio of *mass densities* (i.e., mass per unit volume) should be the same as the ratio of *atomic/molecular masses* (i.e., mass per unit ultimate particle). But then the above numbers would imply that a single molecule of Water Vapor weighs *less* than a single Oxygen atom – a patent absurdity since a Water molecule consists of at least one Oxygen atom *plus some Hydrogen*!

As Dalton summarizes the objection,

> "Though it is probable that the [densities] of different [gases] have some relation to that of their ultimate particles, yet it is certain that they are not the same thing; for the ultimate particles of water or steam are certainly of greater [mass] than those of oxygen, yet [oxygen] is *heavier* than steam."

The same basic point can be understood thinking about the reaction in terms of volumes: two unit volumes of hydrogen combine with one unit volume of oxygen to make two unit volumes of water vapor. I argued above that this made perfect sense from the point of view of Dalton's theory if the number densities for the three gases were in simple ratios (1:2:1 as it turns out). But suppose instead we consider that the number densities of the three gases are *the same*, as Gay-Lussac wanted to assume. Then the 2:1:2 ratio of volumes will imply that the ultimate particles themselves are present in this same ratio, 2:1:2. More concretely, this means that $2N$ hydrogen atoms combine with N oxygen atoms to form $2N$ water molcules.

But that's impossible! There can't be more water molecules produced at the end than there were oxygen atoms at the beginning. Each water molecule must contain at least one *whole* oxygen atom. Yet the numbers above would require either that only half of the water molecules have an oxygen atom (which would mean we wouldn't have a pure substance at all, but rather two distinct sorts of molecules) – or that each water molecule has *half* of an oxygen atom (which violates the whole spirit of the atomic theory).

Thus Dalton passionately rejected the simplest and most alluring possible way of accounting for the law of combining volumes.

Now that we have clarified this controversy between Dalton and Gay-Lussac, let us see how Avogadro proposed to rebut the apparently fatal criticisms made by Dalton. Here is Avogadro's own statement on this point:

> "For instance, the volume of water in the gaseous state is, as M. Gay-Lussac has shown, double that of the oxygen which enters into it, or, what comes to the same thing, equal to that of the hydrogen instead of being equal to that of the oxygen. But a means of explaining facts of this type in conformity with our hypothesis presents itself naturally enough. We suppose that the constituent molecules of any simple gas (i.e., the molecules that are at such a distance from each other that they cannot exert their mutual action) are not formed of only one [atom], but are made up of a certain number of these [atoms] united by attraction to form a single whole. Further, that when such molecules unite with those of another substance to

form a compound molecule, the integral molecule which should result splits up into two or more parts (or [atoms]), each composed of half, quarter, &c., the number of [atoms] forming the constituent molecule of the first substance.... Thus, for example, the [molecule] of water will be composed of a [single atom] of oxygen with one molecule, or, what is the same thing, two [atoms], of hydrogen."

What Avogadro is proposing here is that – in direct contradiction to Dalton's rule of simplicity – the smallest particles or molecules of some (indeed, as it turns out, many) elements are *poly-atomic*.

For example, in the case of hydrogen and oxygen combining to make water, Avogadro proposes the basic reaction be viewed as follows:

$$2\text{⊞–⊞} + \text{O–O} \rightarrow 2\text{⊞–O–⊞}. \tag{10.31}$$

Thus, two "smallest particles" of hydrogen combine with one "smallest particle" of oxygen to produce two "smallest particles" of water. The point is that this is perfectly consistent with the proposed explanation of Guy-Lussac's Law (viz., Avogadro's hypothesis) and yet gets around the objection (based on the tacit assumption that the molecules of hydrogen and oxygen are monatomic) that one can't have a single oxygen atom giving rise to two water molecules since each of the latter must contain at least one whole oxygen atom.

This is a brilliant solution to the puzzle, which at least removes the apparently fatal character of Dalton's criticism. But we must also appreciate that this is, after all, merely a hypothesis at this stage. Avogadro is basically arguing against Dalton's rule of simplicity, on the (somewhat ironic) grounds that greater simplicity can be achieved elsewhere in the system (namely in the rule for how a gas's volume relates to its number of component molecules) by giving this up. This is a very reasonable move to contemplate, but it is by no means obvious *a priori* that Avogadro's hypothesis is true, nor that the molecular structures and relative atomic weights calculated on the basis of it are correct.

One could, after all, simply reject Avogadro's hypothesis – explaining the data pertaining to hydrogen, oxygen, and water as we did above: by supposing that a unit volume of oxygen contains *twice as many* (monatomic) oxygen molecules as a unit volume of hydrogen contains hydrogen atoms. Then there is no problem with Equation (10.29): two hydrogen atoms combine with two oxygen atoms to make two water molecules. And, with this assumption about the respective *number densities* of the two gases, the relative *mass densities* – see Equation (10.30) – would imply atomic weights in the ratio $m_{\text{⊞}} : m_{\text{O}} = 1 : 8$. Which is just what Dalton had said from the beginning.

It turns out, of course, that Avogadro was right and Dalton was wrong. The point here is then to stress that this is not obvious from anything that has been said so far. In fact, the most dramatic physical evidence in support of Avogadro's hypothesis took almost 50 years to appear and be widely accepted.

§ 10.4 The Kinetic Theory

We mentioned above that the caloric theory of heat was widely accepted in the 17th and 18th centuries. Here we explore an alternative theory – of both the nature of heat as such and also of the physics underlying the "spring" and thermal expansion of gases – which gained prominence and eventually widespread acceptance in the 19th century.

10.4.1   The Kinetic Theory of Heat

The alternative to the caloric theory of heat was the so-called kinetic theory, according to which heat was not a distinct physical substance, but was rather simply the *motion* of the atoms/molecules composing ordinary chemical substances. Joseph Black, the 18th century Scottish physicist and chemist, explained that heat, according to this viewpoint, "is a rapid tremor, or vibration.... in the small particles of the heated bodies..." Black, despite being himself ultimately a supporter of the caloric theory, concedes that the evidence is not conclusive:

> "Heat is plainly something extraneous to matter. It is either something superadded to ordinary matter or some alteration of it from its most spontaneous state. Having arrived at this conclusion, it may perhaps be required of me, in the next place, to express more distinctly this something – to give a full description, or definition, of what I mean by the word *heat* in matter. This, however, is a demand that I cannot satisfy entirely. I shall mention, by and by, the supposition relating to this subject that appears to me the most probable. But our knowledge of heat is not brought to that state of perfection that might enable us to propose with confidence a theory of heat or to assign an immediate cause for it."

Black then goes on to review some of the evidence put forward by those who had speculated that heat is motion, discussing for example "the consideration of several means by which heat is produced, or made to appear, in bodies, such as the percussion of iron, the friction of solid bodies, the collision of flint and steel", and the fact that "thick forests are said to have taken fire sometimes because of the branches rubbing one another in stormy weather."

The idea here is simply that it seems possible to produce heat "from scratch" – for example by rubbing two solid objects together. The friction between them produces sensible heat, but without any obvious source (for example, a nearby hot object which cools off during the process, or an exothermic chemical reaction in which, according to the caloric theory, some caloric fluid is released). Indeed, the only thing added to the system appears to be *motion* – in particular, the mechanical *work* required to move the one object, against friction, across the other. Could it then be that what we call "heat" is actually just the kinetic energy produced by this work, but in a hidden form (namely, the random motions of the small particles of the bodies, as opposed to the coherent motion of the bodies as wholes)?

That is precisely the idea of the kinetic theory of heat.

A crucial piece of evidence in support of this idea was produced by Count Rumford (Benjamin Thompson), and summarized in his 1798 paper "An Inquiry Concerning the Source of Heat which is Excited by Friction". Rumford first recounts his belief that profound scientific discoveries can be made if only we pay careful attention to seemingly mundane goings-on:

> "It frequently happens that in the ordinary affairs and occupations of life, opportunities present themselves of contemplating some of the most curious operations of Nature.... I have frequently had occasion to make this observation; and am persuaded that a habit of keeping the eyes open to everything that is going on in the ordinary course of the business of life has oftener led – as it were by accident, or in the playful excursions of the imagination, put into action by contemplating the most common appearances – to useful doubts and sensible schemes for investigation and improvements than all the more intense meditations of philosophers in the hours expressly set apart for study."

We include that passage here only because it seems like generally good advice: don't leave your scientific curiosity at the door when you leave the classroom!

Rumford then describes the specific observations which led him to further subsequent experiments on the nature of heat:

> "Being engaged lately in superintending the boring of cannon in the workshops of the military arsenal in Munich, I was struck with the very considerable degree of heat that a brass gun acquires in a short time in being bored, and with the still higher temperature (much higher than that of boiling water, as I found by experiment) of the metallic chips separated from it

by the borer. The more I meditated on these phenomena, the more they appeared to me to be curious and interesting. A thorough investigation of them seemed even to ... give a farther insight into the hidden nature of heat; and to enable us to form some reasonable conjectures respecting the existence, or nonexistence, of an *igneous fluid* – a subject on which the opinions of philosophers have in all ages been much divided."

The crucial question was:

"From *whence comes* the heat actually produced in the mechanical operation above mentioned? Is it furnished by the metallic chips which are separated by the borer from the solid mass of metal? If this were the case, then, according to the modern doctrines of latent heat and of caloric, the *specific heat* of the parts of the metal, so reduced to chips, ought not only to be changed, but the change undergone by them should be sufficiently large to account for *all* the heat produced.

The "specific heat" c of a material is a measure of the amount of heat that is required, per unit mass, to increase its temperature by one degree:

$$c = \frac{\Delta Q}{m \, \Delta T} \tag{10.32}$$

where ΔQ is the quantity of heat added to an object of mass m and ΔT is its resulting temperature increase.

You already know how the mass and temperature of an object can be measured, but how can we measure the heat ΔQ that flows into an object? The idea is simple: take, as a unit of heat, the "calorie", which we define to be the quantity of heat required to increase the temperature of one gram of water by one degree (Celcius, or equivalently, Kelvin). That is, the specific heat of water can be *defined* to be

$$c_{water} = 1.0 \frac{\text{calories}}{\text{gram K}}. \tag{10.33}$$

To measure the specific heat of some other substance, then – e.g., the metallic chips given off during the cannon-boring process observed by Rumford – we can simply place a mass M of the substance (at some initial temperature T_i) into a mass m bath of water at temperature T_0. Assuming $T_0 < T_i$, heat will flow from the substance into the water until everything equilibrates at some intermediate temperature T_f. The quantity of heat that flowed *out* of the substance will evidently be $\Delta Q = c M (T_i - T_f)$, whereas the quantity of heat that flowed *into* the water is given by $\Delta Q = c_{water} m (T_f - T_0)$. Equating these gives a formula for the specific heat c of the substance in terms of measureable values (and the stipulated value for c_{water}):

$$c = c_{water} \frac{m (T_f - T_0)}{M (T_i - T_f)}. \tag{10.34}$$

Anyway, Rumford's idea was that, if the considerable heat produced from the boring of the cannon had been extracted from the brass chips that were being cut away, then those brass chips would be left in an unusual, caloric-deprived state which should plausibly change the specific heat from its ordinary value.

"But no such change had taken place; for I found, upon taking equal weights of these chips and of thin slips separated from the same block of metal by means of a fine saw, and putting them at the same temperature (that of boiling water) into equal weights of cold water initially at the temperature of $59\frac{1}{2}°$ F, the portion of water into which the chips were put was not, to all appearance, heated either less or more than the other portion in which the slips of metal were put. This experiment being repeated several times, the results were always so nearly the same that I could not determine whether any, or what, change had been produced in the metal, *in regard to its specific heat*, by being reduced to chips by the borer."

This of course strongly suggests that the heat produced by the (rather extreme) friction between the brass cannon stock and the drill bit, is genuinely *produced*, as opposed to merely being released from a latent state. It thus appears to disconfirm the idea, implicit in the caloric theory of heat, that the total quantity of heat in the world is a fixed constant – and in its place to support the idea that heat is just a particular type of ordinary mechanical energy.

In a further experiment, Rumford repeated the boring of the cannon stock, but with now the whole apparatus underwater. This modification was designed to rule out the hypothesis that the heat "produced" was actually being contributed by the surrounding air. With the apparatus underwater, of course, there was no surrounding air. And if the heat was contributed instead by the surrounding water, it should have been easy to observe and measure the resulting *decrease* in the temperature of the water. Of course, no such decrease in the water temperature occured: "At $2\frac{1}{2}$ hours [the water] ACTUALLY BOILED! It would be difficult to describe the surprise and astonishment expressed in the countenances of the bystanders on seeing so large a quantity of cold water heated, and actually made to boil, without any fire."

Rumford summarized his findings as follows:

> "It is hardly necessary to add that anything which any *insulated* body, or system of bodies, can continue to furnish *without limitation*, cannot possibly be *a material substance*; and it appears to me to be extremely difficult, if not quite impossible, to form any distinct idea of anything capable of being excited and communicated in the manner in which heat was excited and communicated in these experiments, except be it MOTION."

In short, the experiments seem to prove that heat *just is* motion.

10.4.2   The Mechanical Equivalent of Heat

The kinetic theory of heat implies the interconvertability (and hence unity) of "heat" and mechanical work or energy. Its establishment is thus a major step toward the assimilation of thermal phenomena under the growing umbrella of the Newtonian, mechanical worldview.

In the decades after Count Rumford's experiments, many physical scientists explored more carefully and quantitatively the conversions between heat and mechanical energy. (This issue was extremely relevant to the subject of steam engines, which was an important developing technology at the time. But more on this next week.) Let us discuss in particular the work of James Joule, whose experiments in the 1840s finalized the evidence for the kinetic theory of heat by determining the exact numerical correspondence between the "calorie" and familiar units of mechanical energy such as the (aptly named) Joule.

Joule designed an apparatus to measure what he called "the mechanical equivalent of heat." The idea was to let some object do a certain, precisely-measurable amount of mechanical work on another system, and then to measure precisely the temperature increase produced. Joule invented a paddle-wheel apparatus, in which a falling weight turned a crank that was connected to a paddle-wheel inside a thermally insulated canister of water. See Figure 10.10. The mechanical work that is done on the water in the canister by the falling weight is then just equal to the decrease in the weight's potential energy: Mgh, where h is the height it falls through.

The heat added to the water (ΔQ) is related to its temperature change (ΔT) according to the formula written earlier:

$$\Delta Q = c_{water}\, m\, \Delta T \tag{10.35}$$

where c_{water} is the specific heat of water and m is the mass of water in the canister. (A more precise calculation would take into account that some of the material of the canister also absorbs some heat, but this gives the basic idea.)

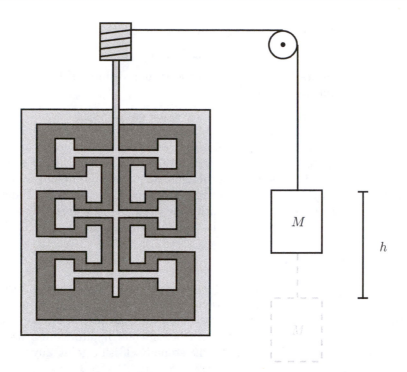

Figure 10.10: Joule's apparatus to measure the "mechanical equivalent of heat." The insulated canister contains water, whose temperature is measured before, and then again after, the mass M is allowed to fall through some vertical height h. The mass is hung from a string which is wound around the axle of a paddlewheel inside the cannister, so as the weight falls, the paddlewheel turns and churns the water, delivering to it a known quantity (Mgh) of mechanical energy. Given the mass m of the water in the canister and its temperature change ΔT during the experiment, the heat delievered to the water can be easily calculated: $\Delta Q = c_{water}\, m\, \Delta T$. One thus determines what is (according to the kinetic theory of heat) one and the same physical quantity in two different units (say, calories and Joules).

According to the kinetic theory of heat, this added heat just *is* the (kinetic) energy produced by the mechanical work Mgh. By measuring the temperature change ΔT produced in a known mass m of water, we can measure the heat produced in calories. And by simply calculating the product Mgh, we measure the heat produced in mechanical energy units such as Joules. Equating these then allows us to discover the number of Joules that corresponds to a calorie – i.e., the "mechanical equivalent of heat."

Another way to view the experiment is that it allows us to calculate the specific heat of water – not in calories/gram-degree, which value we don't *need* to do any experiment to know (because the value is, by definition of "calorie", one calorie/gram-degree) – but rather in Joules/gram-degree. The relevant formula can be found by equating ΔQ from the previous equation, with the expression for the mechanical work done, and solving for the specific heat:

$$c = \frac{Mgh}{m\Delta T}. \tag{10.36}$$

What Joule actually found when he performed the experiment is that the quantity of heat needed to raise one pound of water by one degree Farenheit was equivalent to the mechanical work involved in the fall of 772 pounds through a distance of one foot. Or, restating that in contemporary units: a weight of one Newton (i.e., a mass of about 100 grams) needs to fall through a height of 4.2 meters in order to raise the temperature of one gram of water by one degree Kelvin. In short:

$$1 \text{ calorie} = 4.2 \text{ Joules,} \tag{10.37}$$

which formula provides the quantitative link between the standard units of heat and energy, and thus establishes their in-principle inter-convertability.

The main significance of Joule's work for the further development of physics was that it made clear for the first time that "energy conservation" was a principle with a much wider domain of applicability than previously appreciated. It was not merely a principle of "mechanics" which applied only in so far as there wasn't too much friction. It was rather an apparently universal law, applicable even where friction was present and important, the result of friction (and other similar phenomena) being simply to convert the obvious and potentially useful sort of mechanical energy into a more microscopic, hidden form: heat. As we'll discuss in more detail in the next chapter, the principle that *energy is always conserved when heat is taken into account* is extremely fundamental and will be designated as the first law of thermodynamics.

10.4.3 The Kinetic Theory of Gases

The kinetic theory of heat implies that the atomic (or molecular) constituents of material substances are in some kind of incessant motion, with faster motion evidently corresponding to higher temperatures. The "kind of motion we call heat" (to adopt a phrase used by Clausius, whom we shall meet shortly) must be random and disordered – for example, it can't be that all the atoms in a hot stone in the fire are moving the same direction, because, observably, the stone itself isn't going anywhere.

If we apply these ideas to the gaseous state of matter in particular, it follows that – contrary to the picture, sketched earlier in Figure 10.7, suggested by Newton and developed by Dalton – the individual atoms or molecules in a gas are not at all stationary, but are rather flying, chaotically, every which way. And this picture of the microscopic goings-on in what appears macroscopically to be a quiescent gas, would seem to have further implications for the other macroscopic properties of gases, in particular *pressure*. Could it be, for instance, that gas pressure has nothing to do with repulsive forces between atoms, but instead arises from the "outward push" produced when the moving molecules encounter and reflect from the walls of their containers?

This idea had actually been around for a long time. Boyle himself, for example, believed that "the elastical particles of the air ... may ... be like the springs of watches, coiled up and still endeavouring to fly abroad [and] may have a continual endeavour to stretch themselves out, and thrust away the neighbouring particles ... [and] attain their full liberty." But, he continued, "I will allow you to suspect that there may be sometimes mingled with the particles that are springy, ... some others that owe their elasticity not so much to their structure, as their motion, which ... whirling them about, may make them beat off the neighbouring particles and thereby promote an expansive endeavour in the air..."

The possibility that *all* of the "expansive endeavour in the air" could be due to random translational motions of the constituent particles had been developed as early as 1738 by the physicist Daniel Bernoulli. Bernoulli showed that, assuming a constant average speed for the molecules, one could understand gas pressure – and in particular the relation between pressure and volume known as Boyle's Law – as arising from such collisions. Bernoulli's argument, in essence, was just that, if pressure arises from the collisions of moving gas molecules with the walls of the container, then twice as many molecules (in the same volume and moving with the same speed) should imply twice as many collisions and hence twice the pressure. Pressure, therefore, should be proportional to the *density* of the gas. But density being the total quantity of gas divided by its volume, the proportionality of pressure and density is equivalent to the *inverse* proportionality of pressure and volume, which is Boyle's Law.

Despite containing the essential ideas of what is now known as the kinetic theory of gases, Bernoulli's work on this topic was largely ignored. An obscure English physicist, John Herapath, independently

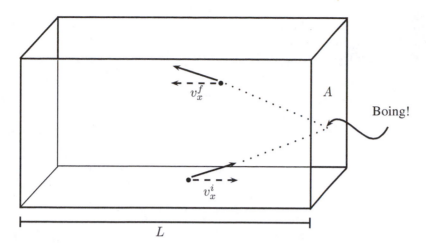

Figure 10.11: A molecule is initially moving to the right, with velocity x-component $v_x^i = v_x$, inside a box of length L and cross-sectional area A. It eventually collides with the wall on the right hand side and bounces back toward the left. During the collision, the x-component of its momentum changes by $\Delta p_x = 2mv_x$.

rediscovered the same ideas in the 1820s, but his ideas, too, were mostly ignored. James Joule recognized the importance of Herapath's work and tried to publicize the ideas in the late 1840s, but this too had little effect. In the mean time, the same ideas had again been independently rediscovered (and developed even more cogently) by J. J. Waterston, a Scottish physicist. But he, too, was ignored by the scientific community. Sometimes it takes a long time before good ideas are widely accepted!

In fact it wasn't until 1856, when the German physicist Rudolf Clausius published his paper on "The Kind of Motion We Call Heat", that the kinetic theory of gases was taken seriously and, then, widely accepted by the scientific community.

Let's go through a more detailed derivation of the ideal gas law from the kinetic model, to bring out some important technical points.

We'll start by considering just a single molecule bouncing around inside a box. Suppose the box has a length L in the x-direction and a cross-sectional area A – and hence a volume $V = LA$. Now suppose the molecule has mass m and is moving such that the x-component of its velocity is v_x. (It might as well be moving at some angle, so the other components of the velocity need not be zero – it just won't matter exactly what they are.) Now consider the collision of the molecule with the right-hand-side wall of the container. Prior to the collision, the x-component of the molecule's momentum will be $p_x = mv_x$. After the collision, which we assume is elastic, it will be headed back the other way with $p_x = -mv_x$. Thus, the x-component of its momentum *changes* by

$$|\Delta p_x| = 2mv_x. \qquad (10.38)$$

How frequently do these collisions occur? Well, after the collision the molecule is headed back to the left. Eventually it will collide with the wall on the left, and bounce back to the right. (There may be other collisions, too, with the other walls, but these will not affect the motion in the x-direction.) The time between two subsequent collisions with the right-hand-side wall should be just the total distance traveled (in the x-direction) between two collisions (that is, $2L$) divided by the x-component of the velocity, v_x:

$$\Delta t = \frac{2L}{v_x}. \qquad (10.39)$$

Now, according to Newton's second law, the momentum change of the molecule upon a collision is a result of a force exerted on it by the wall. The *average magnitude* of this force (averaged, that is, over a long long time during which the particle makes many back-and-forth trips) can be computed this way:

$$F = \frac{\Delta p_x}{\Delta t} = \frac{2mv_x}{2L/v_x} = \frac{mv_x^2}{L}. \tag{10.40}$$

And this average force is, by Newton's *third* law, equal in magnitude to the time-average force the molecule exerts *on the wall*.

But then, the *pressure* exerted on the wall is just the force exerted on the wall divided by the area of the wall:

$$P = \frac{F}{A} = \frac{mv_x^2}{LA} = \frac{mv_x^2}{V} \tag{10.41}$$

where $V = LA$ is just the total volume of the container. Note that we are already starting to see the emergence of Boyle's Law, that is, the inverse proportionality between pressure and volume.

So far we've only considered a single molecule. But a real gas will contain lots and lots of molecules – N of them, say. (Let's assume for simplicity that they all have the same mass m.) Well the above analysis applies separately to each individual particle, so we can write immediately that the *total* pressure exerted on the wall (resulting from collisions of the individual molecules with the wall) is just the sum of the pressures produced by each individual molecule:

$$P = \sum_i \frac{m_i v_{x,i}^2}{V} = \frac{Nm\langle v_x^2 \rangle}{V} \tag{10.42}$$

where the angle-brackets in the last expression denote the average of the indicated quantity over all the different molecules.

Now, in a gas, at any particular instant, the motions of the molecules will be very complicated and random – some flying this way, some that, etc. In particular, there's no apparent reason why the average value of v_x^2 over all the molecules, should be any different from the average value of v_y^2 or v_z^2. Thus, we can write

$$\langle v_x^2 \rangle = \langle v_y^2 \rangle = \langle v_z^2 \rangle = \frac{1}{3} \langle v^2 \rangle \tag{10.43}$$

where $\langle v^2 \rangle = \langle v_x^2 + v_y^2 + v_z^2 \rangle$ is the average squared *speed* of the molecules in the gas.

This then allows us to write

$$m\langle v_x^2 \rangle = \frac{2}{3} \langle KE \rangle \tag{10.44}$$

where $\langle KE \rangle = \langle \frac{1}{2}mv^2 \rangle$ is the average translational kinetic energy for the different molecules in the gas.

Putting this all together, we are left with (moving the V to the other side)

$$PV = \frac{2}{3} N \langle KE \rangle. \tag{10.45}$$

To make the full implications of this result clear, it is helpful to write it this way:

$$PV \sim \langle KE \rangle \tag{10.46}$$

where the proportionality holds for a fixed quantity (e.g., by mass) of the gas. The point of course is that this has the same structure as the empirical ideal gas law – the product of P and V is proportional to the (absolute) temperature – *if we make the further identification of temperature as being proportional to the average kinetic energy of the gas molecules.* Let us then officially endorse this hypothesis and see what it allows us to understand. Formally, let us assume that

$$\langle KE \rangle = \frac{3}{2} k_B T \tag{10.47}$$

where, of course, T is the (absolute) temperature, the factor of 3/2 is introduced for later convenience, and the proportionality constant k_B is proposed as a new fundamental constant of nature which, in effect, plays the role of connecting the microscopic-mechanical units for energy to the macroscopic-thermometric units for temperature. This constant is called "Boltzmann's constant" (hence the subscript B) after the late 19th century physicist Ludwig Boltzmann, whose work we will study in the coming weeks. It is very important to appreciate that, at this stage, the *value* of this new constant is completely unknown!

In any case, though, taking Equation (10.47) as a new postulate, we may rewrite Equation (10.45) as follows:

$$PV = N k_B T. \tag{10.48}$$

Recall that N here refers to the total number of molecules contained in a sample of gas. This number, for a real sample of gas, is of course also totally unknown.

Let us step back and consider the implications of this result.

First, the kinetic theory model – in which gas pressure arises from the collisions of moving molecules and (absolute) temperature is identified with the average translational kinetic energy of the moving molecules – accounts for the behavior summarized in the empirical ideal gas law. In particular, the kinetic theory model predicts that, if the temperature T is held fixed, the pressure of a given sample of gas should be inversely proportional to its volume (Boyle's law), and also that, if the pressure is held fixed, the volume should be proportional to the (absolute) temperature (Charles' law). On those grounds alone, one should consider the kinetic theory as a very serious candidate description of the microscopic physics of gases.

Second, the kinetic theory model actually predicts that Avogadro's hypothesis should be true: with k_B a universal constant, Equation (10.48) implies some particular value of N for a given P, V, and T. That is, equal volumes (V) of different types of gas, under the same pressure (P) and temperature (T) conditions, must contain the same number N of particles, just as Avogadro had hypothesized. So to whatever extent one finds the kinetic theory of gases to be independently compelling, one will regard it as strong evidence for the *truth* of Avogadro's hypothesis.

Let us finally rewrite the ideal gas law in one final form – the form in which you are probably most familiar with it from previous chemistry or physics classes. Recall that, according to Avogadro's hypothesis, the number densities of different chemical species of gas (under the same pressure and temperature conditions) are all the same. And so the mass densities – which can be understood as the products of the number densities (number of molecules per unit volume) and the molecular weights (mass per molecule) – are simply proportional to the molecular weights. So one can infer, for example, from the fact that the mass density of oxygen gas is about 16 times that of hydrogen gas, that the relative atomic weight of oxygen is 16 times that of hydrogen:

$$m_O : m_H = 16 : 1. \tag{10.49}$$

(Strictly speaking, what follows from the 16:1 ratio of mass densities is the 16:1 ratio of *molecular* weights, but since both hydrogen and oxygen are required to be di-atomic molecules, on Avogadro's hypothesis, the relative molecular weights and the relative atomic weights are identical.) A parallel argument establishes, for example, that the atomic weight of nitrogen is 14 times that of hydrogen

$$m_N : m_H = 14 : 1 \tag{10.50}$$

and similarly for many other elements. The point is, it appeared that of all the different types of atoms, Hydrogen atoms were the *lightest*.

Let us then agree to call the number of Hydrogen atoms in a one gram sample of hydrogen gas "Avogadro's number", N_A. Or, equivalently, since hydrogen gas appears to be diatomic, a *two* gram sample of hydrogen gas will be said to contain N_A hydrogen molecules. Of course, we do not yet have any idea what "Avogadro's number" actually *is*. We just know that, for example, two grams of hydrogen gas, or 28 grams of nitrogen gas, or 32 grams of oxygen gas, all contain the same number of molecules, namely N_A (whatever it is!).

In the late 19th century, the term "mole" was introduced to refer to the quantity of a given substance which contains Avogadro's number of molecules. Thus, one mole of hydrogen gas has a mass of two grams, one mole of oxygen gas has a mass of 32 grams, and so on. The number of moles, n, in a given sample can be written as the number N of molecules in that sample, divided by Avogadro's number:

$$n = \frac{N}{N_A}. \tag{10.51}$$

Re-expressing our kinetic theory version of the ideal gas law, Equation (10.48), in terms of n gives

$$PV = n\,(N_A k_B)\,T \tag{10.52}$$

or, giving a new name to the constant product of two other constants,

$$PV = n\,RT. \tag{10.53}$$

Note that, although the "ideal gas constant" R is the product of two other constants whose values remain unknown and cannot straightforwardly be determined from observation, the value of R *can* easily be determined from observation! For example, one mole (i.e., two grams) of hydrogen gas, at room temperature ($T = 300\,\text{K}$) and atmospheric pressure ($P = 1\,\text{atm}$) is observed to occupy a volume of 24.5 Liters. So, evidently,

$$R = \frac{1\,\text{atm}\ 24.5\,\text{L}}{1\,\text{mole}\ 300\,\text{K}} = 0.082\,\frac{\text{L}\cdot\text{atm}}{\text{mole}\cdot\text{K}}. \tag{10.54}$$

So, the two unknown constants – "Boltzmann's constant" k_B, which was involved in the assumed proportionality between temperature and average translational kinetic energy, and "Avogadro's number" N_A – are both unknown, but are intimately linked. If we could determine a value for one of these, we could immediately infer a value for the other using

$$R = k_B N_A \tag{10.55}$$

and the known value of R. Determining this single unknown quantity would be equivalent to determining the absolute size of atoms. For example, if the value of N_A were known, we would know that the mass of an individual Hydrogen atom is $m_H = 1\,\text{gram}/N_A$.

Establishing the absolute scale of atoms in that way was a major goal of scientists in the 19th century. In the remaining weeks of the course, we will trace the path that finally, in the first decade of the 20th century, led to this major achievement.

Questions:

Q1. Here is a description of an experiment performed by Joan-Baptista van Helmont in 1648:

> "That all plants immediately and substantially stem from ... water alone I have learnt from the following experiment. I took an earthern vessel in which I placed two hundred pounds of earth dried in an oven, and watered with rain water. I planted in it the stem of a willow tree weighing five pounds. Five years later it had developed a tree weighing one

hundred and sixty-nine pounds and about three ounces. Nothing but rain (or distilled water) had been added. The large vessel was placed in earth and covered by an iron lid with a tin-surface that was pierced with many holes. I have not weighed the leaves that came off in the four autumn seasons. Finally I dried the earth in the vessel again and found the same two hundred pounds of it diminished by about two ounces. Hence one hundred and sixty-four pounds of wood, bark and roots had come up from water alone."

Do you think Helmont's experiment proves what he says it proves? How would you design an improved experiment to discover the source of the growing tree's increasing mass?

Q2. Atmospheric pressure (at sea level) can support a 34 foot high column of water. The density of air is roughly a thousand times less than the density of water. Roughly how high is the Earth's atmosphere? Torricelli was the first to introduce the concept of a "blanket of air" surrounding the Earth, on precisely these grounds.

Q3. Atmospheric pressure is evidently equal to the weight, per unit area, of a 34 foot high column of water: $P_{atm} = \rho g h$ with $\rho = 1$ gram/cc^3, $g = 9.8$ m/s^2, and $h = 34$ feet. Plugging in and simplifying reveals that $P_{atm} \approx 100,000$ Newtons/m^2. That is, the air around you right now is exerting a force of *a hundred thousand Newtons* (i.e., roughly the weight of ten cars!) on every square meter area that it touches, including, for example, the outside of your body. Can this be right?! What prevents your chest from getting crushed in by this enormous force?

Q4. Why does a helium balloon go up instead of down? Sure, helium is less dense than air. But it still has a positive density, i.e., a positive weight, i.e., it will experience a downward gravitational force (just like a brick or anything else). So there must exist some other (upward) force on the balloon that is at least as big as its weight. What kind of force is this and where does it come from?

Q5. Explain briefly how the law of definite proportions is evidence for the existence of atoms. But does this empirical law by itself *conclusively prove* the existence of atoms? Why or why not? Can you invent a (half-baked) theory which is non-atomic, but which predicts the law of definite proportions? (If you can't, or rather if you're convinced that nobody could, it seems you'd have to believe in atoms *just* on the basis of the law of definite proportions... which somehow doesn't seem very plausible.)

Q6. Suppose elements A and B can combine to form two different compounds, C1 and C2. By weight, C1 is 30 percent A and 70 percent B. C2 is 46 percent A and 54 percent B. Is this consistent with the law of multiple proportions?

Q7. Suppose 1 gram of element A can combine with 3 grams of element B to make a certain compound X, or that same 1 gram of A can combine with 6 grams of B to make a different compound, Y. Based just on this data, which compound (X or Y) do you think Dalton would be more likely to think is "binary"?

Q8. Leaving aside (for the moment) the controversy between Dalton and Gay-Lussac about how to properly determine the relative atomic weights, one of the points emerging from this chapter is that one can in principle determine the weights of atoms relative to one another much more easily than one can determine any of those weights absolutely. For example, one can infer that an oxygen atom weighs 16 (or perhaps 8) times as much as a hydrogen atom, even while having no idea whatsoever about what the mass of a hydrogen atoms *is*. What does this remind you of from the first part of the course? (Hint: that the Sun's mass was about 330,000 times the mass of the Earth was known even before Cavendish performed his experiment.) Can you think of any other similar examples?

Q9. Mendeeleev proposed the periodic table of the elements in 1869. Basically what he did was to list the chemical elements in order by atomic weight, and notice that their other chemical properties (such as molar volume, boiling point, melting point, etc.) displayed certain regular periodicities such that certain groups of elements (the "families") could be identified as sharing similar properties. Mendeeleev importantly left gaps in his table for as-yet-undiscovered elements, many of which were subsequently discovered. Why didn't anybody do this earlier? What specifically had to be sorted out first before the periodic table could be put together?

Q10. The cold soda diet: Suppose you drink two liters of water that is just above the freezing point of water, $T = 0°$ C. The water will eventually be brought into thermal equilibrium with the rest of your body, $T = 37°$ C. The two liters of water has a mass of 2000 grams, and so 74,000 calories of energy are required just to maintain your body temperature after drinking the water. Actually, we might as well add some sugar and flavor to the water. A two liter bottle of soda contains roughly 1,000 Calories, so if you drink an entire two liter bottle of soda (appropriately chilled) your net calorie intake is a *deficit* of 73,000 calories. The typical human diet consists of some 2,000 calories per day. So all you have to do to lose weight is add a cold bottle of fully-sugared soda to your normal diet every couple of weeks. What's wrong with this argument? (Hint: the small-c "calorie" defined in the text is different from the capital-C "Calorie" that is used to measure the energy content of food.) What fraction of the energy content of a 2-liter bottle of soda would you "burn" just by raising its temperature from $0°$C to $37°$ C?

Q11. One argument against the kinetic theory of heat was the following: if heat is merely the motion of ordinary material particles, how does heat from the Sun propagate to the Earth, given that the space between the Sun and Earth is almost entirely empty? What do you make of this argument?

Q12. If the piston (in a cylinder containing some gas) is slowly pulled back, so the volume of the gas increases, the *temperature* of the gas will decrease. Explain why. (Hint: consider the effect of a collision of a gas molecule with the moving piston.) What happens to the temperature if a sample of gas is compressed?

Q13. In rehearsing the kinetic-theory-of-gases-based derivation of the ideal gas law, we argued that the time-average force, exerted by a single particle on one wall of its container, is given by $F = mv_x^2/L$. But this time-average seems to have little to do with what actually occurs: the force exerted on the wall by the particle is actually zero almost all of the time, but is then some enormously large value during the extremely brief periods when the particle is actually in contact with the wall. Why, despite this, is it reasonable to compute the pressure using the time-averaged force F?

Exercises:

E1. Show that the purely-empirical "$PV = gMT$" version of the ideal gas law is compatible with the final "$PV = Nk_BT$" version. In particular, what is the relationship between the empirical proportionality constant g for a given species of gas, and the mass m of a single molecule of that gas? Show explicitly that the two empirical values of g reported in the text (for Hydrogen and Nitrogen) are consistent with the atomic weights for those elements that are implied by Avogadro's hypothesis.

E2. 8 grams of oxygen gas will combine with 1 gram of hydrogen gas to form water (with none of either reactant left over). Also, 14 grams of nitrogen gas will react with 3 grams of hydrogen gas to form ammonia (with none of either reactant left over). In light of this data, what relative atomic weights would Dalton assign to hydrogen, nitrogen, and oxygen?

E3. It was noted in the text that

> 3 grams of Carbon and 8 grams of Oxygen can combine to form one pure gas compound (which plants love but which kills birds and other animals); and, under different condi-

tions, the same 8 grams of Oxygen can combine with 6 grams of Carbon to form a distinct gas compound (which even plants don't like).

Based on this data, what would Dalton say are the relative atomic weights of carbon and oxygen, and what would he say are the molecular formulas of the two different gas compounds? Explain carefully how you apply the "rule of simplicity" here.

E4. Suppose someone adamantly insisted that the exact relative atomic masses of oxygen and hydrogen were: $m_O : m_H = 137 : 19$. Is this possible – i.e., is it logically consistent with the empirical fact that 1 gram of hydrogen combines with 8 grams of oxygen to form 9 grams of pure water? If you say it isn't possible, explain clearly why it isn't; if you say it is possible, explain how and give the corresponding molecular structure of the water molecule.

E5. Suppose that three different compounds (X, Y, and Z) can be made with elements A and B:

- 5 grams A + 1 gram $B \rightarrow$ 6 grams X

- 5 grams A + 4 grams $B \rightarrow$ 9 grams Y

- 5 grams A + 2 grams $B \rightarrow$ 7 grams Z

If X is binary, what do the molecules of Y and Z look like? If Y is binary, what do the molecules of X and Z look like? If Z is binary, what do the molecules of X and Y look like? Which compound would Dalton think is binary, and what would he claim are the relative atomic weights of A and B

E6. One unit volume of nitrogen gas can combine with three unit volumes of hydrogen gas to make two unit volumes of ammonia gas. If, like Dalton, you believe that ammonia is a binary compound of nitrogen and hydrogen, what must be the relative number densities of the three gases? In this same reaction, the ratio of the mass of reacting nitrogen to the mass of reacting hydrogen is 14:3. If somebody suggested that the three gases had equal number densities, how would Dalton reply?

E7. As long as Boltzmann's constant k_B and Avogadro's number N_A are unknown, the absolute mass m and absolute kinetic energy KE of (say) a single gas molecule cannot be determined. However, as first pointed out by Herapath and then again later by Clausius, the kinetic theory of gases does allow the "typical" *speeds* of gas molecules to be calculated. Figure out how this works and then calculate, for example, the RMS (root mean square) speed of nitrogen and hydrogen molecules at room temperature and atmospheric pressure. It is interesting to compare the results to the speed of sound in air (about 330 m/s) as well as the Earth's escape velocity (about 11,000 m/s).

E8. Suppose the tires on your car are supposed to be inflated to a pressure of 30 psi. (A "psi" is one pound per square inch; one atmosphere is about 15 psi.) This actually means that the pressure should be higher than the outside ambient pressure by 30 psi, so the actual pressure of the gas inside the tires should be 45 psi. Suppose you have your tires properly inflated on a nice summer day, when the temperature is about 25 °C. Now suppose after a few months there is a very cold day when the temperature drops to 0 °C. By what fraction will the pressure in your tires decrease? Do you need to stop at the gas station for some air before driving?

E9. Archimedes' principle of *buoyancy* states that an object submerged in a fluid experiences an upward force equal in magnitude to the *weight of the displaced fluid*. Derive this formula by considering a rectangular block (of height h and cross-sectional area A) submerged in a fluid of density ρ.

E10. Work through the kinetic theory derivation of the ideal gas law *without* making the assumption that all the molecules have the same mass. Does it still work? Consider in particular a mixture of different species of gas, e.g., the air. Does the kinetic theory account for the fact that the air obeys the ideal gas law? Suppose some nitrogen gas (whose molecules are moving, on average, at

500 m/s) is mixed with some oxygen gas (whose molecules are also moving at 500 m/s). What will happen?

Projects:

P1. Reproduce a simplified version of Boyle's and Charles' experiments. Is your data consistant with Boyle's Law and Charles' Law? Use your data to determine values for P_{atm} and the temperature (in degrees Celcius) of "absolute zero".

P2. You will be given some information about the masses and, for gaseous substances, volumes of various quantities involved in several (hypothetical) chemical reactions. Your job will be to determine the relative atomic weights of each element involved, as well as the molecular structure of each compound, twice – once using the approach that Dalton would have taken, and then again using the approach that Gay-Lussac and Avogadro would have taken. Leaving aside additional evidence (e.g., the kinetic theory of gases), does either scheme seem inherently more reasonable to you?

P3. Measure the specific heats of some samples of different materials by heating them up to some known initial temperature and then dropping them into a known quantity of water and measuring the temperature change. You will be given a table with information about two different possible sets of relative atomic weights for various materials – one computed using Dalton's approach and one computed using Avogadro's. Can you find any relation between the measured specific heat values and the conjectured relative atomic weight values that would provide evidence for the correctness of either approach?

P4. Perform your own experiment to determine "the mechanical equivalent of heat", i.e., the conversion between calories and Joules. A crude but fun way to do this is to measure the temperature of some water in a thermos bottle, shake the heck out of it for several minutes, and then measure its temperature again. If you estimate the mechanical work (force times distance) done on the water per shake, and count the shakes, you can estimate the total energy added to the water in Joules. Measuring the water's temperature increase (and knowing its mass) allows you to determine the energy added in calories. You may also perform a more rigorous experiment in which energy is provided to the water by a submerged resistor through which a known electrical current is passing.

CHAPTER 11

Thermodynamics

IN Chapter 10 we discussed the rejection of the caloric theory of heat in favor of the kinetic theory of heat – mostly just as a way to set the stage for a discussion of the kinetic theory of gases. This analysis of the micro-physics of gases was seen to provide an almost irresistably elegant explanation of almost everything that was known about the physics of gases by the middle of the 19th century, and to support the truth of Avogadro's hypothesis to boot.

In the present Chapter we will focus on a different but equally important implication of the kinetic theory of heat – the idea, enshrined in the first law of thermodynamics – that when we take into account that heat is a form of energy (or, more precisely, a form of energy transfer), energy can be recognized as a rigorously conserved quantity. In particular, we will work to develop and reconcile two perspectives on certain thermal phenomena: the macroscopic perspective, in which one analyzes things in terms of abstract conservation principles, and the microscopic perspective in which one thinks in more mechanical detail about the motions of the atoms and molecules that constitute the systems in question.

We will also work to develop this same, dual perspective on processes relating to the *second* law of thermodynamics. This will give us occasion to again contemplate the possible existence of an invisible fluid (this one corresponding not to "heat", but to "entropy"). The resolution of this puzzle, described in Chapter 12, will finally lead us to an experimental determination of Avogadro's number.

§ 11.1 The First Law of Thermodynamics

The upshot of Joule's experiments, discussed in the previous chapter, can be put this way: in cases where friction acts (such that the total macroscopic energy of a system appears to decrease), an amount of heat is created that is proportional to the missing energy. Joule's conclusion – the essence of the kinetic theory of heat – is that the new heat *just is* the missing energy. In particular, "temperature" is really a measure of the average translational kinetic energies of the invisible, microscopic particles (atoms and/or molecules) that compose macroscopic objects. On this view, friction is not a process whereby energy is *lost*. Rather, it is a process whereby energy is *hidden* – hidden away, that is, in the invisible microscopic jigglings of atoms and molecules.

The implication is that, contrary to macroscopic appearances, the total energy in any physical process is conserved. It can be helpful to add the following emphasis: the total energy in any physical process is conserved *when heat is taken into account*. This fundamental principle is called the first law of thermodynamics.

In thermodynamics, we are typically concerned with a certain object or sample which is our "system" of interest. The upshot of the first law is then that the total energy of the system can change in two ways.

First, *mechanical work* can be done on the system – for example, we can apply an external force to the system which does work on it by compressing it. And second, *heat* can be allowed to flow into (or out of) the system – for example, the system could be placed next to a hot fire. We summarize this by writing

$$dE = dW + dQ \tag{11.1}$$

where dE represents the change in the energy of a system during some "small" process, and dW and dQ represent respectively the work done *on* the system during that process and the heat which flows *into* the system during that process.

It is important to emphasize that there is no such thing as "the heat contained in" a system, just as there is no such thing as "the work contained in" a system. Heat and work are not the names of specific, distinguishable substances that a thermodynamic system contains. That, at least for "heat", was the view of the caloric theory of heat, but it is essential to understanding the kinetic theory of heat (and so the first law of thermodynamics!) that this is not the case. Instead of thinking of "heat" and "work" as particular kinds of "stuff" that can be added to a system, you should instead think of them as particular *ways* that the same one "stuff" – energy! – can be added to a system. That is what the first law of thermodynamics says.

We will frequently use a sample of gas in a piston as an example system, partly because it is simple and partly just because we've already learned a lot about how gases behave. The overall, macroscopic *state* of such a sample can be characterized by specifying values for various thermodynamic properties such as pressure, temperature, and volume. It is thus convenient to re-write the work done on the gas (dW) in terms of these variables. So: consider a sample of gas in a piston, which initially has pressure P and volume V. Now suppose that an external agent (such as your hand) does some work on the gas by compressing it. In order to move the piston (slowly, with constant speed) your hand must apply a force on the piston equal (but opposite) to the force applied to the piston by the gas itself. This is $F = PA$ where A is the cross-sectional area of the piston. If you slide the piston through a (small) distance dx, the volume of the gas changes by $dV = -A\,dx$ where A is the cross-sectional area of the piston.

The mechanical work you do on the gas during this process is given by $dW = F \cdot dx$, which (using the two expressions from the previous paragraph) can be re-written this way:

$$dW = -P\,dV. \tag{11.2}$$

The minus sign here is necessary because positive work done on the gas (which tends to increase its energy) corresponds to the gas being *compressed*, i.e., reducing its volume, i.e., a negative dV.

With this expression for dW, the first law of thermodynamics can be written this way:

$$dE = dQ - PdV. \tag{11.3}$$

We will use this equation again and again in what follows.

§ 11.2 Heat Capacities at Constant Volume and Constant Pressure

In this section we use the first law of thermodynamics to analyze two example processes that one could perform on a thermodynamic system: heating it while holding the volume constant, and heating it while holding the pressure constant. For each process, some heat flows in and the temperature increases by a certain amount, so we can define an associated "heat capacity" for the system as the ratio of these quantities. These heat capacities can be readily measured, and provide some insight into the underlying molecular nature of gases.

11.2.1   Heating at constant volume

As a first – and simplest – example of a thermodynamic process, let us consider the heating of a sample of gas whose volume is held fixed. Then $dV = 0$ and so $dW = 0$. That is, no mechanical work is done on the system, and so the change in its total energy will equal the quantity of heat added:

$$dE = dQ. \tag{11.4}$$

For simplicity, let us assume that we are talking about *one mole* of gas. Then the heat capacity of our system will be the molar heat capacity for that particular type of gas. (The "molar heat capacity" is just like the "specific heat", except the heat input required per degree of temperature increase is figured per mole of the substance in question, rather than per unit mass.) Formally, we define the heat capacity as the proportionality constant between the heat added (dQ) and the resulting temperature increase dT:

$$dQ = c_V dT \tag{11.5}$$

where we have written c_V for the (remember, molar) heat capacity to remind us that this is the heat capacity associated with a process that occurs at *constant volume*. (We will later be discussing other sorts of processes, such as heating the gas at constant *pressure*, and so there will be distinct associated heat capacities that we'll encounter shortly.)

We can combine the previous two equations by eliminating dQ and solving for c_V:

$$c_V = \frac{dE}{dT} \tag{11.6}$$

which just means that the (molar) heat capacity is the ratio of the change in energy of (one mole of) the gas, to its change in temperature, when a little heat is added. Very often one will see this equation written instead this way

$$c_V = \left(\frac{dE}{dT}\right)_V \tag{11.7}$$

where the subscript V on the derivative reminds us that the ratio pertains to a process in which the volume V is held fixed.

Now let us recall our discussion of the kinetic theory of gases from last week. We posited there that the average translational kinetic energy of the molecules was proportional to the (absolute) temperature:

$$\langle KE \rangle = \frac{3}{2} k_B T. \tag{11.8}$$

If the molecules have *only* translational kinetic energy – i.e., if all of their kinetic energy is translational (as opposed to rotational or vibrational, which would seem to have to be the case at least for *monatomic* gases) – it follows that the *total* kinetic energy of the gas is

$$KE = \frac{3}{2} N k_B T = \frac{3}{2} n R T \tag{11.9}$$

where N is the total number of molecules, and $n = N/N_A$ is the number of moles. (N_A is Avogadro's number and R is of course the gas constant discussed last week: $R = 8.3$ Joules/mole K.)

Let us assume further that the gas is *ideal*, meaning here that the molecules don't exert any long-range forces on each other. This will mean that the system possesses no *potential* energy – i.e., *all* of its energy will be the kinetic energy we have written a formula for above. Then, for one mole of such a gas, the total energy will be

$$E = \frac{3}{2} RT \tag{11.10}$$

Gas	c_V (J / mol K)	c_P (J/mol K)	γ	$k = 2/(\gamma - 1)$
Hydrogen (H_2)	20.4	28.7	1.41	4.9
Helium (He)	12.5	20.8	1.66	3.0
Nitrogen (N_2)	20.6	29.0	1.41	4.9
Oxygen (O_2)	21.1	29.5	1.40	5.0
Chlorine (Cl_2)	24.8	34.1	1.38	5.3
Argon (Ar)	12.5	20.8	1.66	3.0
Carbon Monoxide (CO)	20.7	29.0	1.40	5.0
Carbon Dioxide (CO_2)	28.2	36.8	1.30	6.7
Hydrogen Chloride (HCl)	21.4	29.9	1.40	5.0
Nitric Oxide (NO)	20.9	29.3	1.40	5.0
Nitrous Oxide (N_2O)	28.5	37.0	1.30	6.7

Table 11.1: Molar heat capacities of several gases. (Data is for $T = 288\,K$ and $P = 1$ atm.)

which would mean, from Equation (11.6), that

$$c_V = \frac{dE}{dT} = \frac{3}{2}R = 12.5\,\frac{\text{Joule}}{\text{mol K}}. \tag{11.11}$$

Now the molar specific heats of gases are relatively straightforward to measure in the lab. The values for a number of different chemical species are shown in Table 11.1.

What jumps out immediately is that the prediction is not very accurate for most of the gases. But, as anticipated, it is accurate for the two *monatomic* gases in the table: Helium and Argon. The diatomic gases (such as Nitrogen, Oxygen, Chlorine, Carbon Monoxide, etc.) seem to have molar heat capacities of about $21\,J/mol\,K$, or about $5/2\,R$. And the tri-atomic species (such as Carbon Dioxide and Nitrous Oxide) have molar heat capacities that are even higher.

Though it does raise some questions, this is not a complete mystery. It indicates that some fraction of the total energy of poly-atomic molecules is in some form *other than translational kinetic energy*. As already suggested, this makes perfect sense from the point of view of the atomic theory, assuming we picture the atoms as extremely small points. Then, if the molecules of some gas are just individual atoms – if the gas is monatomic – all the molecules can do is move translationally. But if the molecules consist of two atoms joined together in some way, like a little barbell, then the molecules can not only move translationally, but also *rotate*. And in principle there could be *vibrational* motions as well.

Rudolf Clausius, the German physicist who, starting in the 1850s, formalized the science of thermodynamics and integrated it with the kinetic theory of gases, summarizes this way:

> "the hypothesis of a rotatory as well as a progressive motion of the molecules at once suggests itself; for at every impact of two bodies, unless the same happens to be central and rectilineal, a rotatory as well as a translatory motion ensues.

> "I am also of the opinion that vibrations take place within the several masses in a state of progressive motion. Such vibrations are conceivable in several ways. Even if we limit ourselves to the consideration of the atomic masses solely, and regard these as absolutely rigid, it is still possible that a molecule, which consists of several atoms, may not also constitute an absolutely rigid mass, but that within it the several atoms are to a certain extent moveable, and thus capable of oscillating with respect to each other."

Clausius then pointed out that the heat capacity data allows us to compute the fraction of the total energy of the gas that is translational-kinetic. Or more precisely, when we add some energy to the gas in the form of heat, we can compute (from the measured specific heats) the fraction of this added energy which goes into increasing the translational kinetic energy (which corresponds to the temperature) as opposed to the rotational and vibrational energy. The fraction is just

$$f = \frac{dKE_{trans}}{dE} = \frac{\frac{3}{2}nRdT}{nc_VdT} = \frac{3}{2}\frac{R}{c_V}. \tag{11.12}$$

For the monatomic gases listed in the table (for which $c_V = 3/2R$) the fraction is just one. For the diatomic gases, the fraction appears to be roughly

$$f \approx \frac{3/2}{5/2} = \frac{3}{5} \tag{11.13}$$

whereas for tri-atomic and more complicated molecules, the fraction is even lower.

The value $f \approx 3/5$ for diatomic gases can, to some extent, be understood. Any molecule can move translationally in three independent directions: the x direction, the y direction, and the z direction. Each possibility is said to correspond to one "degree of freedom" for the molecule. And we can understand the relation $E = \frac{3}{2}nRT = \frac{3}{2}Nk_BT$ for monatomic molecules by saying that each degree of freedom of each molecule has, on average, $\frac{1}{2}k_BT$ worth of energy. Now, thinking of a diatomic molecule as a little barbell built from two pointlike atoms connected by a little rigid bar, it is clear that there are two additional rotational degrees of freedom. For example, a molecule whose axis is parallel to the z axis can rotate about the x axis and the y axis.

What about rotation about the z axis? We say either that such rotation can't exist – on the same grounds, whatever they are exactly, that a monatomic molecule can't rotate at all – or that, even if it exists, it doesn't contribute anything to the energy since the moment of inertia about that axis is (approximately?) zero. In any case, whatever the reason exactly, the empirical heat capacity data suggests that for a diatomic molecule, rotation about the symmetry axis is not an "active" degree of freedom. (And neither, apparently, are the degrees of freedom associated with vibration.)

That gives five total "active" degrees of freedom for each diatomic molecule, so we need only assume that the two rotational degrees of freedom acquire, on average, the same energy that we already know the three translational degrees of freedom acquire at a given temperature, namely: $\frac{1}{2}k_BT$ each. This principle (which at this stage is really just an *ad hoc* assumption to account for the empirical heat capacity data) is called the "equipartition of energy" – the idea being that, on average, the total energy of the system is divided evenly ("equally partitioned") among all the "active" degrees of freedom. So then, for gas composed of diatomic molecules with 5 such degrees of freedom, the total energy will be

$$E = \frac{5}{2}Nk_BT = \frac{5}{2}nRT, \tag{11.14}$$

3/5 of which will be translational. And for more complicated molecules, there will evidently be additional "active" degrees of freedom (rotation about the third axis and possibly various sorts of vibrations) so the total energy will be even greater in proportion to Nk_BT, and the fraction of the total energy that is translational kinetic will be lower.

In general, the "equipartition" idea tells us that if the molecules of a gas have k "active" degrees of freedom (each), then we will have $f = 3/k$.

There are some questions, though, about all this. For example, the heat capacities of the diatomic gases are not *exactly* $5/2\,R$. That is, the empirically determined values of f are (for the diatomic gases) generally close to – but not always precisely equal to – 3/5. So evidently the different diatomic gases have "active" degrees of freedom which number close to, but not exactly, five – which doesn't really make

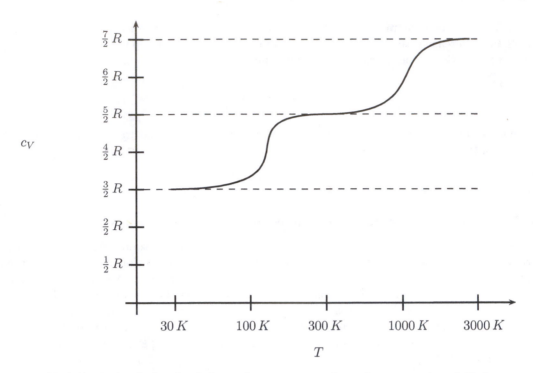

Figure 11.1: Slightly-idealized sketch of the molar constant-volume heat capacity of Hydrogen gas as a function of temperature. At very low temperatures, the heat capacity has the value $c_V = 3/2\ R$ that is expected for gases with only the three translational degrees of freedom being "active". At somewhat warmer (and more ordinary) temperatures, the two rotational degrees of freedom expected for a diatomic molecule gradually "activate". And then, at even higher temperatures, some further (presumably vibrational) degrees of freedom become "active" and so the heat capacity again gradually increases. (At even higher temperatures, the diatomic molecules break apart into their constituent atoms, and so we aren't really any longer talking about the same kind of gas.)

any sense – or the energy is partitioned roughly, but not exactly, according to the equipartition principle, with slight differences for each species of gas. (Or maybe for some of the gases – e.g., Chlorine – there are 5 *fully* "active" degrees of freedom, and then one additional degree of freedom that is only *partially* "active.")

Moreover, it's not at all clear why the vibrational degrees of freedom shouldn't be "active". We suggested above that, in a diatomic molecule, the two atoms can be thought of as being connected by a tiny rigid rod. But probably a more realistic physical model would be that the atoms are connected by a spring (with only some finite stiffness), in which case there is no apparent reason why the two additional vibrational degrees of freedom (one associated with the kinetic energy of vibration, the other associated with the potential energy of the compressed spring) shouldn't be "active" as well.

It also turns out that the heat capacities for various gases depend in an interesting way on the temperature. For example, a (kind of idealized, schematic) graph of the constant-volume molar heat capacity of hydrogen, as a function of temperature, is shown in Figure 11.1.

Despite being a diatomic gas, at very low temperatures, hydrogen acts (in so far as its heat capacity is concerned) like a monatomic gas: apparently only its three translational degrees of freedom are "active". At higher temperatures, the two rotational degrees of freedom become "active" as well (as one would naively expect for a diatomic gas based on the above discussion). Then at very high temperatures, some additional (presumably vibrational) degrees of freedom also become "active". (At even higher temperatures, collisions between the molecules become so violent that the molecules themselves break apart, and the physical properties change dramatically.)

These questions about why certain degrees of freedom (and not others) are "active" (and why this varies with temperature in the way that it seems to) are profound puzzles from the point of view of the kinetic theory of gases and classical physics in general. Historically, they were resolved only in the first part of the 20th century, with the advent of quantum theory. We will discuss this further in Chapter 13.

For now, we turn to the *other* sort of heat capacity that was mentioned at the beginning of this section.

11.2.2   Heating at constant pressure

Let us now consider the process of heating a sample of gas with the pressure P (rather than the volume V) held fixed. Qualitatively, as we add heat to the gas, we increase its energy. But the gas also *expands*, doing positive work on the external atmosphere (or if you prefer, the external atmosphere does *negative* work on the gas), which means that some of the energy we added (in the form of heat) doesn't stay in the gas, but instead escapes into the external environment. We thus expect that it will be necessary to add *more* heat to raise the temperature of the gas by a certain amount dT, than it was necessary to add when we were heating at constant volume. That is, we expect that the heat capacity at constant pressure, c_P, will be *larger* than the heat capacity at constant volume, c_V.

We can develop a more precise relation between these quantities by using the first law of thermodynamics. Rearranging Equation (11.3), we have that

$$dQ = dE + P\,dV. \tag{11.15}$$

We have already shown that the change in the total energy of one mole of gas, dE, is (at least for an ideal gas) proportional to the change in temperature:

$$dE = c_V\,dT. \tag{11.16}$$

The second term on the right hand side of Equation (11.15) can also be simplified, assuming (again) that we are talking about an ideal gas. According to the ideal gas law,

$$PV = n\,RT \tag{11.17}$$

and so, at constant pressure, the volume change dV for $n = 1$ mole of gas can be written

$$dV = \frac{R}{P}\,dT. \tag{11.18}$$

Substituting these results into Equation (11.15) gives

$$dQ = c_V\,dT + R\,dT = (c_V + R)dT \tag{11.19}$$

which means that, for our constant-pressure expansion, the ratio of heat added to temperature increase – i.e., the constant-pressure heat capacity – is (for one mole):

$$c_P = \left(\frac{dQ}{dT}\right)_P = c_V + R. \tag{11.20}$$

Like the constant-volume heat capacity c_V, the constant-pressure heat capacity c_P is relatively easy to measure in the lab. See again Table 1.1 for some actual values.

The results of such measurements for most gases are in very good agreement with Equation (11.20), which confirms that under ordinary conditions most gases are not too different from the hypothetical ideal gas. Conversely, since Equation (11.20) was derived under the assumption that the gas is ideal, the amount by which this relation is *violated* for a given species of real gas is a convenient, empirically accessible way to quantify precisely how un-ideal it is.

We will also later encounter the ratio of specific heats, c_P/c_V, so it is convenient to give that a name here and relate it to the other things we've discussed. Thus, let us define

$$\gamma = \frac{c_P}{c_V} = \frac{c_V + R}{c_V} = 1 + \frac{R}{c_V} = 1 + \frac{2}{3}f \tag{11.21}$$

where f is again the fraction of the total energy (added at a given temperature) that is translational-kinetic. (Note that all of the expressions after the first equation on the left assume that we are talking about a sufficiently ideal gas.)

We may also write γ in terms of the number of "active" degrees of freedom (per molecule), k:

$$\gamma = 1 + \frac{2}{k}. \tag{11.22}$$

Thus, for monatomic gases like Helium, where $f = 1$ and $k = 3$, we expect $\gamma = 1.667$, while for diatomic gases where $f = 3/5$ and $k = 5$, we expect $\gamma = 1.4$. The empirically measured values for γ for several gases are included in Table 11.1. Also included are the values for k that one can compute by plugging in the true, empirical values for γ into Equation (11.22) and solving for k. Note that this is a way of determining, directly from experiment, how many degrees of freedom are "active" in a given type of gas. The important point here is that many of the values turn out to be close to what we have expected: $k = 3$ for monatomic gases, $k = 5$ for diatomic gases, etc. But not all of them are exactly right, and indeed not all of them are even integers. This is yet another perspective on the puzzles we have already encountered.

§ 11.3 More Processes with Gases

In the previous section, we discussed two simple examples of processes one could perform on a gas: adding heat while holding the volume constant, and adding heat while holding the pressure constant. If we wanted to make up fancy-sounding names for these processes, we might call them, respectively, isovolumetric heating and isobaric expansion. And of course we could also imagine the inverse processes, in which, instead of adding heat (at constant volume or pressure, respectively) we arrange for some heat to flow *out* of the system. Such processes would, I guess, be called isovolumetric cooling and isobaric contraction.

That terminology is overkill for those processes, but is a nice warm-up to some terminology we'll need to learn to talk about a host of other possible processes that can be performed on a system. For example, a gas can be expanded or contracted *just* by the application of external forces, *without any heat flow* into or out of the gas. Such processes are called "adiabatic", and we will study them in more detail shortly. It is also possible (easy, in fact) to arrange that some work is done on (or by) the gas while, simultaneously, some heat flows out of (or into) the gas, such that its *temperature remains constant*. Such a process is called "isothermal." And we will also discuss another sort of process called the "free expansion."

Before discussing these new types of processes, however, let us introduce a certain kind of diagram which will provide a nice alternative visual perspective on these concepts, and which will also come in very handy later. The idea is to consider the thermodynamic state of a system (say, our usual sample of gas, with its chemical identity and extent – in grams or moles or whatever – given) as being specified by its

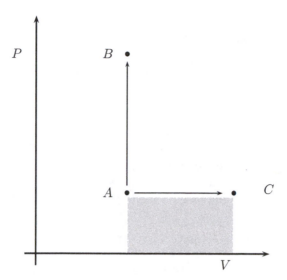

Figure 11.2: A P − V diagram. The process that takes the system from point A to point B is an isovolumetric heating. The process that takes the system from A to C is an isobaric expansion. The mechanical work done by the gas (on the environment) during the isobaric expansion is equal to the area of the shaded rectangle.

pressure P and volume V. Then we'll simply plot that state on a graph whose axes represent P and V. This is called a "$P - V$ Diagram". An example is shown in Figure 11.2.

The points labeled A, B, and C represent three possible thermodynamic states of the gas. The process (indicated by the arrow) which takes the gas from state A to state B is an "isovolumetric heating". The process (indicated by the other arrow) which takes the gas from state A to state C is an "isobaric expansion". Note also that, since in one infinitessimal step of (say) the isobaric expansion from A to C, the work done *by* the gas is $P\,dV$, the total work done *by* the gas during that entire process is $W = \int P\,dV = P\,\Delta V$ (since P is constant) which is equal to the shaded rectangular area in the figure. In general, for any process, the work done *by* the system during that process will be equal to the area under the curve which represents that process in the $P - V$ plane.

A couple of notes before we move on. First, note that we stressed the word "by" several times in the previous paragraph. When we first formulated the first law of thermodynamics, we were mostly interested in the given "system" and the ways that its energy could change. It thus made sense to say that $dE = dQ + dW$, i.e., to interpret dW as the work done *on* the gas by outside agencies. But as we move forward with applications – in particular the heat engine, which we will discuss shortly – we become increasingly interested in the work done *by* the gas on its surroundings. These two works (the work done on the system and the work done by the system) are simply the negatives of each other, and it is easy to get them mixed up. In particular, note that the work represented by the area under the curve in the $P - V$ plane is the work done *by* the gas as its volume increases.

Second, let us mention here an important qualification to everything we've said so far, and much of what is to come. Unless otherwise specified, we assume that all processes are performed *slowly*, such that, at any intermediate step of the process, the gas as a whole is in an equilibrium state. Otherwise, it might not be possible to even assign a particular pressure or volume (or temperature, etc.) to these intermediate states. In thermodynamics, the technical term for a process which is performed slowly enough that the system is always in some kind of equilibrium (such that it is meaningful to talk about its pressure or volume or temperature at any time) is: "quasi-static" and quasi-static processes can be un-done, i.e., they are "reversible". Unless otherwise specified, all processes discussed in this chapter are to be taken as occuring quasi-statically so that they are reversible. But we will also have the opportunity to explore examples of processes which are *not* quasi-static and which are hence irreversible.

11.3.1   Adiabatic processes

Let us first consider adiabatic processes. These are, by definition, processes in which no heat flows into or out of the system. In practice, one arranges for a contraction or expansion to be adiabatic simply by putting some thermal insulation (like styrofoam) around the system. Many "natural" processes are also, to a good approximation, adiabatic, because the systems have poor thermal contact with their surroundings or because the processes happen too rapidly for appreciable heat to flow in or out.

For simplicity, let us again assume we are dealing with an ideal gas. Then the change in the total energy of the gas is proportional to the change in its temperature:

$$dE = n\,c_V\,dT \tag{11.23}$$

where c_V is, of course, the constant-volume molar heat capacity. Since we are talking about an adiabatic process for which $dQ = 0$, we also have that $dE = dW = -P\,dV$. Combining these and using the ideal gas law gives

$$nc_V dT = -P\,dV = -\frac{nRT}{V}dV \tag{11.24}$$

which can be simplified and rearranged into this form:

$$\frac{dT}{T} = -\frac{R}{c_V}\frac{dV}{V}. \tag{11.25}$$

In words, this says that, for an (infinitessimal) adiabatic expansion or compression, the fractional change in temperature is proportional to the fractional change in volume. (The proportionality is negative because, clearly, a decrease in the volume – which involves positive work being done on the gas – will correspond to an increase in the temperature.)

We may now formally integrate both sides, with the result that

$$\log(T) = -\frac{R}{c_V}\log(V) + \text{constant} \tag{11.26}$$

which is mathematically equivalent to

$$TV^{R/c_V} = \text{constant}. \tag{11.27}$$

As an example of the application of this formula, suppose that we are dealing with a sample of Helium, for which $R/c_V = 2/3$. And suppose that we adiabatically compress the Helium to $1/8$ its original volume. Since $(1/8)^{2/3} = 1/4$, this means that the temperature must have increased by a factor of 4 in order that $TV^{2/3}$ remain constant. By contrast, if we performed the same factor-of-8 compression on a diatomic gas like Nitrogen (for which $R/c_V \approx 5/3$), the temperature would have to increase by a somewhat smaller factor $1/(1/8)^{3/5} \approx 3.5$.

We may also use the ideal gas law to convert Equation (11.27) into a pressure-volume relation. Plugging in $T = PV/nR$ (and treating n and R as constants) gives

$$PV^{1+R/c_V} = \text{constant} \tag{11.28}$$

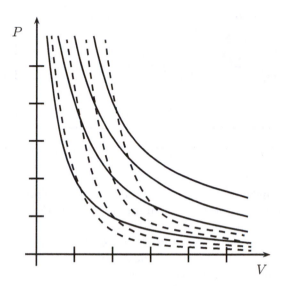

Figure 11.3: P − V diagram showing four isothermal lines (solid) and four adiabatic lines (dashed).

which can be more easily written, using Equation (11.21), as

$$PV^\gamma = \text{constant}. \tag{11.29}$$

(Now you see why we bothered to introduce a special symbol for this otherwise useless-seeming ratio of specific heats!) Again, as an example, suppose we adiabatically compress a diatomic gas for which $\gamma = 1.4$ to half its original volume. Since $(1/2)^{1.4} \approx 0.38$, the pressure of the gas will increase by a factor $1/0.38 \approx 2.64$.

11.3.2 Isothermal processes

Imagine that we have a system (such as a piston full of gas) which is in thermal contact with a large heat reservoir (think: the ocean) whose temperature is T. (The reservoir's being *large* means that, if it exchanges a little energy with the little piston of gas, the temperature of the piston of gas might change – but the temperature of the reservoir won't.) Now suppose we begin to compress the gas, slowly enough that the energy we add to the gas by doing mechanical work on it, has time to escape, in the form of heat, into the heat reservoir, such that the temperature of the gas is maintained at the fixed temperature T. Such a process is called an isothermal compression. (And if we had pulled the piston out instead of pushing it in, it would have been an isothermal expansion.)

In general, there is not too much to say about isothermal processes. But if we restrict our attention (as usual) to ideal gases, we can immediately note that, for a process which occurs at constant temperature, the pressure times the volume is a constant:

$$PV = \text{constant}. \tag{11.30}$$

That, of course, is just Boyle's law, or if you prefer it is the ideal gas law with the T on the right hand side taken as a constant.

Figure 11.3 shows some adiabatic and isothermal lines on a $P − V$ diagram. For the isothermal lines, the pressure is inversely proportional to the volume (to the first power). For the adiabatic lines, the pressure is inversely proportional to the volume to a somewhat higher power, and so the adiabatic lines slope down

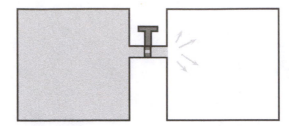

Figure 11.4: Apparatus to produce the free expansion of a gas.

more steeply as one moves to higher volumes, i.e., to the right in the graph. Here is another way to think about this. Consider two processes, one isothermal and one adiabatic, in which one mole of gas begins at the same P and V (and hence also T) values and is then *expanded*. In the isothermal expansion, by definition, the temperature of the gas is constant. The expansion involves doing negative work on the gas, but this is compensated by an equal amount of heat flowing into the gas. In the adiabatic expansion, negative work is also done, but (by definition) no heat is allowed to flow into the gas. So clearly the temperature of the gas will have to decrease during the adiabatic expansion. In terms of the diagram, this means the adiabatic lines will have to be continually crossing isothermal lines (toward lower and lower temperatures) as one moves from left to right in the diagram.

11.3.3   Free Expansions

So far we have discussed four categories of processes that might be performed on a gas: constant volume, constant pressure, isothermal (constant temperature), and adiabatic (no heat flow). As mentioned earlier, for all these processes we have in mind that they are performed *slowly* (i.e., quasi-statically) so that, in the intermediate stages of the process, the gas always *has* some definite pressure, volume, and temperature.

Now (for contrast) we briefly consider a process which is inherently "fast", i.e., which is not quasi-static, and for which it is therefore impossible to plot a particular "path" through the $P-V$ plane corresponding to the process. The particular process we'll consider is called a "free expansion" and can be thought of as follows: take a sample of gas (with some particular initial volume, temperature, and pressure), and then pull the piston back "infinitely fast" so that the gas now has, say, twice as much room. Alternatively, one could imagine a situation like that shown in Figure 11.4, in which a sample of gas, initially confined to one side of a container (with the other side being evacuated), is suddenly allowed to expand into the full container when a wall is removed or a valve is opened.

What makes the free expansion "free" is that, unlike in a quasi-static expansion (adiabatic, isothermal, or constant pressure), there is no moving piston which does (negative) mechanical work on the gas. In addition, since the expansion is relatively sudden, there is no time for heat to flow into or out of the gas. (Alternatively, we can just imagine that the container is thermally insulated.) So, in terms of the first law of thermodynamics, both dW and dQ are zero – and hence dE is also zero, i.e., the total energy of the gas is *unchanged* during the free expansion. For an ideal gas, all of whose energy is kinetic, this means that the temperature will also remain unchanged.

For our current purposes, what is interesting about the free expansion is just that it is not a quasi-static process. If an ideal gas undergoes a free expansion, its temperature will, for the reasons just explained, remain constant. Its initial and final states will thus lie on the same isothermal line in the $P-V$ diagram. But unlike a quasi-static isothermal expansion, there is no smooth "path" in the $P-V$ plane connecting these two points. The reason is that, in the (brief) intermediate stages of this process, the gas is not in an equilibrium state; its pressure varies from one point to another (as does its density and even its temperature) and it has no clearly-defined volume. We will discuss the free expansion process further when we encounter the second law of thermodynamics.

§ 11.4 Heat Engines

We have been discussing various sorts of processes that can be performed on a gas to set the stage for a detailed discussion of a very practical subject which, in fact, developed in parallel with the abstract theoretical science of thermodynamics. That practical subject is "heat engines" – which is a general term for devices which convert heat energy into useful mechanical work. The steam engine developed in the 1760s by James Watt (Scottish mechanical engineer and friend of Joseph Black) is probably the most important historical example, given the role it played in the Industrial Revolution. The internal combustion engine is a more familiar contemporary example.

It is appropriate to mention, here, at the very beginning, the French physicist Sadi Carnot whose 1824 theoretical study of the efficiency of steam engines is a watershed in the historical development of thermo-dynamics. Still working under the assumption of the caloric theory of heat – and writing several decades before the full recognition of the *first* law of thermodynamics – Carnot clearly articulated the *second* law of thermodynamics and its intimate relation to the efficiency of heat engines. One can also appreciate the almost prophetic character of Carnot's statement of the practical importance of the heat engines he was studying:

> "Nature, in providing us with combustibles on all sides, has given us the power to produce, at all times and in all places, heat and the impelling power which is the result of it. To develop this power, to appropriate it to our uses, is the object of heat engines.

> "The study of these engines is of the greatest interest, their importance is enormous, their use is continually increasing, and they seem destined to produce a great revolution in the civilized world.

> "Already the steam-engine works our mines, impels our ships, excavates our ports and our rivers, forges iron, fashions wood, grinds grains, spins and weaves our cloths, transports the heaviest burdens, etc. It appears that it must some day serve as a universal motor, and be substituted for animal power, waterfalls, and air currents."

How right he was.

Let us begin our analysis of heat engines by imagining a simple scenario which will allow us to get the main ideas on the table. Suppose we have some kind of heat source – a burning candle, say – and also some kind of heat sink – say, the cold water flowing in a nearby river. And suppose there is some mechanical task we would like to perform – say, we want to lift a bunch of books from a low shelf up onto a higher shelf. Lifting the books requires us to do mechanical work, i.e., it requires an input of energy. In particular, suppose the books each have mass m and we want to lift them through a vertical height h; then the work required (per book) is just $W = mgh$. The idea is to somehow use the energy contained in the heat source to help us accomplish this task. (At this point it shouldn't be clear why we would need the river at all. We'll come to that shortly.)

A simple scheme for achieving this is illustrated in Figure 11.5. A piston full of an ideal gas is placed next to the shelves. The top of the piston itself constitutes a kind of "elevator". The idea is that we will slide one book at a time from the lower shelf onto the "elevator", then place the candle flame beneath the gas in the piston, allowing heat to flow into the gas. This will cause the gas to expand, lifting the elevator and the book until it can be slid over onto the upper shelf. (We assume that the book-shelf and book-elevator interfaces are frictionless, so no energy needs to be used up to slide the book onto, or off of, the elevator.)

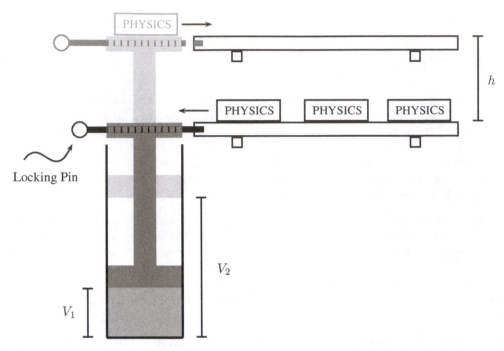

Figure 11.5: A simple scheme for lifting books using heat. The gas in the piston initially occupies volume V_1 and its pressure is just atmospheric pressure, P_{atm}. (We treat the moveable piston/elevator itself as massless.) A locking pin is slid into place, so that a book can be slid onto the elevator without further compressing the gas. Once the book is in place, we heat the gas until its pressure increases enough to support the additional weight of the book at the same initial volume. The required pressure is just $P_{high} = P_{atm} + mg/A$ where A is the cross-sectional area of the piston. Now the locking pin can be removed, and the gas can be heated further (this time at constant pressure) until the elevator's height matches that of the upper shelf. The locking pin is again inserted, and the book is slid onto the upper shelf.

Now that the basic idea is clear, let us be a little bit more precise. Assume that, initially, when the elevator is at the level of the lower shelf, the gas in the piston occupies volume V_1 and is at atmospheric pressure. (Realistically, it would be a little bit higher, because of the weight of the moveable piston itself, but we can for simplicity treat that as massless.) Now we insert a locking pin to temporarily support the weight of the book, which is slid onto the elevator. The locking pin is necessary because, without it, the gas would be compressed by the extra weight of the book, and that would complicate our analysis. In order to increase the pressure of the gas to the point where it can support the additional weight of the book, we now bring in the heat source and allow the gas to absorb heat until its pressure reaches $P_{high} = P_{atm} + mg/A$ where A is the cross-sectional area of the piston. Now the locking pin can be removed without the gas compressing or expanding. Finally, we again allow heat to flow into the gas until its volume reaches $V_2 = V_1 + Ah$, i.e., until the "elevator" reaches the second shelf. The locking pin is then put in place so the book can be slid onto the upper shelf without the piston moving (and the gas expanding).

What we have just described in normal language is actually the following sequence of thermodynamic processes on the gas: a constant volume heating, followed by a constant pressure expansion. The two processes are shown on a $P - V$ diagram in Figure 11.6.

We are going to be interested in the *efficiency* of this process, which we may here define as the quantity

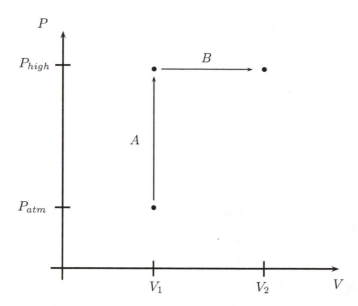

Figure 11.6: Lifting a book using heat, as represented on an abstract $P - V$ diagram. The gas in the cylinder is first (step A) subjected to a constant-volume heating, which increases the pressure from P_{atm} to $P_{high} = P_{atm} + mg/A$. Then (step B) the gas is expanded at constant pressure until the book reaches the upper shelf; this corresponds to the gas occupying volume $V_2 = V_1 + Ah$, as shown in the previous Figure.

of useful work accomplished (W) divided by the quantity of heat consumed (Q_{in}):

$$\eta = \frac{W}{Q_{in}}. \tag{11.31}$$

The useful work accomplished here was obviously just $W = mgh$. To compute the efficiency of our heat engine, though, we'll have to calculate the amount of heat transfered from the burning candle.

Suppose n moles of ideal gas are contained in the piston, whose volume is initially V_1 and whose temperature is initially T_1. Thus, from the ideal gas law, we have that $P_{atm}V_1 = nRT_1$.

Now consider Step A – the constant volume heating (after the locking pin has been put in place and the book has been slid onto the elevator). We need to heat the gas until its pressure increases from P_{atm} to P_{high}. From the ideal gas law, we have

$$\Delta P = P_{high} - P_{atm} = \frac{mg}{A} = \Delta\left(\frac{nRT}{V}\right) = \frac{nR}{V_1}\Delta T \tag{11.32}$$

so that the temperature increase of the gas during Step A is

$$\Delta T_A = \frac{V_1}{nR}\frac{mg}{A}. \tag{11.33}$$

And so the quantity of heat required to produce the needed pressure increase is

$$Q_A = nc_V\Delta T_A = \frac{c_V}{R}\frac{V_1}{A}mg. \tag{11.34}$$

So much for Step A.

Step B was a constant-pressure expansion from initial volume V_1 to final volume $V_2 = V_1 + Ah$. Again from the ideal gas law, we have

$$\Delta V = Ah = \frac{nR}{P_{high}}\Delta T \tag{11.35}$$

so that

$$\Delta T_B = \frac{P_{high}}{nR}Ah. \tag{11.36}$$

And so the quantity of heat required to produce the needed volume increase is

$$Q_B = nc_P\Delta T_B = \frac{c_P}{R}P_{high}Ah. \tag{11.37}$$

Note that, since (by assumption) we are talking about an ideal gas, we have that $c_P = c_V + R$, so that we could also have written

$$Q_B = n(c_V + R)\Delta T_B = nc_V\Delta T_B + P_{high}Ah = nc_V\Delta T_B + P_{atm}Ah + mgh. \tag{11.38}$$

The first term on the right is just the amount by which the energy of the gas itself changes during Step B due to its being heated. The second term represents the work done by the gas in "lifting" the surrounding atmosphere, i.e., the work done by the expanding gas in pushing back the external atmosphere. Only the third term (mgh) represents the useful work done by the engine.

It is then already clear that the process is far less than 100% efficient: the *useful* work done (which we called W before) is only *part* of the total work done by the gas, and a substantial amount of the heat we added goes not into doing work at all, but into heating up the gas and pushing back the atmosphere. The efficiency, evidently, is given by

$$\eta = \frac{mgh}{nc_V\Delta T_A + nc_V\Delta T_B + P_{atm}Ah + mgh}. \tag{11.39}$$

You should, however, be a little suspicious of calculating the efficiency of this particular combination of processes, because we have left the device in a different state than it started in.

What we should do is find a way to return the machine to its initial state, and then consider the efficiency of a complete (repeatable) process. But how exactly should we return the device to its initial state, having already performed processes A and B? Perhaps one's first thought is to retrace, in reverse, those two initial steps. That would mean, evidently, putting the book back onto the elevator and removing the locking pin, then bringing in some of the cold water from the river so that heat flows out of the gas and into the river, allowing the gas to undergo a constant pressure *contraction* until the elevator reaches the lower level. Then we insert the locking pin, slide the book back onto the lower shelf, and again let some heat flow into the river until the gas reaches its original state.

Of course, the problem with *that* strategy is that not only do we return the gas to its original state (which is what we wanted to accomplish), we also returned the *book* to its original state! So that's no good. The net effect of the whole process would simply have been to let some heat flow from the candle into the river, which wouldn't help us at all in our quest to lift books to the upper shelf.

By this point, it's probably occured to you that the solution – or at least a solution – is going to be to "undo" the first two steps *in the opposite order*. That is, after we insert the locking pin and slide the book onto the upper shelf, we bring in the cold river water and cool the gas at constant volume (V_2) until it reaches the lower pressure, P_{atm}. Call that Step C. Now we can remove the locking pin (and the elevator won't spring up or down), and then finally let the gas contract at constant pressure by again transfering some heat to the river water. That's Step D. *This* sequence leaves the gas in its initial state

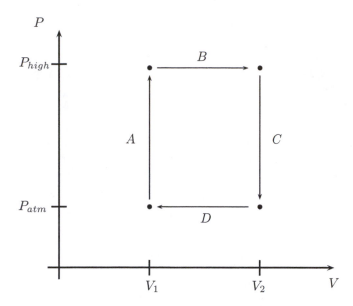

Figure 11.7: A sequence of four steps which (a) leaves the gas back in its initial state, and (b) has raised one book to the upper shelf.

(ready in principle to lift another book in the same fashion), and leaves the first book on the upper shelf, where we wanted to get it. The complete sequence of four Steps is illustrated in Figure 11.7.

How much heat flows out of the gas during the "C" and "D" processes we've just described? Following the calculations above, we can work out that

$$Q_C = -\frac{c_V}{R}\frac{V_2}{A}mg \tag{11.40}$$

and

$$Q_D = -\frac{c_P}{R}P_{atm}Ah \tag{11.41}$$

where the minus signs remind us that here heat is flowing *out* of the gas and into the "heat sink" – the cold river.

It is a useful check at this point to calculate the *total* quantity of heat that flowed into the gas during the entire four step process. This is

$$
\begin{aligned}
Q &= Q_A + Q_B + Q_C + Q_D \\
&= \frac{c_P}{R}Ah(P_{high} - P_{atm}) - \frac{c_V}{R}\frac{mg}{A}(V_2 - V_1) \\
&= \frac{c_P}{R}Ah\left(\frac{mg}{A}\right) - \frac{c_V}{R}\frac{mg}{A}(Ah) \\
&= \frac{c_P - c_V}{R}mgh \\
&= mgh
\end{aligned}
\tag{11.42}
$$

where we have used $V_2 - V_1 = Ah$ and $P_{high} - P_{atm} = mg/A$.

This, of course, is exactly what we should have expected the answer to be. After all, the gas (and the surrounding atmosphere) have been left in their initial states, so the net *work* done *by* the gas during this cycle is just the work done *on the book* – namely, mgh. And since we were everywhere assuming ideal, frictionless processes, it makes sense that the energies precisely balance in accordance with the first law of thermodynamics. As far as the gas in the piston is concerned, the net result of the process is as follows: a quantity of heat Q flowed in, and mechanical work W was done (*by* the gas) on the environment. Since these are numerically equal – that is, since $Q = W = mgh$ – the energy of the gas itself remains unchanged.

That makes it sound a little bit like the process turns out to be 100% efficient, which is definitely *wrong*. The reason is that, from *our* perspective, the heat that flowed into the cold river is just *wasted*. It's *gone*. What *we* care about is what fraction of the heat that *we put into the system* – one might say, what fraction of the heat we *paid for* – gets converted into useful mechanical work. And this is just

$$\eta = \frac{W}{Q_{in}} = \frac{W}{Q_A + Q_B} \tag{11.43}$$

which is just what we calculated before! So, as it turns out, our earlier calculation of the efficiency wasn't premature after all, because we can return the piston to its initial state by letting heat flow into the cold river in a certain way, and (in terms of fuel costs, at least) this is "free".

The book-lifting machine we've been discussing, which uses constant-volume and constant-pressure processes, is just one possible sequence of steps by which thermal energy from a heat source can be used to do useful mechanical work. In a moment, we'll consider an alternative sequence of processes which is, for reasons we will discuss, of greater theoretical interest. Before turning to that, however, let's discuss a few points raised by our warm-up example.

The first is just that, although the specific example above (using a candle to lift books) is a bit contrived and silly, the process captures the essence of what is happening in real power plants. To see this, just note that the overall function of the heat engine we described was to convert (some fraction of) some heat from a heat source into useful mechanical work. We set things up so that this useful mechanical work took the form of lifting books, but it could just as easily have been made, for example, to turn the crank on an electric generator, producing electrical energy which could be transported by wires and used (for example) to power your toaster. Replace the candle from our example with a steady stream of burning coal (or maybe controlled nuclear fission) and replace the ideal gas in the piston with water (a more commonly used "working fluid") and you've got a real power plant. Note that you can now understand why such real power plants are built next to rivers: the heat source is made to boil the water, converting it to high-pressure steam, which then expands through a turbine to produce electricity. The steam is then brought into thermal contact with the cold river water to return it to its initial state so the cycle can continue.

One other point about heat engines that will come up later: in principle, we could always run the cycle for (say) our book-lifting machine *in reverse* instead. The net effect would then evidently be to move one book from the higher shelf down to the lower shelf, and to transfer a certain amount of heat from the low-temperature reservoir (the cold river water) to the high-temperature reservoir.

It's not clear why one would want to try to transfer heat from a river into a candle. But heat engines operating "in reverse" are actually quite useful and common. Your refrigerator does this, for example: it is constantly transfering heat from the (colder) inside to the (warmer) outside – counteracting the inevitable "leaking" of some heat into the refrigerator (for example, whenever you open the door to look for a snack) and keeping the inside at an acceptably low temperature. And note that you have to plug your refrigerator in – that is, you have to pay for this counter-spontaneous heat flow by providing some electrical energy as input to the process. This is the analog of lowering books from our toy example.

Another familiar example of a heat engine running in reverse is an air-conditioning unit. Like a refrigerator, this is a device you have to plug in – i.e., a device that requires the input of some useful energy. It then functions to transfer some heat *opposite* the direction it would flow spontaneously – namely, from the relatively cool inside of your car or house to the relatively warm outside.

§ 11.5 The Carnot Cycle

In the previous section, we considered a heat engine that underwent a *rectangular* cyclic process in the $P - V$ plane. We found that, in order to accomplish a fixed amount of mechanical work, $W = mgh$, we needed to add a quantity of heat given by

$$
\begin{aligned}
Q_{in} &= Q_A + Q_B \\
&= \frac{c_V}{R}\frac{V_1}{A}mg + \frac{c_P}{R}P_{high}Ah \\
&= \frac{c_V}{R}\frac{V_1}{A}mg + \frac{c_P}{R}P_{atm}Ah + \frac{c_P}{R}mgh.
\end{aligned}
\tag{11.44}
$$

Let us think about how we might design the device to make it as efficient as possible. This means *minimizing* Q_{in}. But of course we want to minimize the required input heat subject to some constraints. First and foremost, we want the work done in one cycle to remain mgh, otherwise we won't be able to lift our books. Let's also assume that the low pressure state of the gas must be in equilibrium with atmospheric pressure. And let us also treat c_V and c_P as fixed. So the question is: for a device like that discussed in the previous section, operating with a *given* type of gas, and having a *fixed* lower pressure, how can we maximize the efficiency?

Well what, in principle, is still adjustable once we impose those constraints? Basically, the *shape* of the piston and its initial volume. For example, by adjusting the quantity and/or initial temperature of the gas, we could make the initial volume V_1 as big or as small as we chose. We can also imagine adjusting the cross-sectional area A of the piston. We require that the height of the piston change by h during the process, but this could correspond to any desired $(V_2 - V_1) = Ah$ by simply adjusting A up or down.

Now we can examine Equation (11.44) and see how to minimize the required heat, i.e., maximize the efficiency of our heat engine. The third term represents the work done by the gas on the book. That is not negotiable. But we can make the second term small by making A small, i.e., by using a piston with a small cross-sectional area. Now, making A small tends, all other things equal, to make the first term *big* – but we can still independently choose a value for V_1. So we will say that *both* V_1/A and A are to be made as small as possible. This corresponds to a rectangular process in the $P - V$ plane that is very narrow and very tall and huddled right up next to the P-axis. That, evidently, is the way to design the device if we want the least possible fraction of the heat we "paid for" being dumped into the cold river, i.e., if we want to lift the book as cheaply as possible.

What does this mean in practical terms? Well, it means we need to start off with the gas in a *very cold* initial state. The ideal thing would be to have the gas initially at absolute zero, so then V_1 would be zero (by the ideal gas law) and the first term in Equation (11.44) would be as small as mathematically possible! But then, in order to lift the book to the upper shelf, the gas will have to expand by an astronomical factor in volume – corresponding to a very very *high temperature*.

Remember that the idea from the beginning was to heat the gas in the cylinder using, say, a candle. Well, we're never going to be able to get the gas hotter than the candle flame. So in practice there is a limit to how hot the hot (large volume, high pressure) state of the gas can be. And the same thing on the other side of the cycle. We are supposed to cool the gas down to its initial state by letting heat flow out of it into, say, a cold river. So there is going to be a limit to how cold the cold (small volume, low pressure) state of the gas can be. Of course, we can stretch those limits by using something hotter than a candle flame – maybe a gasoline fire or nuclear fission – to heat the gas during the appropriate parts of

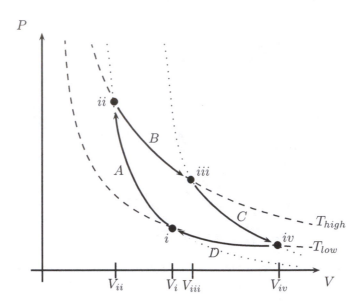

Figure 11.8: The Carnot cycle, operating between temperatures T_{low} and T_{high}. Isothermal lines corresponding to these two temperatures are shown as dashed curves. Two adiabatic lines (dotted curves) are also shown. The cycle consists of the following four processes: (A) an adiabatic compression of the gas in which the temperature increases from T_{low} to T_{high}, (B) an isothermal expansion at T_{high}, (C) an adiabatic expansion in which the temperature is decreased again to T_{low}, and (D) an isothermal compression in which the gas is returned to its initial state. A quantity of heat Q_{in} is added during step B, the isothermal expansion (which could also be called an isothermal heating). Likewise in step D a quantity of heat Q_{out} flows out of the gas and into the low-temperature heat sink. The total work done by the gas during the cycle is $W = Q_{in} - Q_{out}$. This is represented in the figure by the area enclosed by the cycle.

the cycle, and likewise by using something colder – maybe blocks of sub-zero ice or liquid nitrogen – to cool the gas during the appropriate parts of the cycle. But such exotic methods of heating and cooling the gas of course raise their own challenges.

In any case, what matters here is just the qualitative observation that the efficiency of the heat engine appears to have something to do with the minimum and maximum temperatures reached in the cycle. Of course, in quantitative terms, the exact efficiency depends not just on these extremal temperatures, but on the nature of the processes involved – in particular, on the "shape" of the cycle in the $P - V$ plane.

The key insight here, which is due to Sadi Carnot from the 1820s, is to create a special cycle which involves heat flowing in and out of the gas *only* at two particular temperatures. That is, instead of a rectangular cycle in the $P - V$ plane, Carnot envisioned the cycle shown in Figure 11.8 which moves back and forth between two *isotherms* using *adiabatic* compressions and expansions. This means that all the heat flowing into the gas flows at a fixed, high temperature T_{high}, and all the heat flowing out of the gas flows at a fixed, low temperature T_{low}.

The importance of this particular cycle lies in the fact that the efficiency of a heat engine using it is a particularly simple function of the two temperatures. So let us calculate the efficiency.

As usual, for simplicity, assume the working substance is an ideal gas. Suppose that, in each cycle, the engine absorbs a quantity of heat Q_{in} from the heat source at temperature T_{high}, and then dumps a quantity of heat Q_{out} into the heat sink at temperature T_{low}. If the engine is perfectly frictionless and if everything happens slowly and smoothly, the useful mechanical work done – W – will just be the difference between these two quantities of heat. The efficiency of the engine is defined by

$$\eta = \frac{W}{Q_{in}} = \frac{Q_{in} - Q_{out}}{Q_{in}} = 1 - \frac{Q_{out}}{Q_{in}}. \tag{11.45}$$

Since the working fluid is an ideal gas, during the isothermal steps (B and D), the energy of the gas doesn't change. Hence (using the first law) the heat added is equal to the work done by the gas. For example, for step B (which takes us from state ii to state iii as shown in the Figure) we have

$$Q_B = \int_B P\,dV = nRT_{high} \int_B \frac{dV}{V} = nRT_{high} \log\left(\frac{V_{iii}}{V_{ii}}\right) \tag{11.46}$$

where V_{iii} and V_{ii} are respectively the final and initial volumes of the gas along step B.

Similarly, for the heat transfered out of the gas during step D we have

$$Q_D = nRT_{low} \log\left(\frac{V_i}{V_{iv}}\right) = -nRT_{low} \log\left(\frac{V_{iv}}{V_i}\right). \tag{11.47}$$

The minus sign just means that, for step D the heat "added" to the gas is negative, i.e., heat is actually being transferred from the gas to the cold heat sink.

But it turns out that the ratio of volumes in Equation (11.47) is *the same* as the ratio of volumes in Equation (11.46). To see this, consider the adiabatic step A in which the temperature is increased from T_{low} to T_{high}. We have already seen that for adiabatic processes,

$$TV^{\gamma-1} = \text{constant}. \tag{11.48}$$

So for step A we have that

$$T_{low}V_i^{\gamma-1} = T_{high}V_{ii}^{\gamma-1} \tag{11.49}$$

or equivalently

$$\frac{T_{high}}{T_{low}} = \left(\frac{V_i}{V_{ii}}\right)^{\gamma-1}. \tag{11.50}$$

But for the the other adiabatic process, step C, in which the temperature is decreased from T_{high} to T_{low}, we have similarly

$$\frac{T_{high}}{T_{low}} = \left(\frac{V_{iv}}{V_{iii}}\right)^{\gamma-1}. \tag{11.51}$$

Dividing the last two equations reveals that

$$1 = \left(\frac{V_i V_{iii}}{V_{ii} V_{iv}}\right)^{\gamma-1} \tag{11.52}$$

which indeed implies that

$$\frac{V_{iii}}{V_{ii}} = \frac{V_{iv}}{V_i} \tag{11.53}$$

as advertised. And this means, taking the ratio of Equations (11.47) and (11.46), that

$$\frac{Q_{out}}{Q_{in}} = \frac{-Q_D}{Q_B} = \frac{nRT_{low} \log(V_{iv}/V_i)}{nRT_{high} \log(V_{iii}/V_{ii})} = \frac{T_{low}}{T_{high}}. \tag{11.54}$$

And so – finally – we can write the following simple expression for the efficiency of a heat engine using the Carnot cycle and operating between temperatures T_{high} and T_{low}:

$$\eta = 1 - \frac{T_{low}}{T_{high}}. \tag{11.55}$$

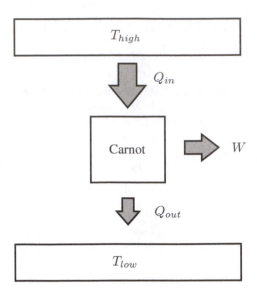

Figure 11.9: Energy flow diagram for a Carnot engine. During each cycle, the engine absorbs a quantity of heat Q_{in} from the high temperature heat source (temperature T_{high}). It converts this heat into a quantity W of useful work, and a quantity Q_{out} of heat, which is dumped into the low temperature heat sink (temperature T_{low}). According to the first law of thermodynamics, and because the engine operates cyclically (and so, in particular, is left in exactly its initial state at the end of one cycle), the total energy in must equal the total energy out: $Q_{in} = Q_{out} + W$. As derived in the main text, for a Carnot engine, there is a maximum amount of useful work W that can be produced for a given input of heat Q_{in}: $\eta = W/Q_{in} = 1 - T_{low}/T_{high}$. (The useful work out will of course be less than this maximum if there are frictional losses within the engine.)

The elegant simplicity of this result suggests that Carnot is on to something profound by having all of the heat flow into the working substance occur at the fixed temperature T_{high} and all of the heat flow out of the working substance occur at the fixed temperature T_{low}. The sense of profundity will only increase when, in the next section, we follow Carnot in arguing further that the formula for the efficiency of a Carnot heat engine is not, for example, an accident of having assumed the working substance to be an ideal gas, but is rather a universal maximum efficiency for any heat engine, no matter how it is constructed, no matter what sort of cycle its working substance follows between the temperatures T_{high} and T_{low}, no matter what the working substance is, etc.

To set the stage for that discussion, however, let us first introduce a convenient abstract representation of a Carnot heat engine, which we might call an "energy flow diagram". See Figure 11.9, which depicts the flow of energy into and out of the Carnot heat engine. A quantity of heat Q_{in} is taken in by the Carnot engine from the high temperature heat source at T_{high}. The engine then delivers a quantity of useful work W, and also dumps a quantity of heat Q_{out} into the low temperature heat sink at T_{low}. The engine operates *cyclically*, which means in particular that after one complete cycle it is left in exactly its initial state. Hence, the total energy *in* during one cycle must equal the total energy *out* during one cycle: $Q_{in} = Q_{out} + W$. That is simply an expression of the first law of thermodynamics.

The calculations made in this section show that, at least for this *particular* type of heat engine, the efficiency is given by $\eta = 1 - T_{low}/T_{high}$. The efficiency is, by definition, the ratio of the useful work obtained (W) to the input heat (Q_{in}). In line with our earlier qualitative observations, the efficiency can be increased by using a hotter heat source (i.e., by increasing T_{high}) and/or by using a colder heat sink (i.e., by decreasing T_{low}). But for any nonzero T_{low} and any finite T_{high}, the efficiency will be *less than 100%*. The energy flow diagram makes it clear why this is the case: some of the input heat is "wasted" – i.e., simply dumped into the low temperature heat sink. One might wonder if, by a sufficiently clever design, some alternative type of heat engine could be designed which would not "waste" any of the input heat in this way. This is precisely the question addressed in the following section.

§ 11.6 The Second Law of Thermodynamics

It is a familiar fact of observation that heat spontaneously flows from hot objects to cold objects, but never vice versa. For example, consider a hot cup of coffee in a cool room. Over time, heat will flow from the coffee into the rest of the room – cooling the coffee and (marginally) warming the room – until the coffee and room both possess a single shared temperature, i.e., until thermal equilibrium is reached.

Note that the opposite process – heat flowing spontaneously from the cool air in the room into the already-hot coffee – would be perfectly consistent with the *first* law of thermodynamics. The inherent *directionality* of spontaneous thermal processes thus requires a new, *second* law, which involves somehow raising to the status of a general postulate the idea that heat spontaneously flows from hot objects to cold objects, but never vice versa.

Of course, it is not true that heat can never be made to flow from cold objects to hot objects. We have already seen a way this can be achieved – namely, by operating a heat engine *in reverse*, as in a refrigerator or air-conditioner. But a crucial point about such devices is that we have to add energy in the form of useful mechanical work – e.g., by lowering books or providing a source of electrical energy. Perhaps, then, the way to formulate our second law of thermodynamics is as follows:

> "It is impossible for a self-acting machine, unaided by any external agency, to convey heat from one body to another at a higher temperature."

This is how Clausius formulated the law. It seems quite reasonable, as a summary of what we've already said: if you want to make heat flow from a cold body to a hotter body, you have to do some work, because it won't happen spontaneously.

It turns out, however, that this rather mundane-sounding statement of the familiar directionality of spontaneous heat flow has some profound consequences. In particular, it turns out to imply that the efficiency of a Carnot heat engine is in some deep sense *maximal* – no other device operating between the same two temperatures T_{high} and T_{low} can have a greater efficiency, regardless of its construction, its mode of operation, what substances are used, etc. Let us run through the argument to support this claim.

It is a proof by contradiction. Thus, suppose we *did* have a heat engine whose efficiency was greater than that of a Carnot engine operating between the same temperatures. Since (as we are in the process of showing) such a device would violate the second law of thermodynamics, it may be termed a "perpetual motion machine of the second kind" or PMM2 for short.[1] The idea is that, if we were given a PMM2, we could join it together in a certain way with a Carnot engine, to produce a hybrid device which would constitute precisely "a self-acting machine, unaided by any external agency" which would "convey heat from one body to another at a higher temperature."

The way to make this hybrid device is simply to connect the "useful work" *output* of the PMM2 to the mechanical work *input* of a Carnot heat engine operating *in reverse*. This is shown schematically in Figure 11.10. Let us run through the argument twice, first qualitatively, and then quantitatively.

---

[1] You might be wondering: why a PMM of the *second* kind? The point is that this type of machine would violate the *second* law of thermodynamics. One can also contemplate – though one cannot build! – a PMM of the first kind. This

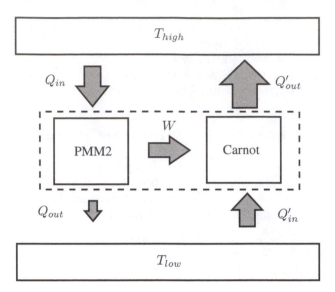

Figure 11.10: Energy flow diagram depicting the hybrid device (enclosed by the dashed line) consisting of a Carnot heat engine (operating in reverse) and the hypothetical perpetual motion machine of the second kind (PMM2). The idea, elaborated in the main text, is that the hybrid device would require no external source of energy, and would result in a net transfer of heat from the low temperature reservoir to the high temperature reservoir, in violation of the second law of thermodynamics.

Qualitatively, the idea is simple. The idea is that the efficiency of the PMM2 is greater than that of the Carnot engine. This means, for a given heat input Q_{in}, it produces more useful work W than could a Carnot engine operating between the same two temperatures. Equivalently, this means that the input heat required by the PMM2 to produce a certain amount of useful work W, is *less* than the input heat that would be required by a Carnot engine (operating between the same two temperatures) to produce that same amount of useful work – and so also *less* than the heat that a Carnot engine operating *in reverse* would *deposit* to the high temperature reservoir for a given input W. And so if, as depicted in the Figure, the useful work output of the PMM2 just *is* the work input to a Carnot engine operating in reverse, the quantity of heat taken in by the PMM2 from the high temperature reservoir (Q_{in}) will be *smaller* than the quantity of heat deposited by the Carnot engine into the high temperature reservoir (Q'_{out}).

It then follows immediately from the first law that the quantity of heat dumped into the low temperature reservoir by the PMM2 (namely Q_{out}) would be *smaller* than the quantity of heat (Q'_{in}) extracted from the low temperature reservoir by the (reverse operating) Carnot engine. And so the *net effect* of one cycle of the hybrid device is to extract some quantity of heat from the low temperature reservoir, and deposit it into the high temperature reservoir. And since, by construction, this hybrid device requires no external energy input, it is *precisely* the kind of device which is forbidden by Clausius' formulation of the second law of thermodynamics. And so, it better be impossible to actually construct such a device – i.e., it better be impossible to construct something like the PMM2 type of heat engine – i.e., it better be impossible for any heat engine (no matter how cleverly it might be constructed!) to have an efficiency greater than a Carnot heat engine operating between the same two temperatures.

Let us run through that argument one more time, in a little more quantitative detail. We have shown, in the previous section, that a Carnot heat engine has efficiency given by $\eta = 1 - T_{low}/T_{high}$. For the case

---

would be a device that produces energy from scratch, in violation of the *first* law of thermodynamics.

of a Carnot engine operating in reverse, as a heat pump, this means

$$\frac{W}{Q'_{out}} = 1 - \frac{T_{low}}{T_{high}} \qquad (11.56)$$

where W is the work input and Q'_{out} is the quantity of heat deposited to the high temperature reservoir, as shown in Figure 11.10.

Now, by assumption, the hypothetical PMM2 device has a *greater* efficiency. This means its ratio of useful work produced to heat input is greater than the Carnot efficiency:

$$\frac{W}{Q_{in}} > 1 - \frac{T_{low}}{T_{high}}. \qquad (11.57)$$

Combining the last two expressions, we then have that

$$\frac{W}{Q_{in}} > \frac{W}{Q'_{out}} \qquad (11.58)$$

which implies

$$Q_{in} < Q'_{out}. \qquad (11.59)$$

This says that a positive net amount of heat is deposited in the high temperature reservoir.

Now, from the first law, we also have that $Q_{in} = W + Q_{out}$ and $Q'_{in} + W = Q'_{out}$. Solving both of these equations for W and setting the two expressions for W equal gives

$$Q_{in} - Q_{out} = Q'_{out} - Q'_{in} \qquad (11.60)$$

which is equivalent to

$$Q_{in} - Q'_{out} = Q_{out} - Q'_{in}. \qquad (11.61)$$

Equation (11.59) implies that the left hand side is negative. Hence, the right hand side must also be negative, i.e.,

$$Q_{out} < Q'_{in} \qquad (11.62)$$

which says that a positive net amount of heat is *extracted* from the low temperature reservoir.

And so, as we argued above qualitatively, the net effect of the cycle is to extract some heat from the low temperature reservoir, and deposit it into the high temperature reservoir. This is just what the second law of thermodynamics (as formulated by Clausius) is supposed to forbid. What got us into trouble, of course, was the assumption that we had a device (the PMM2) whose efficiency exceeded that of the Carnot engine. If the assumed device had had an efficiency *equal* to that of the Carnot engine, the net effect of the hybrid device would have been *nothing*. And if the assumed device had had an effiency *less* than that of the Carnot engine, the net effect of the hybrid device would have been to transfer some heat from the high temperature reservoir to the low temperature reservoir. Both of these would have been perfectly consistent with the second law.

But the assumed PMM2 turned out not to be, so, to the extent we accept Clausius' formulation of the second law as valid – i.e., to the extent we accept that this is really a *law* – we have to conclude that *no* heat engine can have an efficiency greater than that of the Carnot engine – no matter whether it uses an ideal gas or some other substance, no matter what shape its cycle makes when plotted in the $P-V$ plane, etc. This is a remarkable conclusion.

Here is how Carnot put this argument in his 1824 discussion:

"Now if there existed any means of using heat preferable to those which we have employed, that is, if it were possible by any method whatever to make the caloric produce a quantity of motive power greater than we have made it produce by our first series of operations, it would suffice to divert a portion of this power [into a reverse-operating Carnot engine] to make the caloric of the [cold body] return to the [hot body, i.e.,] to restore the initial conditions, and thus to be ready to commence again an operation precisely similar to the former, and so on: this would be not only perpetual motion, but an unlimited creation of motive power without consumption either of caloric or of any agent whatever. Such a creation is entirely contrary to ideas now accepted, to the laws of mechanics and of sound physics. It is inadmissable. We should then conclude that *the maximum of motive power resulting from the employment of steam is also the maximum of motive power realizable by any means whatever.*"

Of course, there is still the question: why should one accept that Clausius' formulation of the "second law" is valid? That is, in Carnot's terms, why should we insist that the "unlimited creation of motive power without consumption either of caloric or of any agent whatever" is "inadmissable"?

In short, why should one accept that the "second law" is genuinely a *law* (on par with the conservation of energy)? Historically, a big part of the evidence was the repeated and seemingly systematic failure of (perfectly clever) engineers to build such perpetual motion machines. As we will see in the next Chapter, however, there is much more to say – and understand – about the second law. It turns out, in fact, that the second law doesn't have quite the same fundamental law status as the first law. But that doesn't mean you should run out and try to build a perpetual motion machine! And the really interesting thing is that understanding all of this will take us back into the atomic theory of matter.

Let us get a head-start on this project here, before closing the present Chapter, by reformulating the second law in terms of a new concept: *entropy*.

§ 11.7 Entropy

In order to summarize and further illuminate the preceding discussion, it is helpful to introduce the concept of "entropy", which was, historically, first introduced (though with a different name) by Clausius.

To understand the motivation for this new concept, consider again the expression we derived earlier for the efficiency of a Carnot heat engine:

$$\frac{W}{Q_{in}} = 1 - \frac{T_{low}}{T_{high}}. \tag{11.63}$$

By the first law, the work done is just the difference between the input and output heats: $W = Q_{in} - Q_{out}$. Plugging this into the previous equation and simplifying gives

$$\frac{Q_{out}}{Q_{in}} = \frac{T_{low}}{T_{high}} \tag{11.64}$$

or, equivalently,

$$\frac{Q_{in}}{T_{high}} = \frac{Q_{out}}{T_{low}}. \tag{11.65}$$

This seems to have the flavor of a kind of conservation law. The idea would be that, when heat flows from the high temperature reservoir into the working fluid of the heat engine, it carries with it a certain quantity of "entropy", given numerically by the quantity of heat divided by the temperature. Then, on the other side of the cycle, some "entropy" is transferred from the heat engine to the low temperature reservoir. The quantity of "entropy" transferred is again equal to the heat transferred divided by the temperature at which that transfer takes place.

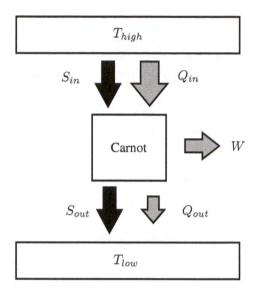

Figure 11.11: An energy-and-entropy flow diagram for the Carnot heat engine.

Let us then just propose that – in addition to the already-familiar physical properties like mass, energy, temperature, pressure, and temperature – physical systems also possess a property called "entropy" whose value increases or decreases whenever some heat flows in or out of the system, according to the following formula:

$$\Delta S = \frac{Q}{T}. \tag{11.66}$$

Here ΔS is the amount by which the entropy changes when a quantity of heat Q flows into the system at temperature T. (If heat flows *out* of the system, then Q is negative, and so the entropy of the system will *decrease*.) Equation (11.65) can then be understood as saying that the entropy *in* equals the entropy *out* during each cycle of the Carnot engine. Which – if there *is* such a thing as entropy – makes perfect sense, since we were after all supposed to leave the working fluid precisely again in its initial state after one complete cycle.

Note that this idea of a new conserved physical quantity allows us to understand one of Carnot's key qualitative insights which has already played a central role in our discussion:

> "...the production of heat alone is not sufficient to give birth to the impelling power: it is necessary that there should also be cold; without it, the heat would be useless."

That is, if, in order to return to its initial state, and having already *increased* its entropy during the first part of the cycle, the working substance in a heat engine has to expel some entropy back out into the environment, then one can see immediately from Equation (11.66) why it is necessary for some of the input heat to be "wasted" – i.e., dumped into a low-temperature reservoir: this is precisely how the working substance expels entropy. And note further that, in order for the "wasted" heat to be smaller than the input heat – i.e., in order for there to be any energy left over to appear as useful mechanical work – the temperature T_{low} at which the waste heat is dumped must indeed be lower than the temperature T_{high} at which the input heat was absorbed from the heat source. Thus, as Carnot already noted (several decades before Clausius introduced the concept of entropy), the existence of a cold heat sink (the river from our original book-lifting example) is really necessary.

Note also that, if we postulate – in addition to the conservation of energy as expressed in the first law of thermodynamics – a conservation law for "entropy", we can re-derive the expression for the efficiency of the Carnot engine (which we now know to be some kind of absolute thermodynamic maximum efficiency) in a very elegant way. Consider the schematic representation of the Carnot engine in Figure 11.11, which is like our earlier "energy flow diagram" but with the addition of indications of "entropy flow". The energy flow part of the diagram, as before, is captured by the following statement of energy conservation:

$$Q_{in} = W + Q_{out}. \tag{11.67}$$

Now we postulate also that, again because the process is supposed to be *cyclical*, entropy (like energy) should not build up in the engine – i.e., the entropy in during one cycle should equal the entropy out:

$$S_{in} = S_{out}. \tag{11.68}$$

Or, re-writing this using the proposed definition of entropy in terms of heat and temperature,

$$\frac{Q_{in}}{T_{high}} = \frac{Q_{out}}{T_{low}}. \tag{11.69}$$

The point is then that Equations (11.67) and (11.69) can easily be solved for the efficiency. Eliminating Q_{out} from the pair of equations immediately gives

$$Q_{in} = W + Q_{in}\frac{T_{low}}{T_{high}} \tag{11.70}$$

which is equivalent to

$$\frac{W}{Q_{in}} = 1 - \frac{T_{low}}{T_{high}} \tag{11.71}$$

which is precisely the Carnot efficiency.

Stepping back, though, we should stress that despite all the talk about "entropy" in the last few pages, it's not entirely clear what, if anything, we're talking about. There is a temptation at this point to regard entropy as some new kind of imponderable fluid, which literally flows into the heat engine when it absorbs energy from the high temperature reservoir, and then flows back out when the heat engine dumps energy into the low temperature reservoir. But if the historical lesson of the caloric theory of heat has taught us anything, it is that we should be rather cautious in leaping too quickly to interpretations of this kind.

Of course, that doesn't mean that it's wrong to *think* of "entropy" as relevantly *like* a fluid which flows into and out of the engine as it transfers energy this way and that. Indeed, despite now knowing that the caloric theory of heat is false, and that, really, "heat" is just a kind of macroscopic perspective on the incessant microscopic motions of the underlying atoms and molecules, we have still been thinking of heat as kind of like a fluid. Just look back at practically any paragraph from the preceding parts of this current chapter, where we repeatedly talk about heat "flowing" this way and that.

Methodologically, the lesson here is that, despite the allure of positing this new thermodynamic concept, "entropy", and (for example) analyzing heat enginges in terms of the "flow" of both energy and entropy, we should remain explicitly-open-minded about how to think about what (if anything) "entropy" really *is* physically – whether, for example, it is some new type of elemental fluid, or (as "heat" turned out to be) just some kind of rough, macroscopic perspective on something the invisible atoms and molecules are doing.

Addressing this is, in essence, the subject of the next chapter, Chapter 12. But, as a kind of teaser, we can already present an argument in favor of the second possibility just mentioned. Actually, the argument is completely parallel to the Count Rumford's argument, discussed in Chapter 10, that "heat" cannot be the manifestation of some new underlying elemental fluid, but must instead be a macroscopic perspective

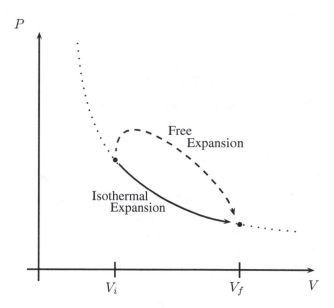

Figure 11.12: *In the free expansion of an ideal gas from volume V_i to volume V_f, no heat flows in, and hence (according to our earlier understanding of entropy) no entropy flows in. Yet the entropy of the gas must increase, as shown by the fact that we could get to the same final state by a reversible isothermal expansion, in which a finite amount of heat – and so a finite amount of entropy – is made to flow into the gas. Thus, in a free expansion, the entropy of the gas increases, even though the entropy didn't come from (didn't "flow in from") anywhere – entropy, in short, was created out of thin air.*

on the random motions of atoms and molecules. The argument, recall, amounted to this: heat is not conserved! You can produce heat "out of thin air".

In the same way, we can argue that "entropy" cannot be an irreducible fluid, because (despite the hints above) it is not conserved. That is, we can create entropy "out of thin air." Here's how. Let some gas undergo a "free expansion" as discussed earlier in this chapter. No heat flows into the gas, so we would expect on the basis of Equation (11.66) that no entropy flows into the gas. And indeed it doesn't. We could imagine that the freely expanding gas is in a perfectly insulated container (or is floating in otherwise empty space) such that there's nowhere for any entropy to have flowed in from.

And yet – as we will now show – the entropy of the gas *does increase*. The way to show this is simply to consider an *alternative* way of getting the gas into the *same final state*. For definiteness, suppose we are dealing with an ideal gas, and suppose the free expansion of the gas is from an initial volume V_i to a final volume V_f. The initial and final states of the gas will share the same total energy (because, for a free expansion, no work is done and no heat flows in, and those are the only two ways the energy of the gas could change), and hence (because the gas is ideal) the initial and final states will share the same temperature. The two states are shown in Figure 11.12, where the free expansion process connecting them is represented by the arching arrow. (Remember that the intermediate states of the gas are not well-defined during this sort of irreversible process, so we cannot represent the process as a well-defined path through the $P - V$ plane.)

We could, however, have gotten to the same final state by expanding the gas isothermally. We have already calculated the amount of heat required to isothermally expand an ideal gas. Since the energy of the gas is constant along an isotherm, we have (for each baby step of the process) that the heat flowing in is equal to the work done by the gas on its environment: $dQ = P \, dV$. And so the total heat added to expand the gas from V_i to V_f is

$$Q = \int_{V_i}^{V_f} P\, dV = nRT \int_{V_i}^{V_f} \frac{dV}{V} = nRT \log\left(\frac{V_f}{V_i}\right). \tag{11.72}$$

And so, by Equation (11.66), the isothermal expansion of the gas involves adding to the gas a quantity of *entropy*

$$\Delta S = \frac{Q}{T} = nR \log\left(\frac{V_f}{V_i}\right) \tag{11.73}$$

which is certainly greater than zero.

The point is, if you believe that the final state of the gas is really *the same* whether we get there by doing a free expansion or by doing a reversible isothermal expansion, you must believe that, in the free expansion, the entropy of the gas *increases* even though there was nowhere for the new entropy to have flowed from. So evidently a free expansion involves the creation, *ex nihilo*, of entropy – just as rubbing a dull drill bit against metal evidently produces heat *ex nihilo*. And so this is evidence that entropy, like heat, is not an irreducible (and so *conserved*) fluid.

Of course, one could question whether the two final states (reached by free expansion and reversible isothermal expansion) are really the same. This is parallel, in the context of Rumford's arguments against the caloric theory, to wondering if the newly-produced heat wasn't, in some non-obvious way, actually extracted from the surroundings. Recall that Rumford compared the specific heat of the metal filings thrown off by the drill bit, to that of pieces of the metal which had not been drilled. He also insulated his experiment from the environment in various ways.

One could do similar experiments here. One could, for example, take two evidently-identical samples of gas, let one undergo a free expansion and the other a reversible isothermal expansion, and then measure and compare various thermodynamic properties of the two samples. We needn't go into details here – it suffices to report that all such evidence supports the original contention that the two final states really are identical. Which in turn supports the argument already presented: in the free expansion of a gas, the entropy of the gas increases, but the new entropy didn't come from anywhere. So entropy was genuinely created from scratch. And that seems to contradict the (otherwise plausible) view that entropy is a conserved substance, and hence to suggest that entropy (like heat) can be understood to be some kind of abstract (collective or average) property of the microscopic state of the gas.

Let us briefly consider one further example of a process that creates entropy. Laying this out on the table will help us clarify some things that have been assumed, in an obscure way, in much of the preceding discussion. Consider two objects, one hot and one cold, which are brought – very briefly – into thermal contact. For example, we dip an ice cube into a cup of hot coffee, and then immediately pull it back out again. Some small amount of heat, dQ, will flow from the coffee to the ice cube – i.e., some heat will flow out of the coffee and a numerically equal quantity of heat will flow into the ice cube. And so, by Equation (11.66), the entropy of the coffee will decrease somewhat and the entropy of the ice cube will increase somewhat.

The key point, however, is that the entropy decrease of the coffee *will not match* the entropy increase of the ice cube. Let's see why. Above, we called dQ the quantity of heat which flows from the coffee to the ice cube. Thus, the heat "added" to the coffee is $dQ_{coffee} = -dQ$ and the heat added to the ice cube is $dQ_{ice} = dQ$. But the heat "added" to the coffee is "added" at a relatively high temperature, T_{coffee}, compared to that of the ice cube, T_{ice}. And so the entropy "added" to the coffee, $dS_{coffee} = dQ_{coffee}/T_{coffee} = -dQ/T_{coffee}$ is not just the negative of the entropy added to the ice cube, $dS_{ice} = dQ_{ice}/T_{ice} = dQ/T_{ice}$. Rather, the entropy of the ice cube increases by *more* than the entropy of the coffee decreases.

Another way to say the same thing is that the total entropy of the whole universe *increases* a little bit during this brief encounter:

$$dS_{universe} = dS_{coffee} + dS_{ice} = dQ \left(\frac{1}{T_{ice}} - \frac{1}{T_{coffee}} \right) \tag{11.74}$$

which is definitely *greater than zero* because $T_{coffee} > T_{ice}$.

The overall lesson here is that it is only in a certain kind of idealized process that entropy is going to be conserved. In our discussion of heat engines and the Carnot cycle, we often said that we were taking the processes involved to be not only frictionless, but also "smooth" or "slow". This was partly just a vague way of capturing something that we can now state more precisely. During an adiabatic expansion of a gas, for example, the piston better be pulled out slowly enough that we don't, even for a brief moment, have something like a free expansion of the gas. And when we, for example, extract heat from the gas during an isothermal compression, it better be that the temperature of the heat reservoir differs only infinitessimally from the temperature of the gas. If we violate these assumptions, we'll be creating entropy from scratch – and so, for example, in order to return the gas to precisely its initial state at the end of a cycle, we'll have to "dump" *more* entropy into the low temperature reservoir than was absorbed from the high temperature reservoir. That means, by Equation (11.66), dumping more *heat* into the low temperature reservoir than would otherwise have been necessary, which in turn means (by the first law) that less of the input heat will be available to do useful mechanical work.

So now we understand a little more clearly the full sense in which the Carnot cycle is an idealization. Not only do we have to imagine that the piston containing the gas is perfectly frictionless, but the heat flow at both the high and low temperatures must occur across an infinitessimally small temperature difference (and so take a really really long time!), and likewise the adiabatic expansions and contractions in the cycle must occur infinitely slowly. Only with this extreme amount of idealization will the efficiency of a Carnot heat engine match our calculated prediction. In other words, real heat engines (where friction and finite speeds and finite temperature differences abound) will always produce some entropy internally, and so will always have efficiencies less than the ideal limit captured by the second law of thermodynamics.

Nevertheless, it should be clear that the ideal Carnot heat engine is of more than mere academic interest. For analyzing it led us (as it led Carnot and Clausius) to the realization that no other type of heat engine (no matter how idealized) can improve on the Carnot efficiency, to the second law of thermodynamics, and to the very concept, "entropy", which helped us understand the nature of the idealization involved in the Carnot engine!

Let us finally summarize and close this chapter by reviewing what we have learned and how it can be formulated in terms of entropy. We began with the first law of thermodynamics, which can be understood as the statement that energy is conserved – as long as heat is recognized as a form of energy and taken into account. This means that the change in energy of a system during some process will be equal to the quantity of heat which flows into the system during the process, plus the quantity of work done on the system during the process. For a *closed system* – one that doesn't interact at all with its environment – the first law of thermodynamics just says that the total energy of the system will be exactly conserved.

This principle was helpful in analyzing and understanding various types of thermodynamic processes as well as in developing quantitative relationships among thermodynamic properties (such as the constant volume and constant pressure specific heats of gases).

We observed, however, that the ubiquitous directionality of spontaneous thermodynamic processes – represented most clearly by the fact that heat flows spontaneously from hot to cold but never vice versa – could not be understood in terms of the first law alone. We thus posited a second law of thermodynamics and found that this entailed a fundamental restriction on the efficiency of heat engines: no heat engine, regardless of its nature, can deliver more useful work from a given input of heat from a high temperature reservoir, than an idealized (reversible) Carnot engine operating between the same two fixed temperatures.

Further analysis of the Carnot cycle led us to posit a new quantity called "entropy" which is conserved in idealized, reversible processes, but which can be created from scratch in irreversible processes of various sorts. This leads to a final re-formulation of the second law of thermodynamics: the entropy of a closed system can never decrease. Formally,

$$\Delta S \geq 0. \tag{11.75}$$

But, to arrive at a clear understanding of the meaning of "entropy" (and hence this final re-formulation of the second law), you will need to continue on to Chapter 12.

Questions:

Q1. Modern experiments reveal that Carbon Dioxide (CO_2) is a *linear* molecule. That is, it looks like this: O–C–O, with the three atoms lying in a straight line. What sort of degrees of freedom, therefore, do you think are responsible for its empirically-determined k-value being 6.7?

Q2. Imagine a room that is perfectly insulated so that no heat can enter or leave. In the room is a refrigerator that is plugged into an electric outlet in the wall. If the door of the refrigerator is left open, what happens to the temperature of the room?

Q3. An ideal gas is one in which the long-range forces the molecules exert on each other precisely vanish. Equivalently, there is no contribution to the potential energy of an ideal gas which varies with the volume. That is why an ideal gas subjected to a free expansion (for which the total energy is necessarily constant) suffers no change in temperature: the potential energy doesn't change when the volume being occupied increases, so neither does the kinetic energy, and that's what temperature *is*. When free expansion experiments are performed with real gases, though, it is usually observed that the temperature changes a tiny bit. In particular, the temperature almost always *decreases* a little bit. What does this imply about the long-range forces that real gas molecules exert on each other? Are the forces attractive or repulsive?

Q4. When a gas is adiabatically compressed, we found that the temperature increases according to Equation (11.27). From the macroscopic perspective, it is clear why the temperature must go up as the volume decreases: compressing the gas means that *work* is being done on it, which increases the energy of the gas, and some fraction of that energy goes into translational kinetic energy of the molecules, which manifests itself as higher temperature. Explain how this can be understood from a purely microscopic perspective. Hint: think about the collisions of the molecules with the *moving* piston.

Q5. Is there an analog to the "free expansion" in which the volume of the gas decreases rather than increasing – a "free compression"? Why or why not?

Q6. The "rectangular" (in the P-V plane) heat engine we began by discussing has heat flowing into (and then later out of) the gas at a whole range of temperatures between some minimum and maximum temperatures reached during the cycle (corresponding to the lower-left and upper-right corners of the rectangle). Is it even possible for such a cycle to be made reversibly? Wouldn't the heat flowing in (from the candle or whatever) necessarily flow across finite temperature differences almost all of the time, and hence create entropy "from scratch" as discussed in the last section above? What modifications to our discussion would be required for the book-lifting machine to truly be able to operate reversibly?

Q7. We have said that a reversible adiabatic expansion or contraction must happen slowly. But – how slowly? Roughly how fast would the piston in a cylinder containing (say) air at room temperature have to be moved before the process became appreciably irreversible? That is, roughly what is the speed of motion for the piston at which (for example) reversible adiabatic expansion becomes (irreversible) free expansion?

Q8. Figure 11.3 shows, in the $P - V$ plane, some isothermal and adiabatic curves for an ideal gas. What do the "isentropic" curves – i.e., the curves corresponding to a constant total entropy of the gas – look like?

Exercises:

E1. You decide to have a heat pump installed on your winter cabin. (A heat pump is just a heat engine operating in reverse.) It will be powered by electricity. You will rarely use the cabin during the summer, so primarily the heat pump will be used as a heater during the winter months. Your neighbor says: "You're stupid! Why not install a (cheaper) electrical heater, which, after all, can convert a full 100% of the incoming electrical energy into heat." How do you respond? (Hint: what is the maximum efficiency of a heat pump allowed by the second law of thermodynamics?)

E2. Consider again the heat engine discussed in Section 9.4, which uses a rectangular cycle in the $P - V$ plane. Write the efficiency of the engine in terms of the four temperatures of the gas at the four corners of the rectangle. Why is the expression complicated compared to the equivalent expression for the Carnot engine?

E3. Suppose, contrary to the second law of thermodynamics, you could generate useful energy simply by extracting it from a system and reducing its temperature. Concretely, suppose you have a device which extracts energy from a small glass of water and powers a 60 Watt light bulb. Suppose the glass contains 300 grams of water initially at room temperature. How long would the light bulb be able to burn before the water turned to ice?

E4. This is the same, in principle, as the previous Exercise, but maybe gives a more interesting perspective on the 2nd law. The total rate at which humanity consumes energy is something on the order of 15 tera-Watts (i.e., 15×10^{12} Joules/second). If we could build a perpetual motion machine of the second kind, and extract this energy from the oceans (hence constantly reducing their temperature), at what rate (in, say, degrees per day or degrees per year or whatever is appropriate) would the temperature of the oceans have to decrease? (The world's oceans have a volume of roughly a billion cubic kilometers.)

E5. As explained in the reading, if Helium is adiabatically compressed to $1/8$ its original volume, its temperature increases by a factor of about 3.5. By what factor would the temperature of a diatomic ideal gas (like Nitrogen, say) increase if it were adiabatically compressed to $1/8$ its original volume? Explain qualitatively why the temperature increase for Nitrogen is different from that for Helium.

E6. Suppose you had the magical ability to snap your fingers and all the molecules in a gas would "spontaneously" move to one side of the container, such that a piston could be quickly slid in (without doing any work on the gas). (Note that this would constitute something like the "free compression" contemplated in Q5.) Explain in detail how you could use your power to create a cyclical engine whose efficiency exceeds the Carnot limit. Hence, show that (in at least this one example) violation of the entropy-formluation of the second law – Equation (11.75) – is equivalent to a violation of the earlier formulations.

E7. Recall that Dulong and Petit found that the molar heat capacities of most heavy metallic solids were about 25 Joules / mole K. Can this be understood on the basis of the equipartition of energy? Which degrees of freedom are "active" for (say) Gold at room temperature?

Projects:

P1. Sound waves are pressure waves in the air in which one region, with a slightly higher-than-average pressure and density, expands – pushing on the next region over so that now it becomes compressed – and so on. A careful analysis reveals that the speed of such a wave should be given by the square root of the derivative of the pressure with respect to the density:

$$c = \sqrt{\frac{dP}{d\rho}} \qquad (11.76)$$

where P is the pressure and ρ is the mass density (mass per unit volume) of the gas. Treating the air as an ideal gas, compute the derivative and calculate the predicted speed of sound (at, say, room temperature). Compare your value to the measured speed of sound. (You may have the opportunity to perform an experiment to measure the speed of sound.) You will probably find that your predicted value is off by about 20 percent. This is probably because you were assuming *isothermal* compressions of the air. As it turns out, though, the compressions are not isothermal, but adiabatic. (Air is not a very good thermal conductor, and the sound waves happen very fast, so there just isn't time for appreciable heat to flow between a compressing section of air and its neighboring sections.) So you should re-do the calculation and show that you get a much more accurate answer.

P2. Invent your own cyclic process that could be used as the heart of a heat engine. For example, your cycle might take a sample of ideal gas back and forth between two isotherms via isobaric (constant pressure) expansions and contractions. Or, instead, via constant volume heating/cooling. Or some other combination. Calculate the work done and the heat flow during each leg of the cycle and develop a formula for the efficiency of a heat engine operating using your cycle.

P3. Derive the formula – Equation (11.27) – relating the temperature and volume for an adiabatic process on an ideal gas, not from macroscopic energy considerations (as was done in the Chapter) but instead from an analysis of the microscopic collisions of gas molecules with the moving piston.

CHAPTER 12

Statistical Physics

By the middle of the 19th century, the atomic theory of matter and the closely-associated kinetic theories of heat and gases had become widely, though still not universally, accepted. The early controversies about the relative atomic weights of different elements had been largely resolved, with Mendeleev's creation of the periodic table of the elements in 1869 serving as a dramatic and conclusive piece of evidence. But the absolute sizes of atoms and molecules remained unknown, and there were also a number of puzzles about, for example, why only certain degrees of freedom appeared to be "active" in gases, and how to understand the laws of thermodynamics and in particular the mysterious concept of "entropy" in microscopic terms.

Our goal this week is to see how the application of *statistical* methods led to progress on a number of these fronts in the late 19th century, culminating in the first accurate measurement of Avogadro's number (and hence the absolute scale of atoms) in the early part of the 20th century.

§ 12.1 The Mean Free Path and The Random Walk

Back in Chapter 10, when we derived the ideal gas law from the kinetic model of gases, we assumed that a given molecule bounces back and forth through the space available to it, moving in straight lines until it bounces, elastically, from a wall. We also saw that the (average) speeds of molecules in a gas could be inferred from empirically measurable quantities. For example, for Nitrogen gas (the most abundant component of ordinary air) at room temperature, the molecules typically move at about 500 m/s. Other molecules, like for example Ammonia (NH_3), which have smaller molecular weights, will typically be moving even *faster* at room temperature.

But, as pointed out in 1858 by the Dutch meteorologist Christophorus Buys-Ballot, this seems to conflict with a phenomenon of fairly common experience: if some noxious gases such as ammonia "be evolved in one corner of a room, entire minutes elapse before they are smelt in another corner, although the particles of gas must have had to traverse the room hundreds of times in a second." Buys-Ballot thus argued that gas molecules could not possibly be moving as fast as the kinetic theory said they were moving. Perhaps, he suggested, there was some fundamental problem with the atomic/molecular model?

In his response to this objection later the same year, Rudolf Clausius stressed that it was never actually an assumption of the kinetic theory of gases that the molecules *never* collide. Instead, the theory required only that the length scale over which the molecules exert appreciable forces on one another – that is, roughly speaking, how close two molecules need to come before they suffer a *collision* – be small compared to the typical distance between molecules. Under these conditions, the molecules would be moving in straight lines, uninfluenced by inter-molecular forces, *most of the time*, and the (empirically correct)

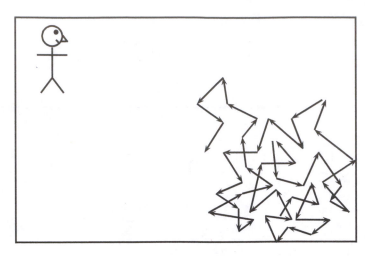

Figure 12.1: An ammonia molecule, released from a bottle of ammonia in one corner of a room, may take an unexpectedly long time to reach the nose of a person in the opposite corner of the room, if the molecule's motion is continually interrupted by collisions with air molecules.

relations between pressure, volume, and temperature would continue to hold.

Actually, as Clausius pointed out, the ideal gas law is "in actually existing gases ... not strictly, but only approximately true". Thus, some deviations from (or additions to) the model used earlier in our derivation of the ideal gas law, are going to be required to make the model fit all the relevant data. In particular, attributing a finite size or "sphere of action" to the molecules can not only bring the theoretically-predicted pressure-volume relation more exactly in line with empirical observations, but it can also provide an answer to the paradox raised by Buys-Ballot.

The key idea here is that, if the molecules have a finite size, each one will only travel some finite average distance – the so-called "mean free path" – before suffering a collision with some other molecule which sends both molecules moving off in some new, random directions. The trajectory of each molecule through space, therefore, will not be a uniform straight line, but a so-called "random walk" such as that illustrated in Figure 12.1. And such a picture obviously helps explain why it takes far more than the expected tiny fraction of a second for molecules moving at hundreds of meters per second, to progress a few meters.

Let us follow Clausius and try to understand the relation between the size of the molecules and the average distance they travel between collisions (i.e., the mean free path).

We start by assuming that two molecules can be considered to have collided if the separation between their centers gets to be as small as some threshhold distance d. If the molecules were hard spheres (like tiny billiard balls), d would just be the sum of the radii of the two molecules. If, on the other hand, there is some short-range repulsive force that gets stronger as the molecules get closer together, d will be the separation corresponding to some appropriate intermediate strength of the force. Of course, we don't know much at this point about which model is true, but the details will not matter for our purposes. All that matters is that a collision will occur if some other molecule enters a given molecule's radius-d "sphere of action".

Now consider some particular molecule, flying off in some random direction through the background of other molecules. The question is: how far, on average, will it get before colliding with another molecule? The answer will clearly depend on how many other molecules are around (per unit volume), and on the size of the given molecule's sphere of influence. An explicit formula can be derived with the help of Figure

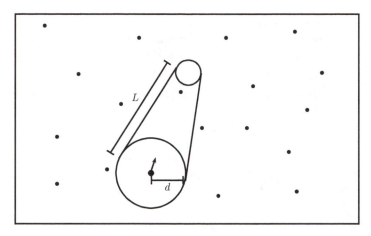

Figure 12.2: A given molecule sweeps out a cylinder of volume $V = L\pi d^2$ as it traverses a distance equal to the mean free path. (d here is the radius of its sphere of influence.) In order that the mean free path should have its intended meaning, there should be on average one target molecule contained in the volume V – i.e., traversing a distance L should, on average, produce one collision.

12.2. The given particle's sphere of action sweeps out a cylinder of radius d as it moves through the gas.

The mean free path L can then be visualized as the length of such a cylinder which contains, on average, one (other, "target") molecule, i.e., it is the length of a path which produces, on average, one collision with another molecule. Mathematically, this means that the *number density* of target molecules (n), multiplied by the volume of the cylinder swept out as the given molecule traverses a mean free path, should equal 1: $n\pi d^2 L = 1$. And so the mean free path is evidently given by:

$$L = \frac{1}{n\pi d^2}. \tag{12.1}$$

Actually this formula is not quite precise, because we have erroneously assumed that all the "target" molecules are stationary. In fact, they, too, will be moving, and this (as it turns out) slightly reduces the mean free path. In any case, for our purposes, it is probably best to regard this simply as an order-of-magnitude estimate, so we may drop the factor of π and write

$$L \sim \frac{1}{nd^2} \tag{12.2}$$

where d is thought of as roughly the diameter of a single molecule.

Unfortunately, neither the absolute number density of a gas, nor the radius of a single molecule, is known, so the formula we just derived for the mean free path is rather useless – if our goal is to compute just how far the mean free path actually *is* in some realistic situation. It didn't take long, however, for Maxwell (yes, the same James Clerk Maxwell who played such a central role in Part 2 of the book) to demonstrate – through a series of sophisticated analyses that we need not go into here – that the mean free path is also related to several more directly measureable properties of gases.

In particular, Maxwell showed that mean free path L can be inferred from experimental measurements of the *thermal conductivity* and also the *viscosity* of gases. Using the results of a colleague's study of the viscosity of air, Maxwell calculated that

> "the value of $[L]$, the mean distance travelled over by a particle between consecutive collisions, = 1/447000th of an inch [or about 5×10^{-8} meters] and each particle makes $8,077,200,000$ collisions per second."

The reported value for the number of collisions per second is just the mean free path L divided by the typical molecular speed of about 500 m/s. It is rather amazing to consider that each molecule in the air suffers a collision, that sends it moving off in some new direction, something like ten billion times each second!

Returning to the objection of Buys-Ballot, we now see that our introductory Figure 12.1 actually understates the point quite significantly. We drew the mean free path there as if it were something of order several centimeters. But it is actually, for typical molecules like those composing air or the ammonia molecule we had in mind for that earlier discussion, many many orders of magnitude smaller than that. And so it is very plausible indeed that the molecules should take a surprisingly long time to random walk their way across the room. We will explore the mathematical details further in the last section of this chapter.

Shortly after Maxwell first estimated the mean free path of air molcules, the Austrian physicist Josef Loschmidt produced the first decent ballpark estimate of Avogadro's number. His calculation was based on our Equation (12.2), which (with L known from viscosity and other similar measurements) can be thought of as one equation in the two unknowns n (the number density in the gas) and d (the diameter of the molecules). What Loschmidt added was a *second* equation in these same two unknowns which then allowed both of them to be algebraically determined. (His second equation was fairly rough and approximate, which is why the value of N_A he came up with can only be thought of as a ballpark estimate. You can reproduce this interesting calculation in the Exercises.)

§ 12.2 Statistics, Equilibrium, and the 2nd Law

In addition to solidifying Clausius' concept of the mean free path by relating it to empirically measurable properties of gases, Maxwell also introduced a number of statistical analyses pertaining to the nature of and the approach to thermal equilibrium. We review some of these analyses in this section and then turn to Maxwell's famous thought experiment in support of the claim that, unlike the 1st Law of Thermodynamics, the 2nd Law must have a non-strict, statistical character.

12.2.1   The Distribution of Molecular Speeds

In our discussion of the kinetic theory of gases back in Chapter 10, we found that the average (translational) kinetic energy of molecules was proportional to the absolute temperature:

$$\langle KE^{\text{trans}} \rangle = \frac{3}{2} k_B T \tag{12.3}$$

where the angle brackets denote the average (over all particles present in the gas), and k_B, Boltzmann's constant, is related to the ideal gas constant R and Avogadro's number N_A as follows: $k_B = R/N_A$. Using the familiar relation for the kinetic energy, this leads to the following expression for the average value of the squared speed of the molecules:

$$\langle v^2 \rangle = \frac{3 k_B T}{m} \tag{12.4}$$

or, equivalently, a "root mean square" (RMS) speed:

$$v_{\text{RMS}} = \sqrt{\frac{3 k_B T}{m}} \tag{12.5}$$

which can be taken as a measure of the typical speed of the molecules in a gas. Multiplying on the right hand side by $N_A/N_A = 1$, one converts the above expression into a formula for the RMS speed of a gas molecule in terms of more readily determinable quantities:

$$v_{\text{RMS}} = \sqrt{\frac{3RT}{\mathcal{M}}}.$$ (12.6)

where $\mathcal{M} = m\,N_A$ is the mass of one mole of a given type of molecules. This, of course, is the formula by which it is known that, for example, Nitrogen molecules in room temperature air move at about 500 meters/second.

It is a near certainty, however, that all the molecules in the gas will *not* simply be moving with precisely this speed. For, as Maxwell explains,

> "it is easy to see that if encounters take place among a great number of molecules, their [speeds], even if originally equal, will become unequal, for, except under conditions which can be only rarely satisfied, two molecules having equal [speeds] before their encounter will acquire unequal [speeds] after the encounter."

Maxwell then argues that, by using statistical methods, we can say something about the *distribution* of molecular speeds:

> "if all the molecules have the same velocity originally, their encounters will produce an inequality of velocity.... If, however, we adopt a statistical view of the system, and distribute the molecules into groups, according to the velocity with which at a given instant they happen to be moving, we shall observe a regularity of a new kind in the proportions of the whole number of molecules which fall into each of these groups."

The reason for assigning molecules to "groups" is that, for example, the x-component of a molecule's velocity could have, at a given moment, any of a continuous infinity of possible values. At some random moment, there are, in all likelihood, precisely zero molecules which have some one *particular* value of v_x from the continuum.

But if we coarse-grain the continuum into "bins", with each "bin" corresponding to a certain small-but-finite *range* of v_x-values, then we can meaningfully discuss how many of (or what fraction of) the molecules have v_xs that lie somewhere in that range.

Consider a sample of gas containing N molecules. Let's call the number of molecules with v_x values lying in a range of width Δv_x and centered at v_x

$$\eta(v_x)\Delta v_x.$$ (12.7)

Note that it is the product – η (the Greek letter "eta") times Δv_x – which is the number; $\eta(v_x)$ itself is not a number, but rather a "number density" (meaning, here, a number-per-unit-velocity).

If $\eta(v_x)\,\Delta v_x$ of the N molecules have their v_x values in some range, then we may regard

$$p(v_x)\Delta v_x = \frac{\eta(v_x)\Delta v_x}{N}.$$ (12.8)

as the *probability* for some randomly-selected molecule to have its v_x in that range. Note again that the function $p(v_x)$ is not a probability, but rather a probability density – a probability per unit velocity.

With that notation in place, we can rehearse Maxwell's clever argument for a specific formula for the distribution of molecular speeds. The key idea is that the x-, y-, and z-components of the velocity should be determined *independently*: the probability, for example, of a given molecule having a v_x in some range, should not depend on what its v_y and v_z are.

Thus, the probability that a molecule has a v_x in some range *and* has a v_y in some range *and* has a v_z in some range, should just be the probability that it has a v_x in the appropriate range *times* the probability that it has a v_y in the appropriate range *times* the probability that it has a v_z in the appropriate range. That is:

$$p(\vec{v})\Delta v_x\,\Delta v_y\,\Delta v_z = p(v_x)\Delta v_x\,p(v_y)\Delta v_y\,p(v_z)\Delta v_z. \tag{12.9}$$

But as Maxwell also points out, in a real gas there should be nothing to pick out any particular direction as unique or special. This has several consequences.

One consequence is that the probability distributions for the x-, y-, and z-components should all be the same. We have actually already assumed this in our notation, which uses the same function "p" for the probability densities associated with all three orthogonal directions. But it is worth noting explicitly.

A second consequence is just that the probability for a molecule to have a certain velocity (i.e, to have v_x, v_y, and v_z values in specific ranges) should be independent of the *direction* of the velocity. That is, the probability density should depend only on the *speed*. Mathematically, this implies that

$$p(v_x)\,p(v_y)\,p(v_z) = \{\text{Some function of } v = |\vec{v}| \text{ only}\}. \tag{12.10}$$

Maxwell's beautiful insight was that it is possible to infer – from this curious relation alone – what the functions actually *are*. The crucial point is that the speed v relates to the individual velocity components this way:

$$v^2 = v_x^2 + v_y^2 + v_z^2. \tag{12.11}$$

Equations (12.10) and (12.11) tell us that we need a function p which has the following property: p of v_x, multiplied by p of v_y, multiplied by p of v_z, gives a function of $v_x^2 + v_y^2 + v_z^2$.

A little reflection reveals that p must be an exponential function. For example:

$$e^{-\alpha v_x^2}e^{-\alpha v_y^2}e^{-\alpha v_z^2} = e^{-\alpha(v_x^2+v_y^2+v_z^2)} = e^{-\alpha v^2}. \tag{12.12}$$

Thus, the probability distribution for each velocity component must be a so-called "Gaussian" function. The constant α can be determined to be

$$\alpha = \frac{m}{2k_BT} \tag{12.13}$$

using the requirement that the RMS speed should match what we said before in Equation (12.5). And the overall normalization factor for the probability density can be determined using the requirement that the total probability (i.e., the probability for a given molecule to have *some* velocity) be one. The result (skipping over the technical details described in the last couple of sentences) is that

$$p(v_x) = \sqrt{\frac{m}{2\pi k_BT}}\,e^{-\frac{1}{2}mv_x^2/k_BT}. \tag{12.14}$$

This Gaussian probability distribution is plotted in Figure 12.3.

This then implies that the probability for a molecule's velocity to be near (in the appropriate sense) a specific value $\vec{v}$ is

$$p(\vec{v})\,\Delta v_x\,\Delta v_y\,\Delta v_z = \left(\frac{m}{2\pi k_BT}\right)^{3/2}e^{-\frac{1}{2}mv^2/k_BT}\,\Delta v_x\,\Delta v_y\,\Delta v_z. \tag{12.15}$$

What, then, is the probability for a molecule to have a *speed* in a certain range of width Δv?

To answer that, we need to add up the probabilities associated with the many different velocities that all share that same magnitude, the speed. We have been dividing the "velocity space" (whose axes are

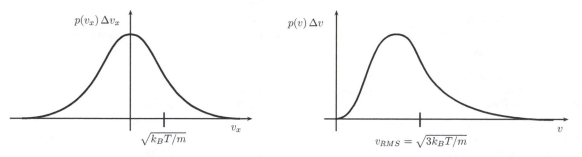

Figure 12.3: The left graph shows the probability distribution for one velocity component (here v_x) for molecules in a gas. It is a Gaussian function with half-width $\sqrt{k_BT/m}$. The right panel shows the probability distribution for molecular speeds. *Note that the root mean square (RMS) speed is slightly to the right of the graph's peak.*

v_x, v_y, and v_z) into little boxes of volume $\Delta v_x \Delta v_y \Delta v_z$. How many such boxes are there in the thin spherical shell of radius v and thickness Δv which corresponds to the speed being in a range of width Δv around v? Evidently the answer is the volume of that shell divided by the volume of a single box, i.e., $4\pi v^2 \Delta v/\Delta v_x \Delta v_y \Delta v_z$. Multiplying the previous equation by this factor thus gives the probability for a molecule to have a speed near (i.e., in a range of width Δv around) v :

$$p(v)\Delta v = 4\,\pi\,v^2 \left(\frac{m}{2\pi k_B T}\right)^{3/2} e^{-\frac{1}{2}mv^2/k_B T}\,\Delta v. \tag{12.16}$$

This function, the Maxwell speed distribution, is also plotted in Figure 12.3.

One perhaps-surprising feature of the graph is that, although each individual velocity *component* is most likely to have a value near zero, essentially none of the particles have zero *speed*. The reason for this is the ratio discussed in the previous paragraph. For larger speeds, there are more distinct velocities which correspond to that same one speed. So although zero is the most probable *velocity* for a molecule to have, essentially no molecules will actually have this velocity because there is only one way to have it. Whereas, for higher speeds, there are many different ways to have it, and their probabilities add. In any case, the most probable speed turns out to be just a bit lower than the RMS speed we calculated before, as indicated in Figure 12.3.

12.2.2   Collisions and the Approach to Thermal Equilibrium

The distribution of velocities, for the molecules in a gas, that we have just been discussing is evidently a kind of equilibrium distribution. Suppose, though, that all the molecules in a certain box of gas were initially (never mind how!) confined to one corner of the box and given all the same exact speed. Evidently the subsequent evolution of the gas (in accordance with Newton's laws of motion) would involve lots of collisions and would result in the gas spreading out in space to occupy the whole box more or less uniformly, and the distribution of speeds evolving toward the equilibrium distribution we found in the previous subsection. The question is: can we learn anything about exactly how these gross changes occur by analyzing the collisions which, evidently, bring them about?

The answer turns out – at least in principle – to be "yes". But the mathematical details are a little advanced for this course. Still, though, it is extremely important to appreciate that the sorts of thermal equilibrium distributions we have just been discussing for the speeds of molecules in a gas, are just that – equilibrium distributions that will arise, dynamically, through collisions, if a system is started off in some other non-equilibrium state. To give the flavor of this sort of dynamical approach to equilibrium, then, we consider instead a simpler example first treated by Maxwell.

Thus, suppose we have two different species of gas (say, one with heavy molecules and one with light molecules) at two different temperatures (say, the gas with heavy molecules is hot, and the gas with light molecules is cold). Now suppose we *mix* them together.

Qualitatively, we know what must happen: the hot gas should transfer some of its energy to the cold gas, until their temperatures – that is, the average translational kinetic energies of their molecules – are equal. And somehow this net transfer of energy from the initially hot gas to the initially cold gas must occur during *collisions* between the two types of molecules. That is, it must be that when a molecule with *lots* of kinetic energy collides with a molecule with *just a little* kinetic energy, the kinetic energies tend to equalize. Of course, they need not *exactly* equalize. Indeed, it will even be possible that, *sometimes*, the difference in the molecules' kinetic energies could be *increased* rather than decreased by the collision. But, in order to explain microscopically what we know must happen macroscopically, it better be that this kind of situation is *rare*, i.e., that "most of the time" or "on average" or "typically" the difference in the two colliding molecules' kinetic energies *decreases* during the collision.

Can we understand how this comes about? In general, no – because unless we know something about the shapes of the molecules and/or the nature of the short-range forces that come into play when they "collide", it will be impossible to analyze the collisions with the necessary sort of detail. But, as Maxwell showed in a beautiful 1860 paper, we can make some progress here if we use an unrealistic but simple model for the molecules and the forces between them.

In particular: let's assume that the molecules are perfect spheres, with radii R_1 (for the one species) and R_2 (for the other species) respectively, and suppose the masses of the molecules are m_1 and m_2 respectively. And suppose that the only forces they exert on each other are the "contact forces" that come into play when the molecules literally touch, i.e., when their centers come to be separated by a distance $d = R_1 + R_2$. (By the way, it isn't actually important here that there be two chemically distinct species of gas. It might as well be that we are mixing two samples, one hot and one cold, of the same chemical species of gas. But it is convenient, mostly for the sake of drawing pictures, to imagine that we are mixing two different types of gas – because then we can draw one of the molecules as bigger than the other, and this will help us remember that one is supposed to have a higher initial kinetic energy than the other.)

We want to analyze a collision between a "hot" molecule and a "cold" molecule. The way we're going to approach this analysis is as follows: we'll take the initial velocities of the two colliding molecules as given; this will allow us to determine the velocity of the center of mass of the two-particle system; we can then proceed to analyze the collision from the point of view of the reference frame in which the center of mass is at rest; this makes certain things a little simpler. And then, finally, we can return to the original frame and write down some expressions for the velocities of the two particles after the collision. Then we'll put all of this together by looking at how the *difference* in the kinetic energies of the two particles *changes* during the course of the collision.

So, to begin with, consider a collision like that illustrated in Figure 12.4. One of the "hot" particles (with mass m_1, radius R_1, and velocity $\vec{v}_1^i$) flies toward one of the "cold" particles (with mass m_2, radius R_2, and velocity $\vec{v}_2^i$). The velocity of the center of mass of the two particle system will be given by

$$\vec{v}_{\text{CM}} = \frac{m_1 \vec{v}_1^i + m_2 \vec{v}_2^i}{m_1 + m_2} \tag{12.17}$$

and the velocities of the two particles *in* the center of mass frame will be given respectively by

$$\vec{u}_1^i = \vec{v}_1^i - \vec{v}_{\text{CM}} \tag{12.18}$$

and

$$\vec{u}_2^i = \vec{v}_2^i - \vec{v}_{\text{CM}}. \tag{12.19}$$

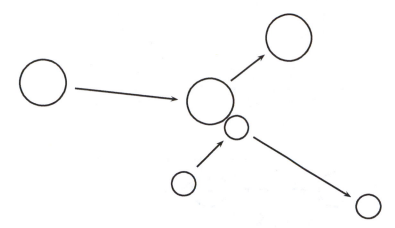

Figure 12.4: A collision between one of the "big" (initially hot) molecules and one of the "small" (initially cold) molecules.

(We use "u" for the velocities relative to the center of mass frame, simply to avoid too many different v's with proliferating subscripts and superscripts!)

Now we're going to have to do some work to analyze the collision in the CM frame. The collision itself is pictured in Figure 12.5. The initial velocities of the particles (in this frame) are in opposite directions. (How else could the total momentum be zero?!) But the collision need not be exactly "head on". That is, in general, the lines representing the trajectories (prior to the collision) of the centers of the two particles will not coincide, but will be separated by some distance s. This is usually called the "impact parameter", and its value (together with the values of the radii for the two particles) will determine the *angle* that the particles are deflected by – the angle θ shown in the Figure.

Our first task is to figure out precisely what the relationship is between the impact parameter s and the scattering angle θ. To see this, it is helpful to introduce the angle ϕ shown in the Figure. On the one hand, considering the right triangle whose hypotenuse is the line segment between the centers of the two particles at the moment they collide, we have

$$s = (R_1 + R_2)\sin(\phi) = d\sin(\phi). \tag{12.20}$$

And on the other hand, since (because of the spherical shape of the particles) the collision will be symmetrical about that same line connecting the centers of the particles, we have that

$$\theta = 2\phi. \tag{12.21}$$

Combining the last two expressions gives the desired relation between s and θ

$$s = d\sin(\theta/2). \tag{12.22}$$

This tells us precisely how the directions the particles will scatter into (relative to their original directions) depend on the impact parameter for the collision.

That's about as far as we can go with the information given. So now we will introduce several additional assumptions of a statistical character. First, we assume that the value of s is in some appropriate sense "random". That is, we assume that there are no particular correlations between the initial positions and velocities of particles which happen to collide, which would make any particular type of collision (characterized by a given s value) more likely than any other. This doesn't mean, however, that all values

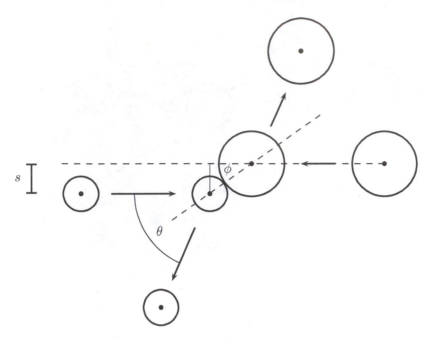

Figure 12.5: The same collision as seen in the center of mass frame.

of s are equally likely. That would be the case if we were talking about the collision of two-dimensional circles, which is (unfortunately) the image conveyed by the pictures! So we have to remember that we are dealing with a three-dimensional collision between spheres. This means that a given range of s values (say, between s and $s+ds$) actually corresponds to an *annulus* (of radius s and thickness ds) in the plane perpendicular to the velocities of the colliding particles.

Our "randomness" assumption then will be that the probability for the impact parameter to lie between s and $s+ds$ is proportional to the area of this annulus:

$$P(s)\,ds = \frac{2\pi s\,ds}{\pi d^2} \tag{12.23}$$

where the denominator is simply the total area which will produce *some* collision, i.e., the area of the circle whose radius $d = R_1 + R_2$ is the maximum possible value of s. Said another way, the assumption here is that a collision, described by some particular point in this circle of radius d, is equally likely for any of the points in that circle.

We now know what scattering angle θ is produced by a given impact parameter s, and how likely we are to have a collision with impact parameter s. So we can combine these to derive a formula for the probability that the post-collision scattering angle will lie between θ and $\theta + d\theta$. We need merely plug Equation (12.22) – and its differential

$$ds = \frac{d}{2}\cos(\theta/2) \tag{12.24}$$

– into the right hand side of Equation (12.23), and interpret the result as $P(\theta)\,d\theta$. That is,

$$P(\theta)\,d\theta = \frac{2\pi d\sin(\theta/2)(d/2)\cos(\theta/2)\,d\theta}{\pi d^2} = \sin(\theta/2)\cos(\theta/2)\,d\theta = \frac{1}{2}\sin(\theta)\,d\theta. \tag{12.25}$$

But this particular probability distribution for the particles to scatter into the various possible angles θ actually means that the particles are equally likely to scatter into *any* direction, i.e., that the scattering is (on average) *isotropic*.

To understand this, it is helpful to think about a large sphere (radius D, say) centered at the collision. Suppose one of the scattered particles is assumed to be equally likely to hit any point on the large sphere – i.e., suppose the scattering is perfectly isotropic. But then, the probability of its hitting a point on the big sphere with polar angle between θ and $\theta + d\theta$ will just be the area of a thin strip at angle θ (which, flattened out, will be a rectangle whose length is $2\pi D \sin(\theta)$ and whose width is $D d\theta$), divided by the total area of the sphere:

$$P_{iso}(\theta)\,d\theta = \frac{2\pi D \sin(\theta) D d\theta}{4\pi D^2} = \frac{1}{2}\sin(\theta)d\theta. \tag{12.26}$$

This shows that, as claimed, the probability distribution for θ indicated in Equation (12.25) corresponds to *isotropic scattering*.

Let's pause to summarize the important point of the last few paragraphs. For colliding spherical particles considered in the center of mass frame, the final velocities of the scattering particles are *equally likely to be in any direction*. It is also true that, in the center of mass frame, the *speeds* of the particles do not change during the collision – the velocities simply change direction. We may summarize this by writing

$$|\vec{u}_1^{\,f}| = |\vec{u}_1^{\,i}| \tag{12.27}$$

and

$$|\vec{u}_2^{\,f}| = |\vec{u}_2^{\,i}| \tag{12.28}$$

and remembering that the *directions* of $\vec{u}_1^{\,f}$ and $\vec{u}_2^{\,f}$ are completely (isotropically) random.

Now let's compare the difference in the kinetic energies of the two particles before and after the collision. Before the collision, the difference is

$$\Delta KE^i = \frac{1}{2}m_1|\vec{v}_1^{\,i}|^2 - \frac{1}{2}m_2|\vec{v}_2^{\,i}|^2. \tag{12.29}$$

After the collision, the difference is

$$\Delta KE^f = \frac{1}{2}m_1|\vec{v}_1^{\,f}|^2 - \frac{1}{2}m_2|\vec{v}_2^{\,f}|^2. \tag{12.30}$$

But the final velocities of the particles in the lab frame are related as follows to the final velocities in the CM frame:

$$\vec{v}_1^{\,f} = \vec{u}_1^{\,f} + v_{CM} \tag{12.31}$$

and

$$\vec{v}_2^{\,f} = \vec{u}_2^{\,f} + v_{CM}. \tag{12.32}$$

So we can plug these expressions into Equation (12.30), and use also Equation (12.17), to rewrite the difference in kinetic energies (after the collision). After a decent bit of algebra, the result can be written as follows:

$$\begin{aligned}
\Delta KE^f &= \left(\frac{m_1 - m_2}{m_1 + m_2}\right)^2 \Delta KE^i + \frac{1}{2}m_1\vec{u}_1^{\,f}\cdot\vec{v}_{CM} \\
&\quad -\frac{1}{2}m_2\vec{u}_2^{\,f}\cdot\vec{v}_{CM} + \frac{2(m_1 - m_2)m_1m_2\vec{v}_1^{\,i}\cdot\vec{v}_2^{\,i}}{(m_1 + m_2)^2}.
\end{aligned} \tag{12.33}$$

This is the result for some one particular collision. If we consider, however, an *average* over all the *possible* collisions, this expression will simplify considerably. In fact, because there are no particular correlations between the initial velocities of colliding particles, the average value of $\vec{v}_1^{\,i}\cdot\vec{v}_2^{\,i}$ should be zero, and so the corresponding term drops out. And because, as we showed just above, the final velocities in the CM frame are equally likely to be in any direction, the terms involving $\vec{u}_1^f \cdot \vec{v}_{CM}$ and $\vec{u}_2^f \cdot \vec{v}_{CM}$ will also drop out when we take the average. We are thus left with

$$\langle \Delta KE^f \rangle = \left(\frac{m_1 - m_2}{m_1 + m_2} \right)^2 \Delta KE^i \tag{12.34}$$

where the angle brackets on the left denote a statistical average over all possible collisions.

The key result here is that the factor in parentheses on the right hand side is definitely less than one – so, overall, the equation says that *on average* the difference in the kinetic energies of the two colliding particles will *decrease*. That is, on average, there is a tendency for collisions to equalize the kinetic energies of the two particles. Of course, it's possible that in some particular collision the difference in kinetic energies will increase – a particle with lots of kinetic energy can have its kinetic energy *increased* at the expense of a particle which initially had only a little kinetic energy. But what Maxwell's analysis shows is that this is, statistically, a relatively uncommon occurrence. Anomalous events like this can and do happen, but they are swamped by collisions in which the particle with lots of kinetic energy loses some of its kinetic energy to the particle with, initially, only a little.

And of course the significance of that is that it allows us to understand – at least for this one admittedly idealized example – how it comes about, dynamically, over some period of time, that the temperatures of two gases, which have at the beginning different temperatures but are allowed to mix, eventually equilibrate. This is an example of the kind of behavior required by the second law of thermodynamics (or perhaps more accurately, it is an example of the kind of observed behavior from which we inferred the second law of thermodynamics). What the example seems to suggest is that the transfer of heat from the relatively hot to the relatively cold body is not some kind of separate or additional constraint, but is something which emerges naturally from the elementary dynamical relations which govern the systems' interaction (namely, Newton's laws) – and the inherent randomness which seems to be implied by the sheer (microscopic) complexity of the systems involved.

12.2.3   Maxwell's Demon

Another way to put the point of the previous subsection is this: the second law of thermodynamics, apparently, isn't a genuine *law* at all! It is more like, say, the "law" according to which you are wasting your money if you play the lottery. In some (important!) senses, this is true: the vast majority of people who play the lottery lose money, and lottery players as a group are guaranteed to be money-losers (because, in the long-term, the total payouts are guaranteed to be less than the total ticket sales). Nevertheless, there's always one lucky person who wins.

What Maxwell's argument from the previous subsection seems to show is that the dynamical approach to thermal equilibrium is like this. That the temperatures of the two gases will, in time, equalize is like the proposition that you are throwing your money away by playing the lottery: it is very likely to be true, but its truth depends on an element of chance, and there is a tiny (but nonzero) probability that the unexpected will happen. In particular, it is apparently *possible* "in principle" that you could mix a hot gas with a cold gas, wait a while, and find that the hot gas had gotten even hotter, and the cold gas even colder – i.e., that heat had been spontaneously transferred, in wanton violation of the second law, from the colder substance to the hotter. The content of the second law of thermodynamics, then, seems to be something like: while possible, this is nevertheless very very very improbable.

Maxwell invented a very interesting thought experiment to make this same point – i.e., "to show that the second law of thermodynamics has only a statistical certainty":

> "the second law of thermodynamics ... is undoubtedly true as long as we can deal with bodies only in mass, and have no power of perceiving or handling the separate molecules of which

they are made up. But if we conceive a being whose faculties are so sharpened that he can follow every molecule in its course, such a being, whose attributes are still as essentially finite as our own, would be able to do what is at present impossible to us. For we have seen that the molecules in a vessel full of air at uniform temperature are moving with velocities by no means uniform, though the mean velocity of any great number of them, arbitrarily selected, is almost exactly uniform. Now let us suppose that such a vessel is divided into two portions, A and B, by a division in which there is a small hole, and that a being, who can see the individual molecules, opens and closes this hole, so as to allow only the swifter molecules to pass from A to B, and only the slower ones to pass from B to A. He will thus, without expenditure of work, raise the temperature of B and lower that of A, in contradiction to the second law of thermodynamics."

This "being" – who came to be known as "Maxwell's Demon" – would be able to effect, by means of his "sharpened" faculties, systematic violations of the second law. Of course, no such being (as far as we know) actually exists. And there is a long history of arguments and discussions of this scenario, with many attempts to prove, for example, that if we only consider the microscopic mechanisms of some such "demon", we will find that, in the process of performing his operations, the demon will expend work, heat up, explode, or otherwise increase the overall entropy of the system in such a way as to render the whole process again consistent with the second law.

Such analyses are interesting and revealing, but in another way miss Maxwell's point. In a letter to a colleague, Maxwell once requested that his imagined creature be called "no more a demon but a valve." This seems to suggest that the demon – whose role is only to *ensure* that heat will flow from cold to hot – was not essential. He could be replaced by a mere "valve", at the price of merely (but greatly) reducing the probability of the violation of the second law. His point, that is, was simply that violations of the second law are *possible*. He didn't mean to commit himself to the further claim that reliable, systematic violations of the second law could be achieved in practice.

In the following section, we will develop this qualitative point in more detail by surveying Ludwig Boltzmann's statistical interpretation of the concept "entropy".

§ 12.3 Boltzmann's Statistical Interpretation of Entropy

Let us begin here by re-considering the "free expansion" process we discussed briefly in Chapter 11. Recall that, in a free expansion, a barrier is suddenly removed or a piston is pulled back (*very quickly!*) so that a portion of gas, initially confined to some volume V_1, can now (irreversibly) expand to occupy the larger volume V_2. Suppose we are dealing with an ideal gas, so that the temperature of the gas, T, is the same before and after the expansion. Then, as we discussed last week, the entropy of the gas increases. We can calculate the entropy change $\Delta S = Q/T$ by calculating the amount of heat Q that would need to flow in to arrive at the same final state via an isothermal expansion, with the result:

$$\Delta S = n\,R\,\log\left(\frac{V_2}{V_1}\right). \tag{12.35}$$

Since we are going to be interested in relating this formula to the microscopic state of the gas before and after the transition, it is helpful to re-write it in terms of the actual number N of molecules (rather than the number n of *moles*). Using $N = nN_A$ and $R = k_B N_A$ (where N_A is Avogadro's number and k_B is Boltzmann's constant), and also massaging the result somewhat, we have that

$$\Delta S = k_B \log(V_2^N) - k_B \log(V_1^N). \tag{12.36}$$

This way of expressing the change in entropy suggests that the first and second terms can be related to the entropy of the final and initial states, respectively. (After all, what we mean by "ΔS" is $S_2 - S_1$.) It

Figure 12.6: Ludwig Boltzmann (1844-1906)

is a little unusual, however, to take the logarithm of something with dimensions (i.e., non-trivial units). And here both terms apparently involve taking the logarithm of a quantity with the dimensions of a volume, raised to the power N. Let us therefore introduce a (somewhat arbitrary) constant, V_0, with the dimensions of a volume. Its physical interpretation will emerge shortly.

We may introduce the new constant V_0 into the expression for ΔS (without changing its value – in effect we are adding something and then subtracting the same thing away) as follows:

$$\Delta S = k_B \log\left[\left(\frac{V_2}{V_0}\right)^N\right] - k_B \log\left[\left(\frac{V_1}{V_0}\right)^N\right]. \tag{12.37}$$

So far we have just been playing mathematical games, the point of which is to have massaged the expression for ΔS into a form that makes a certain interpretation very natural. Here, finally, is that interpretation. Suppose we take V_0 to be a very tiny fixed volume, introduced essentially to make it possible count the number of distinct "places" in the full volume, V_2 or V_1, available to the gas molecules. The idea here is to imagine chopping up the big volume (V_2 or V_1) into a whole bunch of discrete (and finite-sized, but small) "regions" or "places" of volume V_0. Then the "number of distinct places" in volume V_2 is just the ratio V_2/V_0 (and likewise for V_1).

This in turn suggests the following way to read Equation (12.37). Before the expansion, the entropy of the gas was (Boltzmann's constant, k_B, times) the logarithm of the following: the "number of distinct places" available to the gas, raised to the power N (the number of molecules composing the gas). And after the expansion, the entropy was (Boltzmann's constant times) the logarithm of: the (new, larger) "number of distinct places" available to the gas molecules, raised to the power N. That is, in general, the hypothesis is that the entropy is always just: Boltzmann's constant times the logarithm of the "number of distinct places" available to the gas, raised to the power N.

But now: what is the "number of distinct places" available to each gas molecule, raised to the power N? It is nothing but a count of *the total number of ways the gas molecules could be arranged among those distinct possible places*. That is, it is a count of the number of distinct *configurations* the N gas molecules might be in. Think of it this way. Suppose all the molecules are pulled to the side and it is your job to put them all back into the box. How many different ways could you do that? Well, the number of different possible places you could put the first particle is just (V/V_0). And then you have to make a separate, independent choice from among those same (V/V_0) possibilities for where to put the second particle. So the total number of distinct configurations for the first two particles will be $(V/V_0)^2$. And so on.

A detailed specification of the precise location of each individual particle – that is, a detailed specification of the configuration of the whole N-particle system – is called a *microstate* of the system. Let us use the letter W to denote the number of such microstates of a system that are consistent with whatever is known about its macroscopic state. Then the idea we are developing about the meaning of entropy can be written this way:

$$S = k_B \log(W) \tag{12.38}$$

where the idea is to think of W as the number of microstates that are consistent with some coarser, macroscopic specification of the state of the system – like that its N particles are distributed over some volume V.

Equation (12.38) captures Boltzmann's statistical interpretation of entropy. The basic idea is that, because of the enormous difference in scale between macroscopic and microscopic descriptions of a thermodynamic system, a specified macrostate will correspond to a very large number of distinct microstates. The *entropy* of the system in that macrostate is a measure of the number of such corresponding microstates.

Given the way we have set things up here, the actual number of microstates corresponding to a given macrostate (and hence the entropy) is to some degree arbitrary – it depends on the arbitrarily chosen quantity V_0 which we used to divide volumes into "distinct places." Note, however, that while changing the value of V_0 does change W and hence S, changing V_0 merely shifts S by a constant. Thus, although there is (given what we've said so far) some vagueness in talking about "the entropy" associated with a given macrostate, there is no vagueness at all in talking about the amount by which the entropy *changes* as a system evolves from one macrostate to another.

A key part of Boltzmann's approach to understanding entropy is the assumption that all microstates – at least, all microstates consistent with whatever is known about the macroscopic state of the system in question – are *equally likely*. This is where statistical considerations enter the picture, and allow us to understand in microscopic terms the nature of *equilibrium*. As an example, consider again the gas in a box where (for simplicity) we take the left and right sides of the box as our two distinct places. But let us suppose that the number N of molecules composing the gas is very large. It is a fact of experience that, in such a situation, the gas will distribute itself uniformly throughout the volume – in particular, the number of particles on the left side of the box should be very very close to $N/2$.

The question is: what guarantees this? What makes the molecules distribute themselves more-or-less evenly between the two sides of the box? Boltzmann's answer is: nothing. Each molecule is equally likely to be found on either side of the box, and (consequently) any specific arrangement of particles between the left and right sides is as likely to occur as as any other. But where there is *just one microstate* in which (for example) all the particles are found on the left side of the box, there are *many, many microstates* in which more-or-less half of them are on the left. For example, it could be the first, third, seventh, eighth, eleventh, ... particles which are on the left (with the others all on the right) – or it could be the first, second, fourth, ninth, eleventh, ... particles which are on the left (with the others all on the right) – or it could be the second, third, seventh, ninth, tenth, ... on the left (with the others on the right) – or ... and so on. The point is, there are going to be *lots and lots of ways* in which the particles can be more-or-less evenly distributed between the two sides, and just a relatively few ways in which the particles can be un-evenly distributed. So no special assumptions or laws or coincidences or forces are required to explain why, in practice, we always observe a uniform distribution. We just need to accept that macroscopic arrangements which can happen in many different ways, will tend to appear more often than arrangements which can only happen in a small number of ways.

So far we have been talking as if entropy (or the underlying counting of microstates) was merely an issue of the *spatial* configuration of the atoms or molecules composing the system. This, as it turns out, is an artifact of the example we used to motivate this whole discussion – namely, the free expansion of an ideal gas, in which the temperature is the same before and after the expansion. We know, though, that the precise microstate of a gas consists, in principle, of more than just the *positions* of all the particles. It should include, in addition, a specification of the *velocities* of all the particles. And so we expect a corresponding contribution to the entropy associated with the distribution of particles with respect to velocity.

To make this precise, it is helpful to use the idea of a "velocity space" that we introduced before. This is just a "space" whose axes represent, not the coordinates in space of a particle's location, but the components of its velocity. Thus, for example, a particle which has velocity zero would be represented by a point at the origin of the velocity space. Then, just as we introduced the quantity V_0 to allow us to count "distinct places" above, so we can introduce another new quantity – call it V_0', say – which represents the volume of a certain little "cell" in the velocity space and hence allows us to again count "distinct places" (in the velocity space).

What kind of macroscopic information do we typically have about a system which is relevant to the velocity part of its entropy? We might know, for example, its temperature, which (as we have discussed) is equivalent to knowing the *average* translational kinetic energy of the gas. So the question arises: for (say) a bunch of molecules composing a gas, what is the equilibrium distribution of molecular speeds – i.e., what distribution of speeds corresponds to the greatest number of distinct velocity microstates? The answer, of course, is just the Maxwell speed distribution we already discussed, though understanding in rigorous mathematical terms *why* this is true is tricky. (We will say more about why it's tricky shortly.)

So far we have discussed Boltzmann's idea that entropy is a measure of the number of distinct microscopic states which correspond to a given macroscopic description, and how this idea can be used to understand why a certain sort of equilibrium behavior arises (for example, the uniform distribution of gas molecules in physical space, or their Maxwell distribution in velocity space). Let us now apply Boltzmann's ideas to the approach to equilibrium.

For the most part, there isn't much new to say here. For example, we might consider again the gas in a box, and suppose that the gas was initially confined to the left hand side of the box by a barrier which is suddenly removed at some moment. Then, just after the barrier is removed, the gas can be thought of as being in a non-equilibrium state: it is free, in principle, to fill the entire volume, but (because we have gone out of our way to arrange things this way by first compressing it to one side and then suddenly removing the partition) happens to be confined exclusively on the left hand side of the box. What happens next?

Well, the system evolves toward equilibrium – for example, if we wait a certain amount of time, the left/right distribution will go from 100/0 to 90/10, then 80/20, and so on, until it eventually reaches 50/50 – and *stays* there. Microscopically, what is driving this approach to equilibrium? Again, nothing – nothing, that is, but statistics. We can simulate this process in class.

A similar kind of analysis applies to the approach to equilibrium with respect to the velocities of the molecules composing a gas. If, as Maxwell explained in the quote from the previous section, every one of the molecules in a gas was somehow made to be moving initially with exactly the same speed, collisions between molecules would result in their speeds becoming dis-equalized, with some equilibrium distribution of speeds (the Maxwell distribution!) eventually coming about. The point here is just to explain that this dynamical approach to equilibrium can be understood in purely statistical terms. There are simply *overwhelmingly more distinct ways* the gas as a whole can possess a Maxwell speed distribution, than ways in which all of the particles can possess the same speed. (There are actually lots of microstates corresponding to all the particles having the same speed, because there are lots of possible directions for the velocity to be in, even if its magnitude is fixed. But there are still *enormously more* microstates corresponding to the Maxwell distribution.) So there is no mystery about how or why the Maxwell distribution comes about: we start in a state that is relatively unusual, and then we mix things around and get into some new microstate. It is overwhelmingly likely that this new state will exhibit the Maxwell speed distribution, even though it is selected at random from all possible microstates, because the overwhelming majority of all possible microstates exhibit this distribution!

Boltzmann explains this point as follows:

> "From an urn, in which many black and an equal number of white but otherwise identical spheres are placed, let 20 purely random drawings be made. The case that only black balls are drawn is not a hair less probable than the case that on the first draw one gets a black sphere, on the second a white, on the third a black, etc. The fact that one is more likely to get 10 black spheres and 10 white spheres in 20 drawings than one is to get 20 black spheres is due to the fact that the former event can come about in many more ways than the latter. The relative probability of the former event as compared to the latter is the number $20!/10!10!$ $[= 184,756]$, which indicates how many permutations one can make of the terms in the series of 10 white and 10 black spheres, treating the different white spheres as identical, and the different black spheres as identical. Each one of these permutations represents an event that has the same probability as the event of all black spheres...."

> "Just as in this simple example, the event that all molecules in a gas have exactly the same velocity in the same direction is not a hair less probable than the event that each molecule has exactly the velocity and direction of motion that it actually has at a particular instant in the gas. But if we compare the first event with the event that the Maxwell velocity distribution holds in the gas, we find there are very many more equiprobable configurations to be counted as belonging to the latter."

Let's see if we can actually calculate the contribution to the entropy of a gas (of N particles with total energy E) from the scatter of its particles in the velocity space. Recall that the velocity (at some instant) of some particular molecule composing the gas, can be represented as a dot at the appropriate location in the velocity space. The velocity part of the microstate of the whole N-particle gas, then, can be represented by N dots in the velocity space (each dot being somehow "labelled" as to which particular particle it represents the velocity of).

So far, this is exactly analogous to the *spatial* distribution of particles that we used here to motivate Boltzmann's statistical (micro-state counting) approach. But while in principle *any* distribution of the particles *in (regular, physical) space* is possible, there is a *constraint* on the possible distributions of particles in the velocity space: different "positions" in the velocity space correspond to different kinetic energy, and (usually) we want to consider a situation where the total energy of the gas (or equivalently the average energy per molecule, the temperature) is fixed.

A further difference between the spatial and velocity distributions is that, at least for a bunch of molecules in, say, a box, there is a hard spatial boundary outside of which the particles cannot go. This is not really the case in the velocity space – a given molecule might, in principle at least, have any arbitrarily large speed, i.e., be an arbitrarily large "distance" from the origin of the velocity space. Thus, if there were no constraint on the total energy, each individual particle would have an infinity of different possible "positions" (in the velocity space) to occupy, the number of different configurations (in the velocity space) for the N particles would be infinity raised to the power N, and the velocity-distribution part of the entropy would fail to be well-defined.

One can see how a constraint on the total energy prevents *this* disaster. But one can also see how *counting* the total number of possible microstates – the total number of distinct distributions of N labelled dots in the velocity space – subject to the constraint that the total energy E of the N particles is fixed, is going to be a mathematically difficult problem. There isn't going to be an obvious and simple formula like the $W = (V/V_0)^N$ we had in the context of the spatial part of the entropy.

Actually, though, we can get what we need here by using a kind of weird mathematical trick. Instead of thinking in terms of a three-dimensional velocity space in which we place N labeled dots to represent the (velocity part of the) microstate of the gas, let us instead introduce a single $3N$-dimensional velocity space, whose axes are the x-, y-, and z-components of the velocities of *every single one of the N particles*. Then we can specify the (velocity part of the) microstate of the entire N-particle system by placing *a single dot* somewhere in this space. Moreover, since the total (translational kinetic) energy of the gas is given by

$$E = \sum_{i=1}^{N} \frac{1}{2} m \left((v_x^i)^2 + (v_y^i)^2 + (v_z^i)^2 \right) \tag{12.39}$$

all of the possible microstates (which correspond to the total energy being E) lie on the surface of a $3N$-dimensional "sphere" in this space. The radius of this constant-energy "sphere" is given by

$$R = \sqrt{\frac{2E}{m}}. \tag{12.40}$$

Now the key point is that the number of distinct microstates (considering just the velocities, of course) which correspond to the total energy being E is going to be proportional to the *surface area* of this "sphere" in this crazy space. Now that is impossible to draw or visualize, but we can at least say that the area is proportional to R^{3N-1}, just as the surface area of a regular 3-dimensional sphere is proportional to $R^{3-1} = R^2$. We thus have that

$$W \sim R^{3N-1} = \left(\frac{2E}{m} \right)^{(3N-1)/2}. \tag{12.41}$$

Since we already have a mere proportionality anyway, and are only concerned here about the E-dependence of the result, we might as well just write this as follows:

$$W \sim E^{(3N-1)/2}. \tag{12.42}$$

Now we can calculate the entropy by following Boltzmann's general formula:

$$S = k_B \log(W) = k_B \frac{3N-1}{2} \log(E) + \text{constant} \tag{12.43}$$

where we need to add an arbitrary constant because we only knew W up to some kind of proportionality factor, and the log of a product is the sum of the logs.

Since N is (for any reasonable, macroscopic system) *much, much* greater than 1, we might as well simplify the above result to

$$S = \frac{3}{2} N k_B \log(E) + \text{constant}. \tag{12.44}$$

Of course, this is just the entropy associated with the distribution of particles in velocity-space. The *total* entropy of a given sample of gas will be the sum of the energy- and volume-dependent terms:

$$S = \frac{3}{2} N k_B \log(E) + N k_B \log(V) + \text{constant} \tag{12.45}$$

where here the "constant" term depends on neither the total energy E nor the volume V.

Just to connect this formula up to something familiar, consider a process in which some heat dQ is allowed to flow into an ideal monatomic gas of N particles whose volume V is fixed. The total energy of the gas will increase by $dE = dQ$ according to the first law, since no work is done. Thus, according to the above expression for the entropy, the entropy will increase by

$$dS = \frac{3}{2} N k_B \log(E_f / E_i) = \frac{3}{2} N k_B \log(1 + dQ/E) = \frac{3}{2} N k_B \frac{dQ}{E} \tag{12.46}$$

Figure 12.7: A typical (velocity) micro-state for a gas consisting of $N = 100$ (monatomic) ideal gas molecules. Actually, what's depicted is not a single micro-state, because the dots are not labelled. To commit to a particular micro-state, we'd have to write, near each dot, the name or number of the particle whose velocity that particular dot is supposed to represent. Note that, for $N = 100$, there are 100! (one hundred factorial) $\approx 10^{158}$ distinct ways of labeling the dots – that is, the arrangement of dots in the Figure actually corresponds to about 10^{158} distinct micro-states! And of course there are way more than 100 particles in a macroscopic sample of gas, so even this incomprehensibly large number is a ridiculous underestimate for the number of possible micro-states in a real gas. And of course there are lots of (subtly, or dramatically) different ways we could have arranged the dots!

where we have used the fact that $\log(1 + x) = x$ for small x, and here E is the initial total energy of the gas. But (for a monatomic ideal gas) $E = \frac{3}{2}Nk_BT$, so the last expression simplifies to

$$dS = \frac{dQ}{T}. \tag{12.47}$$

This, of course, is just Clausius' original expression for the entropy change produced by a certain heat flow at a certain temperature. So we have shown that Boltzmann's statistical account of entropy is consistent with the entropy concept as originally formulated by Clausius.

Let's also not lose sight of the key claim, here, in regard to velocity distributions. We have just attempted to *count* the number of distinct velocity micro-states (consistent with the total energy constraint), and used this to express the (velocity-, or energy-dependent part of the) entropy of a monatomic ideal gas. But the more interesting qualitative claim in the background is something we haven't really shown: that the *overwhelming majority* of these possible (velocity) micro-states have *distributions* that look, more or less, like the Maxwell equilibrium distribution. (That, in this account, it what it *means* for the Maxwell distribution to *be* an equilibrium distribution.)

The claim, that is, is that the overwhelming majority of possible micro-states (subject to the energy constraint) look basically like the micro-state illustrated in Figure 12.7. This is not something we've proved, though neither did we really prove the analogous claim for the alleged equilibrium *spatial* distributions. (We'll play with some toy models in class to at least try make these claims plausible.) In any case, what we want to be able to say is that we can, in a purely statistical way, understand or *explain* the approach of a non-equilibrium velocity distribution toward the equilibrium distribution. That is, as in Maxwell's example, the claim is that we can explain why a sample of gas in which, somehow, all the molecules were initially given the same speed will eventually come around to the Maxwell speed distribution, in purely statistical terms: the initial condition, here, is one of a group of "special" or a-typical microstates, and

Figure 12.8: Boltzmann's tomb in Vienna, with his formula, "$S = k_B \log(W)$", summarizing the statistical interpretation of entropy.

so, purely because there are relatively so many "typical" microstates compared to the "a-typical" ones, the odds are very good that the system will end up in one of the typical ones. So, as with the tendency for gases to expand to fill the volumes available to them with a quite uniform density, no special "force" need be postulated to explain the final state. The process is just random in the sense that all of the possible micro-states are equally likely to be realized at the end of the experiment. It's just that the overwhelming majority of those microstates have something in common (namely, the equilibrium distribution of particles in space, or velocity space, as appropriate), and so – despite the ultimate randomness – we are overwhelmingly likely to see, at the end of the experiment, a state that has the equilibrium character.

§ 12.4 Entropy and Statistics: Some Formal Developments

The previous section was mostly qualitative, attempting to explain the basic idea of Boltzmann's statistical concept of entropy, and his associated statistical understanding of the second law of thermodynamics. In this section we will be a little bit more formal and try to flesh out several important mathematical consequence of the qualitative ideas introduced above.

12.4.1 Temperature and Thermal Equilibrium

Let us begin by trying to understand how temperature can be understood in terms of the statistical concept of entropy. Suppose that some small-ish thermodynamic system is placed into thermal contact with a much larger system – a "heat reservoir". For simplicity, assume that the system and the reservoir together constitute a closed system, i.e., they can transfer energy (in the form of heat) back and forth between them, but the total energy E of the system and reservoir together is fixed. We can then ask: what is the most probable (i.e., equilibrium) distribution of this total energy between the system and reservoir?

Suppose the system takes a portion E_S of the total available energy, with the remainder $E_R = E - E_S$ taken by the reservoir. Now the key statistical insight is that the number of available microstates for both the system and the reservoir will be a (rapidly) increasing function of their energy. (See, for example, Equation (12.42).) Let us give these functions names: $W_S(E_S)$ to denote the number of available system microstates when the system has energy E_S, and $W_R(E_R)$ to denote the number of available reservoir microstates when the reservoir has energy E_R.

In order to answer the question of how the energy is most likely to be divided between the system and reservoir, however, we must consider the system and reservoir together as a single meta-system. The crucial point is then to consider the total number of microstates of this meta-system, as a function of (say) the amount of energy taken by (just) the system. This will be the *product*

$$W(E_S) = W_S(E_S) \times W_R(E - E_S) \tag{12.48}$$

since, for *each* of the $W_S(E_S)$ possible microstates of the system, there are $W_R(E - E_S)$ distinct possible microstates of the reservoir. Now the important point is that the function $W(E_S)$ will have a sharp peak – there will be a specific value of E_S which maximizes W, i.e., which corresponds to the majority of all possible meta-system microstates. We can find out where this peak is – i.e., we can find a condition expressing the equilibrium condition for the partioning of energy between the system and reservoir – by the standard calculus technique of differentiating Equation (12.48) with respect to E_S and setting the result equal to zero.

Actually, however, it is mathematically equivalent – and more convenient and illuminating – to first take the logarithm of Equation (12.48) and then differentiate *that*. (It's mathematically equivalent because the logarithm of a function peaks at the same place the function itself peaks – the logarithm being after all a monotonically increasing function.) Carrying out the indicated operations gives

$$0 = \frac{d\log(W_S(E_S))}{dE_S} + \frac{d\log(W_R(E - E_S))}{dE_S} \tag{12.49}$$

which can be simplified into

$$\frac{d\log(W_S(E_S))}{dE_S} = \frac{d\log(W_R(E_R))}{dE_R}. \tag{12.50}$$

But, recalling the statistical definition of entropy, this is equivalent to the following:

$$\frac{dS_S(E_S)}{dE_S} = \frac{dS_R(E_R)}{dE_R}, \tag{12.51}$$

i.e., the two systems in thermal contact are overwhelmingly likely to distribute the energy so that they share the same "marginal entropy", i.e., so that they share the same value for the derivative of their entropy with respect to their energy.

This makes perfect sense: if the two derivatives, of entropy with respect to energy, are not equal, then a transfer of some energy between the two subsystems will increase the entropy of one of the subsystems more than it decreases the entropy of the other. That is, some further energy transfer will increase the total entropy of the meta-system, i.e., will increase the total number of accessible micro-states. And this, therefore, will tend to happen spontaneously. The point is that thermal equilibrium is reached when the total number of accessible micro-states is maximized, i.e., when further energy transfer can no longer increase the total entropy.

Of course, qualitatively we know that two systems in thermal contact will – in equilibrium – come to share the same *temperature*. So there must be some simple relationship between the temperature of a system and the derivative of its entropy with respect to its energy. It can't be that these are just equal, because the one goes up while the other goes down, and anyway they don't have the same units. But we can identify

$$T = \frac{1}{dS/dE}, \tag{12.52}$$

a relationship which is called the thermodynamic definition of temperature.

Actually, in some ways this isn't a new relationship at all, but merely a rearrangement of Clausius' earlier definition of the entropy (change) associated with a small influx of heat dQ:

$$dS = \frac{dQ}{T}. \tag{12.53}$$

What we have done, in effect, is to use the fact that now – after Boltzmann – we have an independent grounding for the concept "entropy", so we can convert what was previously considered a definition of entropy *in terms of temperature*, into a new definition of temperature. This new formulation of temperature is important because, unlike the definitions in terms of Mercury thermometers or even ideal gases, this formulation does not rely on the properties of any arbitrarily selected substance, but instead relies on the purely statistical fact that increasing the energy of a system increases the number of accessible microstates.

12.4.2   The Boltzmann Factor

In the setup for the previous subsection we described the "heat reservoir" as large compared to the system. Nothing we said in the previous section actually used this assumption, but let us bring it in now. The upshot is that the heat reservoir will contain the vast majority of the total energy E of the system: $E_S \ll E$. That will allow certain mathematical simplifications.

The main conceptual question of interest here is the following: having established which particular value of E_S is most probably realized, can we go further and say something about the relative *probabilities* for the various system microstates which might possibly be realized? These of course will be those whose energy is not too far from E_S, i.e., with the assumption from the previous paragraph, those whose energy is small.

The key point here is that the probability of a given system microstate being realized is (because of the fundamental "all microstates are equally likely" assumption) just proportional to the number of distinct ways this microstate could occur. This is not just "one" because the system is in thermal contact with the reservoir, and a single system microstate corresponds to *many* microstates for the combined meta-system. How many exactly? Well, just the number of distinct microstates *of the reservoir* when the reservoir has all but E_S of the total meta-system energy.

Let's say all of that again with equations. The probability for a certain particular system microstate (with energy E_S) is proportional to the number of distinct microstates *of the reservoir* when the reservoir has energy $E_R = E - E_S$:

$$P(E_S) \sim W_R(E - E_S). \tag{12.54}$$

But the right hand side can be written in terms of the reservoir entropy, using Equation (12.38). The resulting expression for the probability of the one system microstate is

$$P(E_S) \sim e^{S_R(E-E_S)/k_B}. \tag{12.55}$$

But now, since $E_S \ll E$, let us approximate $S_R(E - E_S)$ using a Taylor expansion:

$$S_R(E - E_S) \approx S_R(E) - E_S \frac{dS_R}{dE_R}. \tag{12.56}$$

But – as explained in the previous subsection – the derivative in the last expression is just (one over) the *temperature* of the reservoir. So, putting this all together we have that

$$P(E_S) \sim e^{S_R(E)/k_B} \, e^{-E_S/k_B T} \tag{12.57}$$

which can be more simply expressed as

$$P(E_S) \sim e^{-E_S/k_B T} \tag{12.58}$$

since the other factor is just some additional proportionality constant that doesn't depend on E_S.

The factor on the right hand side is usually called "the Boltzmann factor". It expresses the exponentially-decreasing probability for increasingly energetic possible system microstates when the system is in thermal contact with a large heat reservoir at temperature T.

We have already seen an important example of this behavior. We may consider, as a kind of funny abstract "system", (say) the x-component of the motion of some particular one molecule in a box of gas – taking all of the *other* particles as a kind of "heat reservoir" with which our "system" can exchange energy. What we have just shown, then, is that the probability for our system to take on a particular microstate with energy E is given by the Boltzmann factor. But for our system, the microstates are labeled simply by the value of the x-component of the particle's velocity – v_x – and the corresponding energy is just the associated kinetic energy, $mv_x^2/2$. And so the Boltzmann factor tells us that the probability for the x-component of the velocity of this molecule to take on the value v_x is given by

$$P(v_x) \sim e^{-\frac{1}{2}mv_x^2/k_BT}. \tag{12.59}$$

And that is precisely what we showed earlier was required by Maxwell's (rather strange) argument for the velocity distribution in the gas. So we have re-derived that result now from a much deeper, purely statistical perspective.

12.4.3   The Equipartition Theorem

We just used the example of a particular degree of freedom for a gas molecule – the x-component of its velocity – as an example to illustrate the Boltzmann factor. Recall, though, that there was some puzzle about the rotational degrees of freedom of poly-atomic molecules. It appeared, from data about the specific heats, that only certain of their degrees of freedom were fully "activated". We are now going to argue that, on the basis of the ideas we've been developing in this chapter, this is inexplicable and deeply problematic.

Consider a degree of freedom for a molecule in a gas. Call it "f" – a deliberately non-committal letter, so that we can imagine f standing for a translational velocity component, or a rotational velocity component (for an extended molecule), or even (say) the amplitude of certain type of vibration that a poly-atomic molecule might undergo. We will just assume that the energy for the microstate (for this particular degree of freedom) labeled by the particular value f is a quadratic function of f:

$$E_f = \alpha f^2 \tag{12.60}$$

where, for example, if f is a translational velocity component, then α will be one half of the mass – or if f is a rotational velocity component, then α will be one half of the associated moment of inertia – etc.

Now the point is that, without knowing anything about the details here – without knowing what *sort* of degree of freedom f refers to – we can calculate the average value of E_f by using the Boltzmann factor. The possible values of f are, by assumption, all of the continuum of values between $-\infty$ and ∞, and the probability of each of those possible values is given by the Boltzmann factor:

$$P_f \sim e^{-E_f/k_BT} = e^{-\alpha f^2/k_BT}. \tag{12.61}$$

And so the *average* value of the energy is just the average of all the possible values, each weighted by its corresponding probability of occurrence:

$$\langle E_f \rangle = \frac{\int df\, \alpha f^2 e^{-\alpha f^2/k_BT}}{\int df\, e^{-\alpha f^2/k_BT}}. \tag{12.62}$$

Both integrals can be done. You can learn how in some of this week's Exercises, but for now we just quote the result:

$$\langle E_f \rangle = \frac{1}{2}\, k_B T. \tag{12.63}$$

This represents an explicit and formal proof of what was earlier called the assumption of "equipartition of energy". We have now raised it to the level of a theorem, which is a serious problem given that, on the one hand, the assumptions of the theorem seem to be otherwise very strongly motivated and, on the other hand, the empirical data pertaining to the specific heats of poly-atomic gases indicate that the consequence of the theorem – that all degrees of freedom whose energy depends quadratically on the parameter specifying the microstate should possess on average $k_BT/2$ of energy in thermal equilibrium – are not true! That is, the data on the specific heats of gases indicates that there are some degrees of freedom (certain rotational degrees of freedom of certain gases, and the vibrational degrees of freedom of most gases) which simply do not contain this much energy on average. And that is simply incomprehensible from the point of view of everything that's been developed here.

Maxwell discussed the seriousness of this puzzle as follows:

> "...here we are brought face to face with the greatest difficulty which the molecular theory has yet encountered....

> "If we suppose that the molecules are atoms – mere material points, incapable of rotatory energy or internal motion – then [the number of degrees of freedom] n is 3 ... and the ratio of the specific heats [at constant pressure and constant volume] is 1.66, which is too great for any real gas. [Helium and other monatomic gases had not yet been discovered!]

> "But we learn from the spectroscope that a molecule can execute vibrations of constant period. It cannot therefore be a mere material point, but a system capable of changing its form. Such a system cannot have less than six [degrees of freedom]. This would make the ... ratio of the specific heats 1.33, which is too small for hydrogen, oxygen, nitrogen, carbonic acid, nitrous oxide, and hydrochloric acid.... Every additional variable ... increases the specific heat... So does any capacity which the molecule may have for storing up energy in the potential form. But the calculated specific heat is already too great when we suppose the molecule to consist of two atoms only. Hence every additional degree of complexity which we attribute to the molecule can only increase the difficulty of reconciling the observed with the calculated value of the specific heat.

> "I have now put before you what I consider to be the greatest difficulty yet encountered by the molecular theory."

We will not be able to resolve this puzzle here. Its resolution came only with the advent of quantum mechanics in the 20th century (and even that remains somewhat unsettled since there remain controversies about how to understand quantum mechanics in physical terms). We will get a glimpse, but only a glimpse, of the way that quantum mechanics resolves the puzzle in Chapter 13. But mostly we include this here just to show that, and why, there remained some room for legitimate skepticism about the statistical understanding of entropy and thermodynamics – and so the atomic theory of matter more generally – even into the late 1800s.

At the same time, though, Boltzmann's statistical interpretation of thermodynamics was very compelling, and the atomic theory of matter had received powerful support from many independent but converging lines of evidence. Let us then finally turn to one last piece of evidence, which dramatically brought the microscopic world of atoms into the realm of direct observation and quantitative measurement.

§ 12.5 Brownian Motion and Avogadro's Number

Starting in 1827, the biologist Robert Brown noticed something that puzzled him and, eventually, scientists from several other disciplines including physics:

"Extremely minute particles of solid matter, whether obtained from organic or inorganic substances, when suspended in pure water, or in some other aqueous fluids, exhibit motions for which I am unable to account and which, from their irregularity and seeming independence, resemble in a remarkable degree the less rapid motions of some of the simplest animalcules of infusions."

That is, tiny little particles of matter (such as pollen grains) were, when viewed through a microscope, observed to jiggle and dance around at random, in a way that was qualitatively similar to the (much larger) microscopic organisms that had been observed swimming around in water since the first invention of the microscope.

But further investigation made the idea that these particles were living, swimming creatures, seem less plausible:

"Having found motion in the particles of the pollen of all the living plants which I had examined, I was led next to inquire whether this property continued after the death of the plant, and for what length of time it was retained. In plants, either dried or immersed in spirit for a few days only, the particles of pollen of both kinds were found in motion equally evident with that observed in the living plant; specimens of several plants, some of which had been dried and preserved in an herbarium for upwards of twenty years, and others not less than a century, still exhibited the molecules or smaller spherical particles in considerable numbers, and in evident motion."

Indeed, "[r]ocks of all ages, including those in which organic remains have never been found, [also] yielded the [incessantly moving particles] in abundance."

Historian Mary Jo Nye summarizes the controversy, during subsequent decades, surrounding this curious "Brownian motion":

"In the next forty years, biologists and physicists prepared a variety of organic and inorganic particles – such as sulphur, mastic, cinnabar, pulverised coal, india ink, and gamboge – in organic and inorganic solvents, studying the influence of alkalis, acids, salts, and even narcotics upon the movement. They subjected preparations of Brownian particles to sunlight and darkness, magnetism and electricity, red and blue light, and heat and cold in hopes of discovering the cause of the motion. In 1886, for example, the *New York Microscopical Society Journal* recorded that one experimenter observed that the Brownian movement of globules in freshly drawn human milk varied, being most active at the birth of the offspring and diminishing in rapidity with an elapse of time."

Of course, one possible explanation, as you have perhaps already guessed, was that the observable Brownian motion of these tiny particles was being caused by the incessant collisions the particles suffered with the (invisible, and even smaller) *molecules* of the surrounding water. On this interpretation, the Brownian motion was a kind of visible revelation of the kinetic theory of heat and, more broadly, the atomic-molecular theory of matter. Indeed, the apparently random trajectories traced out in real time by the Brownian particles, could be thought of as visible, large-scale versions of the "random walks" undertaken by gas molecules in the analysis of Clausius with which we began this chapter.

This interpretation seemed to imply that the average kinetic energy of the Brownian particles should be the same as that of the background water molecules. Indeed, the essence of this interpretation of the Brownian motion was that the pollen grains (etc.) were simply, at least as far as physics was concerned, nothing but very large molecules! And so, for example, the particles should be in thermal equilibrium with the surrounding water and the equipartition theorem should apply to them. One would therefore expect

$$\frac{1}{2}m\langle v^2 \rangle = \frac{3}{2}k_B T \qquad (12.64)$$

where $\langle v^2 \rangle$ is the mean square speed of the particles and m is their mass.

Now, although the particles we're talking about here are small enough that one needs a microscope to observe them, they are also large enough that it was possible to estimate their sizes and densities – and hence the masses of the individual particles, m. And of course it was easy to measure the temperature of the surrounding fluid. So it was realized that, by simply measuring the average translational speed of the particles one could, through the above formula, determine empirically the value of Boltzmann's constant, k_B – and hence of Avogadro's number, N_A, through the relation $k_B = R/N_A$ (where R is the ideal gas constant).

So people set out to measure the speeds at which these incessantly jiggling Brownian particles moved. The problem was, these experiments gave nonsense – they all seemed to give different numbers that didn't agree with each other. Eventually it was realized that the average speed over a certain time period was a strong function of the time period in question – precisely what one would expect for a particle undergoing a "random walk" – and that even what *looked* to the eye like the "instantaneous speed" through a microscope was in fact an average over a time period which was itself large compared to the time between subsequent collisions (between the particle and some adjacent water molecule). Indeed, for such a particle immersed in water, such collisions are for all practical purposes happening continuously, and so there is no hope of observing, through a microscope, the true instantaneous speed of the particle. Thus, even while these observations seemed to establish the correctness of the kinetic theory interpretation of Brownian motion, they also dashed the hope of using the observable Brownian motion to get a handle on k_B and N_A – i.e., on the absolute scale of atoms and molecules.

12.5.1   Einstein and Perrin

Or so everyone thought until a brilliant theoretical physicist and a brilliant experimental physicist shed new light on the issue in the first decade of the 20th century. The theorist was Albert Einstein, and his brilliance can perhaps be indicated by the fact that his 1905 paper on Brownian motion was the third most important contribution he made to physics during that (so-called "miraculous") year. (His other two papers basically created relativity theory and quantum theory, respectively!)

Einstein's contribution to this particular topic was, in essence, to express the typical character of a "random walk" (like that undergone by both gas molecules and Brownian motion particles) not in terms of the mean free path (which is effectively zero for the Brownian motion particles) but in terms of the RMS speed of the particle (which is simply related to the temperature according to the kinetic theory).

Let us first present a derivation of the random walk formula based on the mean free path idea, to set up a foil for Einstein's alternative formula. Suppose a particle takes a sequence of steps whose length is always L, but whose directions are (for each step independently) completely random. Then we can show, as follows, that the mean square distance travelled after N such steps, is proportional to N.

Suppose that the particle's position, after n steps, is $\vec{r}_n$. Its position after $n + 1$ steps will then be

$$\vec{r}_{n+1} = \vec{r}_n + \vec{L} \qquad (12.65)$$

where $\vec{L}$ has magnitude L and a random direction. Squaring both sides of this equation, we have that

$$|\vec{r}_{n+1}|^2 = |\vec{r}_n|^2 + L^2 + 2\vec{r}_n \cdot \vec{L}. \qquad (12.66)$$

But since the direction of $\vec{L}$ is random, the last term will – on average – be zero. We thus have that

$$\langle r_{n+1}^2 \rangle = \langle r_n^2 \rangle + L^2 \qquad (12.67)$$

and so, by mathematical induction,

$$\langle r_n^2 \rangle = nL^2 \tag{12.68}$$

since clearly the *first* step will produce a mean square distance from the starting point precisely equal to L^2. The "typical" – or, more precisely, *root mean square* – distance of the particle from its starting point, after n steps, is therefore proportional to the square root of n:

$$r_n^{\text{RMS}} = L\sqrt{n}. \tag{12.69}$$

Note, by the way, that if the particle moves with a constant speed v, each distance-L step will take a time $\tau = L/v$, and so the number of steps taken in time t will be $n = t/\tau$. Plugging this in, one has that the "typical" (RMS) distance traveled by a particle during time t is proportional to the square root of t:

$$r_t^{\text{RMS}} = L\sqrt{\frac{t}{\tau}}. \tag{12.70}$$

Dividing through by t so that the left hand side becomes the average (in two senses!) velocity of a particle, we get

$$v = \frac{r_t^{\text{RMS}}}{t} = \frac{L}{\sqrt{\tau}}\frac{1}{\sqrt{t}} \tag{12.71}$$

which is itself a function of t. This explains mathematically why the speed of the Brownian motion particles, as observed through the microscope, didn't give any kind of sensible or consistent result.

For our immediate purposes here, however, the problem with this last formula is not just that the right hand side depends on t. Given the nature of the random walk, that much is expected. The problem, rather, is that we want a formula which relates the *observable* speed of the particle (that is, the average speed over some reasonable, finite, *observable* period of time) to the *average instantaneous* speed of the particle – because that is the quantity that the kinetic theory, by virtue of Equation (12.64), tells us something about.

But Equation (12.71) is of no use in that regard. For the Brownian particle, we expect both L (the mean free path) and τ (the time between collisions) to be vanishingly small. But that gives a notoriously ill-defined factor of zero divided by zero, leaving us with no well-defined relationship at all between the distance progressed by the particle and the time.

So what we need is an entirely new formula which, like that developed above, describes the particle's random walk, but in terms of the (RMS) *instantaneous speed* instead of the distance and time between collisions. The formula will, of course, still have the overall structure

$$r_t^{\text{RMS}} = \alpha\sqrt{t} \tag{12.72}$$

because that's just how random walks work. But the idea is to work out an expression for the constant α in terms of the instantaneous RMS speed – i.e., in terms of the temperature – instead of the totally unknown quantities L and τ.

Such a formula is precisely what Einstein worked out in 1905. We present here a version of Einstein's derivation provided just a few years later by the physicist Paul Langevin.

We start with Newton's second law for the Brownian particle, including a drag force produced by the particle's motion through the (viscous) background fluid:

$$m\frac{d^2x}{dt^2} = F_x - 6\pi R\eta\frac{dx}{dt}. \tag{12.73}$$

(For simplicity we analyze only the x-component of the motion.) The first term on the right hand side is the force exerted on the particle by the surrounding fluid's molecules. The second term on the right is the drag force, proportional to the speed of the particle through the fluid, and in a direction so as to tend to slow it down. The specific form of this term arises from the assumption that the particle is spherical, with radius R. The viscosity of the background fluid is denoted by η.

Einstein's key insight can be understood roughly as follows: *on average*, the acceleration of the particle will be zero. And so, on average, must be the whole right hand side of the last equation. This means, surprisingly, that the effect of what amounts to *friction* (the second term) is *just exactly as important* as the effect of what we'd naively think of as the exclusive cause of the Brownian motion – namely, the forces produced by the constant bombardment of the particle by molecules of the surrounding fluid.

One can think of this in energy terms as follows: in order to maintain a roughly constant average speed, the Brownian particle must (on average) be losing as much energy to friction (viscous drag) as it is gaining from the collisions from the surrounding water molecules. But whereas we know virtually nothing about the rate or intensity of these collisions (this, remember, was the failing of the previous formula), the rate of energy loss to viscous drag can be very simply related to the size of the particle (which is known) and the viscosity of the water (which is also known).

With that qualitative understanding in place, let us see how the derivation proceeds mathematically. First multiply the previous equation through by x – the position of the particle. This gives

$$mx\frac{d^2x}{dt^2} = xF_x - 6\pi R\eta\, x\frac{dx}{dt} \tag{12.74}$$

which can be re-written as follows:

$$m\frac{d}{dt}\left[x\frac{dx}{dt}\right] - m\left(\frac{dx}{dt}\right)^2 = xF_x - 6\pi R\eta\, x\frac{dx}{dt}. \tag{12.75}$$

Now we imagine taking the time average of this equation. The first term on the left vanishes: the motion is random, so there is no correlation between the position x and the velocity dx/dt at a given instant; the product is just as likely to be increasing as decreasing. The first term on the right hand side is also going to average to zero, because the force F_x is supposed to be random: so there is no reason it should be correlated with the position x of the particle. That leaves us with

$$\langle m\left(\frac{dx}{dt}\right)^2\rangle = 6\pi R\eta\langle x\frac{dx}{dt}\rangle. \tag{12.76}$$

But the right hand side can be re-written by noticing that $d(x^2)/dt = 2xdx/dt$:

$$\langle m\left(\frac{dx}{dt}\right)^2\rangle = 3\pi R\eta\frac{d}{dt}\langle x^2\rangle. \tag{12.77}$$

And finally, of course, the left hand side is just proportional to the temperature. So our final result is

$$\frac{d}{dt}\langle x^2\rangle = \frac{k_BT}{3\pi R\eta} \tag{12.78}$$

or equivalently

$$\langle x^2\rangle = \frac{k_BT}{3\pi R\eta}t \tag{12.79}$$

which, as advertised, gives us an explicit expression for the mean square distance traveled by a Brownian motion particle (of radius R) in time t, in terms of the temperature T and the viscosity η of the background fluid.

Einstein's derivation of Equation (12.79) set the stage for the brilliant experimental physicist, Jean Perrin, to undertake a systematic empirical study of Brownian motion. The notable experimental achievement – which really made the whole study possible and worthwhile – was his tireless effort to produce microscopic spherical particles with consistent, and accurately known, radii.

Perrin then observed, literally by watching them through a microscope, *many* of these particles undergoing Brownian motion for fixed periods of time t. By measuring how far each particle got during this time period, squaring, and then averaging, the results, Perrin could determine $\langle x^2 \rangle / t$ empirically. He of course also measured the temperature T of the sample, and knew the radius R of the Brownian particles that he had so painstakingly created for the purpose. And the viscosity η of water was known. In short, everything in Equation (12.79) was known except for Boltzmann's constant, k_B, so that – finally – could be determined with some accuracy.

The result was

$$k_B = 1.4 \times 10^{-23} \, J/K \tag{12.80}$$

implying, through $k_B = R/N_A$, the following value for Avogadro's number:

$$N_A = 6 \times 10^{23}. \tag{12.81}$$

This was not the first experimental estimate of Avogadro's number. We mentioned, earlier in this chapter, that the same quantity had been estimated, several decades earlier, by Loschmidt. And there were other similar estimates in the intervening years. But Perrin's measurement was the first to cleanly dispense with arbitrary and dubious assumptions about, for example, the shapes of molecules. And so it was the first really convincing – and really convincingly accurate – measurement of the absolute scale of atoms.

12.5.2   Perrin's Second Experiment

Despite being the *first* convincingly accurate measurement of k_B and hence N_A, the experiments described in the previous subsection was not the only such measurement which appeared around the turn of the 20th century. There were, actually, about a dozen such measurements, using a wide variety of different techniques. By independently arriving at the same value for N_A, this group of experiments proved beyond the shadow of a doubt that the value arrived at in Perrin's first experiment, and more generally the whole atomic-molecular-kinetic-statistical approach to physics, was correct.

To give at least an indication of the variety of these other experiments, we will discuss here one of them which happens to have a close connection to the ideas developed earlier in this chapter – and which happens also to have been undertaken by Jean Perrin.

Perrin's second experiment involves the same, painstakingly manufactured spherical Brownian particles whose random walks he observed in the first experiment. The second experiment, however, concerns not the trajectory followed by the individual particles, but instead the spatial distribution of the particles as a group.

In our previous discussions of the spatial distribution of molecules, we have always found that the equilibrium distribution of particles was *uniform*. For molecules, though, that is only true on small scales. The molecules in a meter-sized box of gas will, in equilibrium, be distributed uniformly throughout the box. But the molecules in, say, the whole atmosphere of the Earth are *not* of uniform density. As any mountain-climber will tell you, the density drops off rather dramatically with increasing altitude. In fact, the density falls off (roughly) *exponentially* with altitude h:

$$\rho(h) = \rho(0) \, e^{-\alpha h} \tag{12.82}$$

where α is a constant. We can understand this exponential density fall-off – and in addition understand the value of the constant α – with a little analysis.

Assume the atmosphere is made of molecules of mass m. Then the (mass) density ρ is related to the (number) density n through $\rho = mn$. Now consider a thin horizontal layer of the atmosphere, lying between altitude h and $h + dh$, and having area A in the horizontal plane. Its volume is thus $A\,dh$, and its total mass is therefore $dM = A\rho\,dh = Amn\,dh$.

This layer of atmosphere has a non-zero mass, and so is pulled downward by a gravitational force of magnitude $F_g = dM\,g$, where $g = 9.8\,m/s^2$ is the acceleration of gravity near the Earth's surface. What prevents this layer of atmosphere from simply falling to the ground under the influence of this force? There is another force – equal in magnitude, but *upward* – that cancels it out and leaves the layer in equilibrium. This other force is the *buoyancy* force, i.e., the net force of the surrounding atmosphere pushing "in" on it. This is a net upward force precisely because the density of the atmosphere – and hence its *pressure* – decreases with altitude. This (upward) buoyancy force will be equal to

$$dF_b = A\left(P(h) - P(h + dh)\right) \tag{12.83}$$

or, using the ideal gas law to rewrite the pressure P in terms of the density ρ,

$$dF_b = Ak_BT\left(n(h) - n(h + dh)\right) = -Ak_BTdn \tag{12.84}$$

where $dn = n(h + dh) - n(h)$.

Equating this last expression for the buoyancy force with our earlier expression for the gravitational force gives

$$Amgn\,dh = -Ak_BT\,dn \tag{12.85}$$

or, simplifying,

$$\frac{dn}{dh} = -\frac{mg}{k_BT}n. \tag{12.86}$$

This first-order differential equation can be easily solved for $n(h)$:

$$n(h) = n(0)e^{-mgh/k_BT} \tag{12.87}$$

where $n(0)$ is the number density at $h = 0$. From here we need merely multiply through by the molecular mass m to get, as advertised, Equation (12.82). Note that we can now identify the constant α from that earlier equation as follows:

$$\alpha = \frac{mg}{k_BT}. \tag{12.88}$$

One might think that, since one can surely measure α by simply measuring how the density (or pressure) of air varies with altitude, one can in turn use this formula to empirically determine Boltzmann's constant, k_B, and hence Avogadro's number. The problem is, the mass m of a single air molecule cannot be known until one knows Avogadro's number – so one is trapped in a circle. This can be seen also by multiplying both the numerator and denominator on the right hand side of the previous formula by N_A. The result is

$$\alpha = \frac{\mathcal{M}g}{RT} \tag{12.89}$$

where $\mathcal{M}$ here is the mass of a *mole* of air, and R is the usual gas constant. So one cannot, after all, determine the absolute scale of atoms by seeing how the density of air varies with altitude.

Before turning to Perrin's second experiment, let us pause briefly for a microscopic and statistical perspective on Equation (12.87). This result, if you think about it, is *exactly* what one should have expected on the basis of the so-called *Boltzmann factor* developed in the previous section. From a microscopic and statistical point of view, the position of any particular air molecule is *random*. But because we're interested in considering potentially large variations in the altitude of the molecule, we should take into

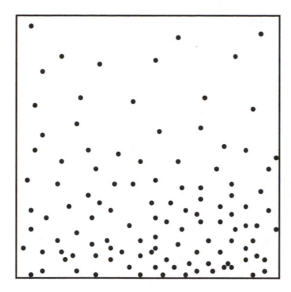

Figure 12.9: Sketch of the expected spatial arrangement of Brownian motion particles in Perrin's second experiment.

account the existence of the gravitational potential energy, which of course varies with height as mgh. Thus, a completely straightforward application of the Boltzmann factor tells us right away that the *probability* for some particular molecule to be found at altitude h is proportional to $e^{-mgh/k_B T}$.

And so what is going to happen if we have an enormous collection of molecules, with each one's position being randomly assigned according to that probability formula? We will of course find – with exceedingly high probability – that the number density of particles varies in that same way with altidue. That is, we'll find Equation (12.87).

Now we're in a position to understand the key idea that made Perrin's second experiment possible. Understanding already that the tiny Brownian motion particles are, as far as physics is concerned, just very large and very heavy molecules, it is clear that, in principle, the statistical analysis of the previous two paragraphs should apply to them just as much as it applies to the air molecules in the atmosphere. That is, in a tiny cell filled (say) with water and a bunch of these Brownian motion particles, we should expect *not* that the particles will all just sink to the bottom, and *not* that they will distribute themselves uniformly throughout the volume, but rather that they will distribute themselves in accordance with Equation (12.87). That is, the number density of particles will be greatest at the bottom of the cell, and will decrease exponentially with "altitude" – i.e., with increasing height from the bottom of the cell. This expectation is sketched in Figure 12.9.

So the idea is to build such a cell, and then count (or in some way approximate) the number of Brownian particles at different heights h. This can be accomplished by looking from above through a standard microscope, which brings some particular horizontal layer in the cell into focus. One can then scan through several different such layers by simply rotating the focus knob of the microscope, keeping track, somehow, of the number of particles observed at each different height. One can then plot the number of particles observed as a function of the height, and fit an exponential curve to find an empirical value for what was (in the context of the discussion of the atmosphere) called α.

Now the crucial point is that – unlike the air molecules from our earlier discussion – the mass m of these Brownian motion particles can be known rather precisely. They are, after all, spheres of a certain known radius R, made from a material (Perrin used something called "gamboge", a yellow-ish latex made from vegetables) of a known mass density ρ. One thus has

$$m = \frac{4}{3}\pi R^3 \rho \tag{12.90}$$

Actually, though, the mass that should enter Equation (12.88) in the context of this experiment is not the absolute mass m given by the last formula, but rather the "effective mass" which takes into account the buoyancy of the gamboge spheres in the background water (or whatever liquid they are immersed in). Remember, the mg on the right hand side of Equation (12.88) is supposed to represent the magnitude of the force which pulls the particles in question down toward the ground; for a single air molecule moving through the vacuum, this is just the gravitational force, but for a particle immersed in a fluid, it is the (downward) gravitational force *minus* the (upward) buoyancy force. Another way to understand this same point is in terms of energy. From the point of view of the Boltzmann factor, what matter is the change in the overall potential energy of the system as the particle in question is raised or lowered. But when we raise one of the Brownian motion particles from "the ground" to a certain height h, we must also (to make room for it) lower an equal volume of water from height h back to "the ground". So the net change in potential energy is not mgh, but $mgh - \frac{4}{3}\pi R^3 \rho_w$, where ρ_w is the density of the water (or other background fluid).

In any case, the upshot is that the correct expression for m in Equation (12.88) is

$$m = \frac{4}{3}\pi R^3 (\rho - \rho_w). \tag{12.91}$$

To repeat, though, what's really important is that this (effective) mass is *known*. And so by running the experiment and determining α (and of course knowing g and T) one can determine an empirical value for Boltzmann's constant k_B, and so also N_A.

Even in the eyes of the most skeptical scientists, Perrin's measurements of Boltzmann's constant k_B – and hence also Avogadro's number N_A – removed any remaining doubts about the reality of atoms. The German chemist Wilhelm Ostwald, for example, had said in 1906 that "atoms are only hypothetical things." But he changed his tune in 1908: "I have satisfied myself that we arrived a short time ago at the possession of experimental proof for the discrete or particulate nature of matter – proof which the atomic hypothesis has vainly sought for a hundred years, even a thousand years."

Or, as Henri Poincarè summarized the scientific consensus: "atoms are no longer a useful fiction; things seem to us in favour of saying that we see them since we know how to count them... The brilliant determinations of the number of atoms made by M. Perrin have completed this triumph of atomism.... The atom of the chemist is now a reality."

Questions:

Q1. In poker, a royal flush (consisting, say, of the 10, J, Q, K and A, all of spades) is a way better hand than, say, the 2 of hearts, the 5 of diamonds, the 6 of spades, the 9 of hearts, and the J of clubs. Yet the probabilities of dealing out these two hands, from a well-shuffled deck, are exactly equal. Indeed, the probabilities of getting any of the $52 \times 51 \times 50 \times 49 \times 48 = 311,875,200$ possible poker hands, are all the same, namely, one in 311,875,200. So why, then, are some hands (like the royal flush) considered better than others?

Q2. Suppose you start with the royal flush described in the previous question. Then you randomly pick one of the cards and replace it with the next card from the (well-shuffled) deck. Then you keep doing this. What is likely to happen to the quality of your poker hand over time? Explain how this relates to Boltzmann's account of the approach to equilibrium.

Q3. If a collection of particles in outer space starts out uniformly distributed over some region, but ends up clumping together (due to the attractive gravitational forces that the particles exert on each other) into, say, a star, has the entropy of that collection of particles increased or decreased? (Make sure you think about both the spatial distribution and also the distribution in velocity space!)

Q4. The typical (RMS) speed of Helium atoms, at room temperature, is about 1300 m/s. (Helium atoms move faster than Nitrogen molecules, at the same temperature, because their average kinetic energies are the same and Helium atoms are less massive!) This is nearly an entire order of magnitude below the Earth's escape velocity. Yet, it would seem natural to explain why there is no Helium in Earth's atmosphere by saying that any Helium that was here, would escape. Can this natural explanation be correct even though $v_{RMS}^{He} < v_{esc}$?

Q5. When a liquid *evaporates*, it is the fastest-moving molecules which are able to break free from the liquid surface. Just based on this fact, what would you expect to happen to the temperature of the remaining liquid? Is this a real phenomenon?

Q6. Consider the gas, initially confined to the left half of a box, with the barrier then pulled out so the gas is now free, in principle, to fill the entire box. And, of course, this is exactly what will happen for the overwhelming majority of possible initial microstates of the gas. But very special states, which produce unusual behavior, are always possible. Give an example of a specific possible initial microstate for which the gas will *not* expand to fill the entire box, but will instead remain confined on the left hand side indefinitely.

Q7. We have discussed how, in a free expansion, the gas molecules are overwhelming likely to end up uniformly distributed over the new, larger volume. Consider the precise micro-state (including both positions and velocities for all the molecules) of a gas at the end of such a free expansion process. And now imagine the different (but closely related) micro-state that is the same, spatially, but has all of the velocities exactly reversed. If the gas evolved from this microstate, it would be like watching the movie of its free exapansion in reverse – the molecules would all end up, later, on one side of the box! But if, for each microstate that has the gas doing the ordinary and expected thing (namely, spreading out uniformly over the volume available to it) there is exactly one other microstate which generates the bizarre behavior of spontaneously confining itself to one side of the box, how can it be the case that the ordinary behavior is overwhelmingly more likely to occur?

Q8. In a passage quoted in the reading, Maxwell notes that his "demon" could arrange for heat to flow from the cooler gas on one side to the warmer gas on the other side, in violation of the second law of thermodynamics. Could the demon also undo a free expansion? How?

Q9. Explain how Figure 12.7 is consistent with the graphs in Figure 12.3. In particular, explain why, even though most of the dots in Figure 12.7 are clustered near the origin, the number of particles with $v = 0$ is vanishingly small. (Hint: how do you divide up the velocity space shown in Figure 12.7 into regions corresponding to roughly constant speed?)

Exercises:

E1. Here is the argument put forward by Loschmidt that was mentioned in the text as the first approximate calculation of Avogadro's number. The idea is to combine Equation (12.2) with some other relation between the number density n and the molecular diameter d. Now, it is obvious that the number density (for, say, exactly one mole of Nitrogen) can be written this way: $n = N_A/V_{gas}$, where N_A is Avogadro's number and V_{gas} is the volume occupied by the Nitrogen gas (roughly 23 liters). What Loschmidt pointed out, however, was that Nitrogen can also be *liquified* (by cooling it down to about 77 Kelvin), and the volume of that same one mole of Nitrogen can be measured in the liquid state. That volume turns out to be $V_{liquid} = 34.7\ cm^3$. Then Loschmidt argued that V_{liquid} should be approximately equal to N_A times the volume of a single molecule, which is roughly d^3, since in the liquid state there is supposed to be hardly any empty space at all between adjacent molecules. This allows one to re-write the right hand side of Equation (12.2) purely in terms of V_{gas}, V_{liquid}, and N_A, so one can in turn solve for N_A in terms of empirically measured quantities. Work through this and determine a value for N_A based on the numbers given here (and in the main text for the mean free path in air).

E2. Suppose two molecules, each of mass m and moving with speed v, collide. Provide a concrete example showing that there exist possible collisions (i.e., collisions in which both momentum and kinetic energy are conserved) for which the speeds of the molecules after the collision are different. (A good way to specify a concrete example of such a collision is to specify the x- and y- coordinates of the velocity of both particles, both before and after the collision.)

E3. Use Equation (12.70) and the values for L and τ from the discussion of ammonia in the first section, to estimate the time t it should take a typical ammonia molecule to random walk its way, through the air, a distance of one meter. (You will probably find a time that is longer than you expected, and it is worth thinking about why the time you calculate this way is unrealistically long.)

E4. Let's learn how to compute Gaussian integrals of the form

$$I_0 = \int_{-\infty}^{\infty} e^{-cx^2}\, dx. \tag{12.92}$$

The trick is to consider the square of this integral, expressed as follows:

$$I_0^2 = \left(\int_{-\infty}^{\infty} e^{-cx^2}\, dx \right) \left(\int_{\infty}^{\infty} e^{-cy^2}\, dy \right). \tag{12.93}$$

We can think of I_0^2 as a double integral over the x-y-plane, which we can then convert to polar coordinates:

$$I_0^2 = \int_{r=0}^{\infty} e^{-cr^2}\, 2\pi r\, dr. \tag{12.94}$$

Do this integral and hence argue that $I_0 = \sqrt{\pi/c}$. Finally, use this result to show that Equation (12.14) is a valid probability density, in the sense that $\int p(v_x)\, dv_x = 1$.

E5. Integrals of the form $I_n = \int_{-\infty}^{\infty} x^n e^{-cx^2}\, dx$ can be done using integration by parts. But the following observation suggests a cute alternative way to do them:

$$I_2 = \int x^2 e^{-cx^2}\, dx = -\frac{d}{dc} \int e^{-cx^2}\, dx = -\frac{d}{dc} \sqrt{\frac{\pi}{c}}. \tag{12.95}$$

Use this trick to work out a formula for I_2 and then calculate both integrals in Equation (12.62) and confirm the result reported in Equation (12.63).

E6. One of the apparently "active" degrees of freedom for a (diatomic) Nitrogen molecule is associated with rotation about an axis perpendicular to the line through the two Nitrogen atoms, with $E = \frac{1}{2}I\omega^2$. Using the now-known value for Avogadro's number $N_A = 6.0 \times 10^{23}$, we can say that the mass of each Nitrogen atom is $m = 14\,\text{grams}/N_A$. And the separation distance between the two atoms turns out to be about $d = 1.5 \times 10^{-10}$ meters. What is a typical angular velocity ω with which Nitrogen molecules rotate in air at room temperature?

E7. One mole of liquid water has a mass of 18 grams and hence (since the density is 1 g/cc) occupies a volume of 18 cubic centimeters. Use the now-known value of N_A to compute the volume V of a single water molecule and then also its "diameter" d (on the assumption that $d^3 \approx V$).

E8. Having heard that violations of the 2nd law are possible (just unusual/rare), your neighbor decides to try to extract energy from the stagnant pond in his backyard by installing a paddle wheel hooked up to an electrical generator. He is patient, and plans to just wait for an unusual statistical fluctuation to cause the water to push the paddle wheel around and generate some useful energy. Suppose the temperature of the pond is $20°C = 293$ K, and he would be satisfied to extract a mere 1.0 Joule of energy from the pond this way. Note that this implies that the entropy of the pond should spontaneously decrease by only $\Delta S = -1J/293K$, which really doesn't seem like much, so how improbable could it be, really? Answer this question using Boltzmann's statistical interpretation of entropy, Equation (12.38). In particular, you can argue that the probability of this kind of fluctuation should be equal to the number of micro-states in which it happens, divided by the number of micro-states associated with whatever the pond is doing normally:

$$P = \frac{W_{fluctuation}}{W_{normal}} = \frac{e^{S_{fluctuation}/k_B}}{e^{S_{normal}/k_B}} = e^{\Delta S/k_B}. \tag{12.96}$$

Projects:

P1. Use dice (or perhaps a computer random number generator) to assign v_x, v_y, and v_z values (each of these just being 1, 2, 3, 4, 5, or 6) to a number of particles. Then, for each particle, compute the speed v of the particle. Make a histogram showing how many of the particles have speeds in various ranges and explain why the graph has the shape that it does.

P2. Simulate a one-dimensional, N-step random walk using coin flips. Do this a number of times and calculate the RMS displacement. Now vary N (or pool data with other students/groups who were assigned a different N) and make a plot of the RMS displacement as a function of N.

P3. Use three dice to simulate the interaction of a system with a heat bath. One of the dice can be the system, and the other two can be the reservoir. The energy of a given die will be represented by the number it is showing. Suppose the total energy of the combined system/reservoir meta-system is $E = 3$. Then there is just one possible microstate (namely, all three dice show "1") so there is not much interesting to say. Now suppose that $E = 4$. Now there are two possible microstates for the system to be in ($E_S = 1$ and $E_S = 2$). How many microstates of the combined system/reservoir are there for each of these? Which is more probable, if all microstates of the meta-system are equally probable? Continue exploring the $E = 5$ and $E = 6$ cases and summarize your findings with appropriate graphs. It is also interesting to similarly explore the case in which the system is one die and the reservoir is three dice.

P4. Write a computer program to simulate a simple model of a free expansion. Suppose there are N particles and all of them are initially on the left hand side of a box. Then, in each subsequent time step, we pick one of the N molecules at random, and change its location (from left to right if it's on the left, or from right to left if it's on the right). Make a plot showing what happens to the number of particles on the left over time and also what happens to the entropy of the system over time. (You might start with, say, $N = 10$ and then explore larger values as well.)

P5. Reproduce Perrin's second experiment and determine your own value for Avogadro's number.

CHAPTER 13

A Glimpse Beyond Classical Physics

A s we saw last week, the atomic theory of matter became firmly established in the first decade of the 20th century with the experiments of Jean Perrin and others. But once it was clear that ordinary matter really was made of atoms, of something like the sort first suggested just a century earlier by Dalton, further questions about the composition and sub-structure of atoms came into sharper focus. In addition, the long-standing puzzle about the specific heats of gases remained unresolved.

This last (and significantly shorter) chapter attempts to bring a temporary closure to the story of atomism (and provide a nice book-end to the course as a whole) by briefly indicating how a more detailed picture of sub-atomic structure was experimentally inferred and how the novel concept of "quantization" clarified several puzzles even while introducing new ones that, despite launching the massively important quantum revolution, remain, in profound ways, unresolved to this day.

§ 13.1 The Copernican Atom

Perrin's experiments had revealed that Avogadro's number, N_A, was approximately 6×10^{23}. This in turn implied that a typical atom had a diameter of roughly an Angstrom, i.e., 10^{-10} meters, and hence a radius of roughly

$$R = \frac{1}{2} \times 10^{-10}\,\mathrm{m} = \frac{1}{2}\,\text{Å}. \tag{13.1}$$

It was also known from early studies of cathode rays (streams of negatively-charged particles, eventually named "electrons", which materials would emit under certain conditions) as well as radioactivity (the process in which certain unstable atoms emit various types of radiation, some of it electrically charged) that atoms are built from smaller, "sub-atomic" particles and are largely held together by electrical forces. The details, however, remained unclear and several speculative models had been proposed.

In one of these models (often called the "plum pudding" model, but only British people know what that means) atoms were pictured as blobs of positively-charged "jelly" – with a radius of order R – with smaller, pointlike negatively-charged electrons embedded within. In the standard configuration of such an atom, the positive charge of the "jelly" was perfectly balanced by the negative charge of the embedded electrons so that the atom as a whole was electrically neutral. But atoms could be "ionized" by adding or removing one or more electrons, such that the atom possessed a net non-zero electric charge. And even a neutral atom could have a non-zero electric field near its surface – and hence exert electrical forces on nearby charged particles – if the distribution of its internal electrons was not perfectly symmetrical.

To get a ballpark sense of the size of the atomic electric fields implied by this particular model, consider a spherical, radius-R blob of jelly with a positive charge equal to the absolute value e of the charge carried

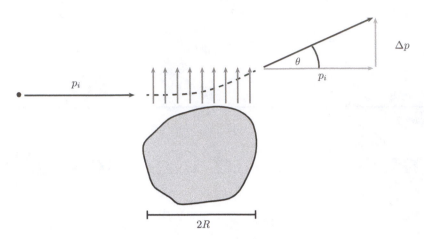

Figure 13.1: Schematic diagram of an alpha particle (represented by the black dot) being deflected by the electric field produced by blob-model atom (of diameter $2R$) it passes near.

by a single electron. (Such a blob would, according to this model, be the positive component of the simplest atom, Hydrogen.) At the surface of the blob, the electric field would have magnitude

$$E = \frac{1}{4\pi\epsilon_0}\frac{e}{R^2} \approx 6 \times 10^{11}\,\text{N/C} \tag{13.2}$$

where we have used the value $e = 1.6 \times 10^{-19}\,\text{C}$ for the elementary unit of charge. The electric field outside an asymmetric but electrically neutral atom – and also the electric field *inside* even this positive Hydrogen ion – will be somewhat smaller than this, whereas the field just outside a heavier atom (with more positively-charged jelly) might be larger than this. But this gives us a rough, order-of-magnitue sense of the electric field strengths that are likely to exist in the vicinity of atoms according to the "blob model".

The strength of atomic electric fields is of particular interest in relation to a series of scattering experiments undertaken by Ernest Rutherford and his colleague Hans Geiger in the first decade of the 20th century. Rutherford and Geiger had noticed that if a beam of $\alpha-$ ("alpha-") particles – these being tiny, positively-charged particles emitted in certain forms of radioactivity – was shot through a thin metal foil, the beam downstream of the foil became somewhat diffused. Some fraction of the alpha particles were evidently being deflected as they passed near (or through) the atoms in the foil.

The α-particles were known to have a mass

$$m_\alpha = 6.6 \times 10^{-27}\,\text{kg} \tag{13.3}$$

(about four times the mass of a single Hydrogen atom) and had a positive electric charge of two elementary units:

$$q_\alpha = 2e = 3.2 \times 10^{-19}\,\text{C}. \tag{13.4}$$

The alpha particles used in these early experiments, emitted by radioactive radium atoms, typically had kinetic energies of $KE = 8 \times 10^{-13}$ Joules and hence moved with a speed of roughly

$$v_0 = \sqrt{2KE/m_\alpha} = 1.6 \times 10^7\,\text{m/s}. \tag{13.5}$$

Let's consider an α-particle flying merrily along at speed v_0 and then passing by an atom in a piece of metal foil, as shown in Figure 13.1. The alpha particle initially has a momentum of magnitude $p_i = m_\alpha v_0$.

But upon passing near an atom in the foil, it traverses an electric field, produced by that atom, which (for simplicity) we approximate as having the uniform magnitude E (estimated above) and direction perpendicular to the alpha particle's initial direction of travel. The sideways electric force causes the trajectory of the alpha particle to deflect by an angle θ which, using the impulse-momentum theorem and the small-angle approximation, should be given by

$$\theta \approx \frac{\Delta p}{p_i} = \frac{q_\alpha E \, \Delta t}{m_\alpha v_0} \tag{13.6}$$

where $\Delta t = 2R/v_0$ is roughly the amount of time during which the sideways force, of magnitude $q_\alpha E$, is being exerted.

Plugging in the known values, we find

$$\theta \approx 10^{-5} \text{ radians} \tag{13.7}$$

which is a tiny fraction of one degree. (Note that the fact that we get a very small angle justifies the use of the small-angle approximation, $\theta = \sin(\theta) = \tan(\theta)$, earlier.)

To summarize, the expected deflection of an alpha particle that passes near an atom is – according to the "blob model" of the atom – extremely small... so small, in fact, that it is hard to understand how Rutherford and Geiger could have even noticed their alpha particle beam becoming diffuse downstream of the foil. Of course, our calculation used the electric field strength that would be associated with a Hydrogen ion. But an atom of (for example) gold – with atomic number 79 – would have nearly 100 times more positive charge than a Hydrogen atom, and so might conceivably produce electric fields up to 100 times stronger. That's a bit implausible, but it's not out of the realm of possibility. In addition, a given alpha particle would probably encounter many atoms as it traverses even a very thin piece of gold foil. For the same reason that it is exceedingly improbable for all of the air molecules in a room to be found, at some moment, on one side of the room, it is exceedingly improbable that a given alpha particle would be deflected in the same direction ("up" in the previous Figure, say) by every atom it encounters. But still, the alpha particles may be deflected one way a little more often than they are deflected the other way, and thus end up deflecting by an angle that is (at least) several times larger than the deflection angle associated with a single atomic encounter.

This combination of effects – larger electric fields associated with heavier metal atoms, and multiple deflections – was probably enough to explain the roughly one-degree deflections that Rutherford and Geiger saw in their preliminary study. Surely, though, alpha particle deflections of greater than a few degrees were out of the question.

But, in science, it's good to to empirically check even ideas that seem impossible. So, as Rutherford would tell the story some years later, he assigned an undergraduate student, Ernest Marsden, to work with Geiger and check carefully for large-angle deflections:

> "One day Geiger came to me and said, 'Don't you think that young Marsden, whom I am training in radioactive methods, ought to begin a small research?' Now I had thought that too, so I said, 'Why not let him see if any α-particles can be scattered through a large angle?' I may tell you in cofindence that I did not believe they would be, since we knew that the α-particle was a very fast massive particle, with a great deal of energy, and you could show that if the scattering was due to the accumulated effect of a number of small scatterings the chance of an α-particle's being scattered backwards was very small."

Of course, despite Rutherford's expectations, Geiger and young Marsden did see something interesting. As he recounted it later:

> "... I remember ... Geiger coming to me in great excitement and saying 'We have been able to get some of the α-particles coming backwards...'

"It was quite the most incredible event that has ever happened to me in my life. It was almost as incredible as if you fired a 15-inch shell at a piece of tissue paper and it came back and hit you."

Contemplating the surprising result, Rutherford slowly realized that the only way to account for the observed large deflections – not just several degrees, but up to 180 degrees! – was for the electric fields exerting forces on the alpha particles to have magnitudes something like 5 orders of magnitude larger than our estimate above. And the only way to produce electric fields that strong is for the positive charge of the atom to be, Rutherford's words, "concentrated in a minute nucleus" at least 5 orders of magnitue smaller than the atom itself.

Atoms, it seemed, were not balls of positively-charged jelly with negatively-charged electrons embedded within, but were instead incomprehensibly dense, positively-charged nuclei being orbited by negatively-charged electrons, much as the planets in the solar system turn out to orbit the massive, central Sun.

Between 1911 and 1913, Rutherford undertook a detailed theoretical analysis of the probability for an alpha particle to be deflected by angle θ in a scattering experiment, and showed in particular that the probability for a fixed-area alpha-particle detector to "click", when situated so as to detect particles that had scattered through an angle θ, was proportional to the fourth power of the co-secant of $\theta/2$:

$$P(\theta) \sim \csc^4(\theta/2) = \frac{1}{\sin^4(\theta/2)}. \tag{13.8}$$

You can work through this same calculation in one of the Projects.

Subsequent experimental tests showed that Rutherford's predictions based on the "nuclear model" – which is also sometimes called the "Rutherford model", although my favorite name for it given the overall structure of this course is the "Copernican model" – matched the data perfectly.

§ 13.2 Stability of the Copernican Atom

In the nuclear/Rutherford/Copernican picture, the simplest atom – Hydrogen – evidently consists of a single electron orbiting a nucleus (containing virtually all of the atom's mass) which was eventually recognized to consist of a single positively-charged particle called a proton. This simple Copernican model of the Hydrogen atom is sketched in Figure 13.2.

Assuming the electron orbits the proton in a circle of radius R, we can find the period T of the electron's orbit by applying Newton's second law:

$$\frac{ke^2}{R^2} = m_e \frac{4\pi^2 R}{T^2} \tag{13.9}$$

so that

$$T = \frac{2\pi}{e} \sqrt{\frac{m_e}{k}} \, R^{3/2} \tag{13.10}$$

which can of course be recognized as an atomic analog to Kepler's third law! Anyway, plugging in the values of e, m_e, k, and R we find that the period of the electron's orbit is 1.4×10^{-16} seconds. The electron is moving fast!

Equivalently, the electron in this Hydrogen atom can be thought of as oscillating sinusoidally (simultaneously in both the horizontal and vertical directions) with an angular frequency $\omega = 2\pi/T = 4.5 \times 10^{16}$ rad/sec. But this raises a question that relates to ideas we developed way back in Chapter 10. In particular, we showed there that a system with an electric dipole moment of magnitude p_0

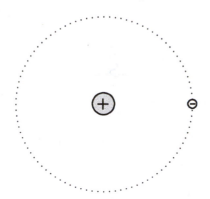

Figure 13.2: A single Hydrogen atom in the nuclear/Rutherford/Copernican picture: a negatively-charged electron orbits around a (much more massive and hence approximately stationary) proton.

which oscillates with angular frequency ω will radiate electromagnetic energy at a rate given by the Larmor formula:

$$P = \frac{1}{6\pi} \frac{p_0^2 \omega^4}{\epsilon_0 c^3}. \tag{13.11}$$

Our "Copernican" Hydrogen atom is just such a system, and so it should evidently be radiating energy in the form of electromagnetic radiation.

Where would this radiated energy come from? Evidently the energy of the atom itself would have to decrease as energy is carried away, which implies that the radius R of the electron's orbit must decrease in time. But a decreasing R implies a faster orbit (with a shorter T and a higher ω) which in turn implies a faster radiation rate P which means that R must decrease even more rapidly than before...

The laws of physics we've developed up to this point thus imply that our Copernican atom should be radically unstable. In particular, it can be shown that our humble Hydrogen atom, starting with an electron in a half-Angstrom-radius circular orbit about the proton, should emit a blinding and indeed catastrophic burst of radiation in which, literally, an infinite amount of increasingly-high-frequency electromagnetic energy is emitted in a time of order 10^{-10} seconds. The analysis involved in making this prediction uses essentially the same techniques you used several weeks ago to understand the famous "chirp" that LIGO recently observed from a binary black hole merger. Another of this chapter's Projects will walk you through this calculation.

Other types of atoms, with two or more electrons orbiting a heavier nucleus, are predicted to suffer the same qualitative fate as Hydrogen.

But all of the atoms in the world exploding in an infinite burst of high-frequency radiation – in a tiny fraction of a second – seems like the kind of thing we would have noticed if it had happened. So it should be clear that there is some fundamental problem with applyling our now-familiar Newtonian and Maxwellian laws to the sub-atomic realm. Novel physics is required to account for the apparent stability of atoms.

§ 13.3 Quantization to the Rescue

The conceptual core of that novel physics was the notion of "quantization" that was first introduced by Max Planck, and then later significantly clarified by Albert Einstein, in the first decade of the 20th century. Planck and Einstein had begun to appreciate that, in order to explain the observed spectrum

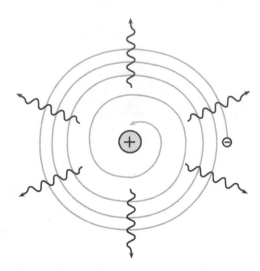

Figure 13.3: According to the known laws of classical physics, the electron in a Hydrogen atom should emit energy in the form of electromagnetic radiation as it orbits the nucleus. Its orbit, therefore, should systematically spiral in toward the nucleus and indeed it turns out that the electron should emit an infinite burst of infinitely high-frequency radiation in about a tenth of a nanosecond!

of thermal radiation emitted by hot objects (e.g., the reddish light emitted by a "red hot" coal in a fire), it was necessary to assume that the amount of energy that could be carried by a ray of light with a certain frequency was not just any old value from the continuum (as one would expect on the basis of the formulas developed in Chapter 10, e.g., that the energy density is proportional to the square of the field strength) but was instead restricted to certain discrete possible values. In particular, Planck and Einstein found agreement with the observed spectrum only if they assumed that the possible values for the energy E in a ray of light were given by

$$E = n\epsilon \tag{13.12}$$

where n is an integer (0, 1, 2, 3, etc.) and ϵ is some basic quantity of energy related to the frequency of the light ray. The allowed energies of a ray of light were not continuous, but were rather "quantized".

Einstein suggested that this strange requirement could be understood if we abandoned – or supplemented? – the idea, which was of course one of the triumphs of 19th century physics, that light consisted of electromagnetic waves. In particular, Einstein's idea was that Equation (13.12) could be understood if we took the energy associated with a light wave of a given frequency not to be carried by the electromagnetic wave itself, but instead to be carried by *particles* – particles which came in time to be called "photons" – that were somehow associated with the waves. This made sense of Equation (13.12) just because the number of particles associated with a given wave would obviously have to be an integer. It is in the very nature of particles that they come in integer-sized quantities only!

But the nature of the "association" between the waves and the particles remained unclear, and despite the massive progress that has been made in the last 100 years in learning how to manipulate photons, the nature of the so-called "wave-particle duality" of light remains shrouded in mystery.

It soon became clear, though, that the mysterious notion of "quantization" applied to more than just light. Just a few years after Rutherford argued, on the basis of his scattering experiments, for the Copernican model of the atom, Niels Bohr introduced the idea that the electron in a Hydrogen atom could not orbit with any old radius R from the continuum of possible values, but must instead orbit with one of a set of discrete possible radii. Electron orbits, in short, were also quantized.

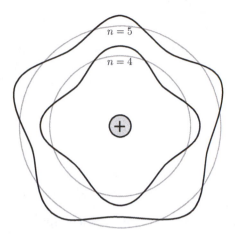

Figure 13.4: In de Broglie's model of the Hydrogen atom, the electron orbits are quantized because electrons have an associated wave, an integer number of whose wavelengths must fit around a complete orbit to avoid discontinuities. The possible orbits with $n = 4$ and $n = 5$ full wavelengths are shown. Note that the idea is not that the electron particle follows the wiggly black curves shown here; rather, the wiggly black curves are meant to (very schematically!) indicate a sort of standing wave, associated with the electron particle, which guides it on a circular path like the ones shown in gray.

Bohr introduced this idea in order to account for the discrete set of frequencies that had been observed in the electromagnetic radiation emitted by Hydrogen. If only certain electron orbits are allowed, and each allowed orbit has its own specific energy, then a certain amount of energy must be released when the electron "jumps" from a certain higher-energy orbit to a lower-energy orbit. This energy will evidently be carried off by a single photon whose energy is related, as mentioned above, to its frequency. So a discrete set of allowed orbits – "energy levels" – in the atom will imply a discrete set of frequencies in the emission spectrum. Bohr cleverly worked out, in effect, a list of which discrete allowed orbits for the electron would be sufficient to account for all of the observed frequencies in the emission spectrum.

Note in particular that Bohr's model implies that there is a *smallest*, lowest-energy allowed orbit for the electron. Because there is no allowed lower-energy state for it to go into, an electron in this smallest orbit would evidently be prevented from radiating energy away, and the observed stability of matter would be accounted for.

But, at the time, Bohr had no idea *why* the orbits of the electron in a Hydrogen atom should be restricted to his list of allowed possibilities. Progress on this front was made in the early 1920s by Louis de Broglie, who suggested that since light had turned out to be not just a wave but also, somehow, a particle, maybe electrons could turn out to be not just particles, but also, somehow, waves? In particular, de Broglie showed that Bohr's list of allowed orbits for the electron in a Hydrogen atom could be understood on the assumption that the electron was accompanied by an associated wave, an integer number of whose wavelengths would need to fit around an orbit to avoid discontinuities. This idea is sketched in Figure 13.4.

Soon after de Broglie's suggestion, direct experimental evidence for the wave-character of electrons was observed, Erwin Schrödinger developed the precise wave equation obeyed by "matter waves", and the quantum revolution was up and running. You will (hopefully) get to learn all about that fascinating story in a future physics course.

For now, though, and to close out this course, let's see how the qualitative idea of "quantization" – that is, the discrete, rather than continuous, character of allowed energies (and other properties) – can resolve the lingering puzzle about the specific heats of gases.

In Chapter 12 we discussed the "equipartition theorem" and showed explicitly that any quantity f (which could be the distance that a certain spring is stretched, or one velocity component of a moving particle, or the angular velocity of a molecule rotating about some axis, etc.) with associated energy

$$E = \alpha f^2 \tag{13.13}$$

will, using the Boltzmann factor to assign probabilities to the different possible values, have an *average energy*

$$\langle E \rangle = \frac{1}{2} k_B T. \tag{13.14}$$

For a single pointlike monatomic molecule – for example a Helium atom – there are precisely three such degrees of freedom: the $x-$, $y-$, and $z-$ components of its velocity. A single such particle, in contact with a "bath" (perhaps consisting of many other Helium atoms) of temperature T, should thus have, on average, $\frac{3}{2} k_B T$ worth of energy. An entire mole of Helium gas will therefore have energy $\frac{3}{2} k_B T \cdot N_A = \frac{3}{2} RT$. And its (constant volume) molar heat capacity will be $c_V = \frac{3}{2} R = 12.5\,\mathrm{J/molK}$, just as is in fact observed experimentally.

A diatomic molecule like, say, Nitrogen has not just 3 but 6 (or perhaps even 7) such properties on which the energy depends quadratically: the three velocity components associated with the center of mass point, the angular velocities associated with rotation about the two orthogonal axes associated with relatively large moments of inertia, and the length of the bond between the two atoms, which can be thought of as a tiny spring. (The possible 7th degree of freedom is associated with rotation about the diatomic molecule's axis. If the moment of inertia about this axis is precisely zero, the associated energy would be precisely zero for any angular velocity. Whereas if the moment of inertia is some small but nonzero value, the energy should be a nonzero value proportional to the square of the angular velocity, and the equipartition theorem should apply.)

And so we should expect the constant-volume molar heat capacity of Nitrogen gas to be $c_V = 3R = 24.9\,\mathrm{J/molK}$ (or perhaps $3.5\,R = 29.1\,\mathrm{J/molK}$).

But, as noted in Chapter 11, the actual experimentally-observed constant-volume heat capacity is only about $2.5\,R = 20.6\,\mathrm{J/molK}$. So it seems that only five of the six (or seven) degrees of freedom are fully "active" in the way one would expect on the basis of the equipartition theorem.

To see how quantization helps resolve this puzzle, let's develop (something like) the equipartition theorem again, but in a slightly different way. Before (both in Chapter 12 and in our quick recap just above) we thought about, and calculated the average energy explicitly in terms of, the quantity "f" on which the energy was supposed to depend quadratically. This required us, in the course of computing the average energy, to evaluate a couple of Gaussian integrals – not the hardest thing in the world, but not the simplest either.

We can simplify the math and gain some insight (while sacrificing only a bit of quantitative accuracy) by instead just thinking and working directly in terms of the energy. So, consider some degree of freedom (the stretch of a spring, a velocity component of a particle, etc.) which can possess any energy E between zero and infinity. The Boltzmann factor tells us that, at temperature T, the probability for the energy to be E is given by

$$P(E) = \frac{1}{A} e^{-E/k_B T} \tag{13.15}$$

where A is a normalization constant. To ensure that the total probability, for the energy to be *any* of the possible values between zero and infinity, is one, we need

$$A = \int_0^\infty e^{-E/k_B T}\, dE = k_B T. \tag{13.16}$$

We can then calculate the average energy by adding up, for all possible energy values, the energy value times the probability for that energy value to be realized. The result is

$$\langle E \rangle = \int_0^\infty E \, P(E) \, dE = k_B T. \tag{13.17}$$

Note that this is missing the factor of 1/2 that we found earlier, when we did this properly, in terms of some f with energy $E \sim f^2$.

The reason this result turns out to be off by a factor of 1/2 is a little bit subtle. It has to do with the fact that if the possible f values are thought of as "evenly spaced", the associated energy values will *not* be evenly spaced, and vice versa. Our earlier calculation – the one that gave the right answer, with the 1/2 – assumed evenly spaced f values and hence non-evenly spaced E values. Our quick and dirty calculation just now tacitly assumes evenly spaced E values and hence isn't quite right. (The subtle thing is: what in the world does it mean for a continuous infinity of possible values to be "evenly spaced" or not?! Unfortunately we don't have time to get into that here.)

Since our aim here is not quantitative accuracy but instead conceptual understanding, let's call this result close enough, not worry about the factor of 1/2, and just say: if all of the E values, from the continuum between zero and infinity, are possible, the average energy value will be (of order) $k_B T$.

It is worth taking a moment to think about this result graphically. The average energy is the sum (over all possible E values) of the E value multiplied by the probability of that E value. That is, the average energy is the integral of $E \cdot P(E)$, as indicated in Equation (13.17). We can visualize this integral as the area under the curve when we plot the product $E \cdot P(E)$. This function of E is plotted in Figure 13.5. It has the shape it does because E starts at zero and goes up, while $P(E)$ starts at a large value ($1/k_B T$) and goes down. So their product goes up and then comes back down.

It is easy enough to show that $E \cdot P(E)$ reaches its maximum value at $E = k_B T$ and that its value at that point – the height of the peak – is e^{-1}. So as a very rough approximation, we can think of the graph of $E \cdot P(E)$ as a bump of width $k_B T$ and height $1/e$, so the area under the curve should be of order $k_B T$.

The question now is this: what if, instead of any energy value from the whole continuum being allowed, energy is quantized? That is, what if the possible energy values constitute only a discrete set, such as the one implied by Equation (13.12)?

We can calculate the new average energy and will do so in a moment. But let's try to first think through what to expect, qualitatively. If the allowed values of E are all the possible integer multiples of some elementary unit of energy, ϵ, we should expect that in the $\epsilon \to 0$ limit we'll get the same average energy as before, namely $k_B T$, just on the grounds that the set of evenly-spaced energy values just *is* the whole continuum if the spacing between them goes to zero.

But if the elementary energy unit ϵ (which represents the spacing between adjacent allowed energy values) is *large*, we expect the average energy to be considerably smaller. Why? Well, think about how we calculate the average energy. We add up, for each allowed energy value, the product of the energy value and the probability that the system actually has that energy value. The first term in that sum, $E = 0$, never contributes anything: the *probability* for E to be zero is quite high (higher than for any other value!), but the energy itself is zero, so when we multiply them we get zero.

But if the spacing ϵ is large, the *next* highest allowed value of E, namely $E = 1\epsilon$, may also contribute very little to the sum because, although E itself will be fairly large, $P(E) \sim e^{-E/k_B T}$ will be exponentially small. And then the rest of the allowed values, $E = 2\epsilon$, $E = 3\epsilon$, etc., will contribute even less for the same reason: the probability $P(E)$ is just too small when $E \gg k_B T$. In effect, if the spacing ϵ between adjacent energy levels is large compared to $k_B T$ – which remember is roughly the width of the peak in

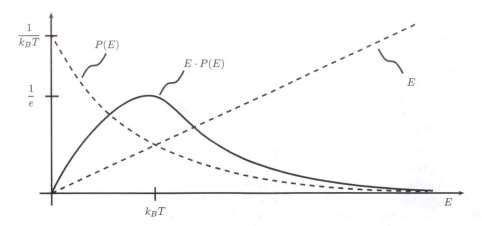

Figure 13.5: To calculate the average energy of a degree of freedom we need to find the area under the curve for the function $E \cdot P(E)$. Because E starts at zero and increases, and $P(E)$ starts at $1/k_BT$ and decreases, the product has the peaked shape shown. The width of the peak is of order k_BT whereas the height of the peak is of order one, so it makes sense that the total area under the curve is of order k_BT. (Note that the three quantities being plotted here – E, $P(E)$, and their product – all have different units, so it is a little confusing and misleading to plot them all on the same graph. The labels I've put on the vertical axis are particularly problematic! But you won't get confused if you understand that the $1/e$ label is just marking the height of the peak in the $E \cdot P(E)$ graph, while the $1/k_BT$ label is just marking the value of $P(E)$ at $E = 0$.)

the Figure 13.5 graph of $E \cdot P(E)$ – then the first nonzero allowed value of E can already be past the peak, and the average energy can come out much smaller than k_BT.

It does indeed work out exactly that way, but let's work through the actual calculation to see. So, suppose the allowed energy values are $E = n\epsilon$ for $n = 0, 1, 2, 3, \ldots$ The Boltzmann factor tells us, as usual, that

$$P(E) = \frac{1}{B} e^{-E/k_BT} \tag{13.18}$$

but with a somewhat different normalization constant than before:

$$B = \sum_{n=0}^{\infty} e^{-n\epsilon/k_BT} = \sum_{n=0}^{\infty} \left(e^{-\epsilon/k_BT} \right)^n. \tag{13.19}$$

This sum is of the form

$$S = \sum_{n=0}^{\infty} X^n = 1 + X + X^2 + X^3 + \cdots \tag{13.20}$$

with $X = e^{-\epsilon/k_BT}$. How to evaluate it? There is a beautiful trick, which I learned from the original Mathemagician, Arthur Benjamin. Multiply Equation (13.20) on both sides by X, giving

$$XS = \sum_{n=0}^{\infty} X^{n+1} = X + X^2 + X^3 + X^4 + \cdots. \tag{13.21}$$

Now let's subtract Equation (13.21) from Equation (13.20) – or, in words, let's "take away the XS"! Almost everything cancels on the right hand side, and we are left with

$$S - XS = 1 \tag{13.22}$$

which is readily solved to give

$$S = \frac{1}{1 - X}. \tag{13.23}$$

Of course, that only works if $X < 1$, but that will be the case here for us since $X = e^{-\epsilon/k_B T}$.

OK, so using that general formula to evaluate the normalization coefficient, B, we have that the probability for our degree of freedom to have n of the ϵ-sized units of energy is

$$P(E) = \left(1 - e^{-\epsilon/k_B T}\right) e^{-E/k_B T}. \tag{13.24}$$

We can then proceed to calculate the average value of the energy:

$$\langle E \rangle = \sum E \cdot P(E) = \sum_{n=0}^{\infty} n\epsilon \left(1 - e^{-\epsilon/k_B T}\right) e^{-n\epsilon/k_B T}. \tag{13.25}$$

The sum here can be calculated using a variation of our "take away the XS" formula. I'll leave it as an Exercise, though, and just quote here the result, which is that

$$\langle E \rangle = \frac{\epsilon}{e^{\epsilon/k_B T} - 1}. \tag{13.26}$$

This result has exactly the qualitative features we anticipated. For $\epsilon \ll k_B T$, the exponential in the denominator can be approximated as $e^{\epsilon/k_B T} \approx 1 + \epsilon/k_B T$, the $+1$ and the -1 in the denominator then cancel, and we are left with $\langle E \rangle \approx k_B T$, as expected.

On the other hand, for $\epsilon \gg k_B T$, the exponential in the denominator is so large that subtracting one from it doesn't change things much, and so the average energy comes out exponentially small: $\langle E \rangle \approx \epsilon\, e^{-\epsilon/k_B T}$.

Equation (13.26) is not, in all of its details, applicable to all of the degrees of freedom of (for example) diatomic gas molecules. For instance, while we assumed in our derivation that the allowed energy values have an equal spacing (ϵ), a full quantum-mechanical analysis reveals that the spacing between adjacent allowed energies for the molecules' rotational degrees of freedom should increase with energy. (Our formula is not totally inapplicable, though: it happens to apply perfectly to the *vibrational* degrees of freedom associated with the bond length, and is also exactly the right formula for the average energy contained in a ray of light with a certain frequency. So you will encounter it again in a future physics course where you study the work of Planck and Einstein in more detail.)

But the qualitative features of Equation (13.26) are right, even for the rotational degrees of freedom, and allow us to explain the otherwise-puzzling data for the heat capacities of different kinds of gases. The idea is simple: The spacing ϵ between adjacent allowed energies will be different for different degrees of freedom. For rotational degrees of freedom, it depends on the moment of inertia, and for the vibrational degrees of freedom, it depends on the stiffness of the spring-like bond holding the two atoms together. And it turns out that, for example, for Nitrogen gas at room temperature, the ϵ appropriate for rotations about the two axes with relatively large moments of inertia is small compared to $k_B T$ (which explains why those two degrees of freedom are fully "active", i.e., have the full average energy we predicted on the basis of the equipartition theorem); whereas the ϵ appropriate for rotations about the other axis (with a moment of inertia that is tiny or perhaps even zero), as well as the ϵ appropriate for vibrations, are quite large compared to $k_B T$ (which explains why those degrees of freedom are "in-active", i.e., why those degrees of freedom do not possess anywhere near the amount of energy that we would have expected based on the equipartition theorem).

We've been thinking of the temperature T as fixed (at, say, room temperature) and discussing how some degrees of freedom are fully active, while others are almost completely inactive, depending on the appropriate value of the quantized energy spacing, ϵ.

But it can be helpful to think about the same relationship from another perspective. For a given degree of freedom with a given ϵ, that degree of freedom will transition from being relatively inactive, to being fully active, as the temperature T (multiplied by Boltzmann's constant, k_B) increases from below ϵ to above ϵ. So as the temperature is steadily increased, from some very low value through to some very high value, we would expect that additional degrees of freedom become active – "come online" – at various intermediate temperatures. And, of course, the coming online of a new degree of freedom implies a fairly sudden increase in the heat capacity of the gas.

The idea of energy quantization, that is, allows us to understand why the heat capacity of a gas should vary with temperature in the way that was indicated, for di-atomic Hydrogen gas, back in Figure 11.1. Below a temperature of about 100 K, only the translational degrees of freedom (associated with the center-of-mass motion of the molecule as a whole) are active, and the molar heat capacity is roughly the value, $\frac{3}{2}R$, we'd expect for a monatomic gas. By the time the temperature reaches room temperature, about 300 K, the two additional rotational degrees of freedom (associated with rotations about axes perpendicular to the symmetry axis of the molecule, i.e., associated with rotations about axes with relatively large moments of inertia) have come online, so there are basically 5 active degrees of freedom and the molar heat capacity has increased to $\frac{5}{2}R$. And if we increase the temperature even further, eventually the remaining degrees of freedom come online and the molar heat capacity increases to about $\frac{7}{2}R$.

And so that one remaining loose end – which remember Maxwell described toward the end of the 19th century as "the greatest difficulty yet encountered by the molecular theory" – has been fairly well tied up... or at least hidden beneath the mountain of new puzzles and worries raised by the advent of quantization. But tackling those puzzles and worries will have to wait for another course in another semester.

* * *

Our goal in this course this semester was to provide a rich and technically-detailed perspective on the grand forest of (at least "classical", pre-20th-century) physics. I think we have succeeded, and I hope seeing this beautiful forest helps motivate you to always remain, in Einstein's words, "a real seeker after truth" – in physics, or wherever your passions lead you.

Questions:

Q1. In describing the "blob" (aka "plum pudding") model of the atom, we said in passing that the magnitude of the electric field inside the blob would be even less than our estimate, Equation (13.2). This is not quite true: the electric field could have an arbitrarily large magnitude close to one of the embedded pointlike electrons. But alpha particles could not be made to scatter into large angles by colliding with electrons, no matter how big the electric fields near them are. Explain why. (Hint: an alpha particle is thousands of times more massive than an electron.)

Q2. The text mentions the idea that energy quantization – and in particular the existince of a lowest-allowed-energy state for the electron in the Hydrogen atom – accounts for the observed stability of atoms. But this raises questions. For example, if the electron in that lowest-allowed-energy state is orbiting in some smallest-allowed circular orbit, isn't it still accelerating, and so shouldn't it still radiate energy? Try to brainstorm some possible ways out of this conundrum. (There are several, but of course it shouldn't be obvious at this point which, if any, turn out to be correct.)

Q3. Summarize in your own words why the average energy stored in a degree of freedom can be significantly smaller than $\sim k_B T$ if the allowed energies are quantized rather than continuous.

Exercises:

E1. Show, as claimed in the text, that the function $E \cdot P(E)$ – where $P(E)$ is given by Equations (13.15) and (13.16) – reaches its maximum value at $E = k_B T$ and that this maximum value is $1/e$.

E2. In our calculation of the average energy value under the restriction of energy quantization, we encountered a sum of the form

$$T = \sum_{n=0}^{\infty} n X^n.$$

Argue that $T = X \frac{dS}{dX}$ and then use the previously-established fact that $S = 1/(1 - X)$ to evaluate T. Confirm that the average energy comes out as claimed in Equation (13.26).

Projects:

P1. Analyze the scattering of alpha particles from a positively-charged nucleus and show that, as claimed in Equation (13.8), the probability for an alpha particle to be detected, having scattered through angle θ, is proportional to $1/\sin^4(\theta/2)$.

P2. Work through the analysis of the (in)stability of the Copernican Hydrogen atom described in the text. In particular: (i) write an expression for the total (kinetic plus electrical potential) energy of the atom in terms of the radius R of the electron's orbit; (ii) rewrite the Larmor formula, Equation (13.11), in terms of the radius R; (iii) use the idea that the energy of the atom must decrease at the same rate energy is being radiated away to write a first-order differential equation for the radius R; and (iv) solve the differential equation and find in particular the time it takes for R to go from half an Angstrom to zero.

Index